Immortal Ideas

Arts and Science Masterpiece:
"A brilliant piece of work for those who appreciate the history and culture of arts and science. No need to be a scientist here! If you have ever enjoyed David Suzuki's "Nature of Things," or Jay Ingram's Discovery Channel, then this is for you! This would also be a great gift for those readers on your Christmas list who are difficult to buy for. Exceptionally well illustrated, the book is a pleasure to hold."
— Carlo Lisi, BSc, MRT(N), Medical Technologist

An Intriguing Story:
"I shall never again look at a sunflower in quite the same way."
— Giselle Whyte, Brock University

The Author Succeeds Brilliantly:
"Despite the mathematical foundations of the ideas discussed, which are thus rock-solid, it is possible to skip the equations and enjoy the ideas. These are demonstrated through a wealth of beautiful illustrations of animals and plants, water-drops and fractals, classical paintings and the heavens, and why the Parthenon looks just right."
— Jennie Carter, former Ontario cabinet minister

Engaging:
"Lavishly illustrated . . . an enjoyable romp through the mysteries of nature and mathematics . . . engaging."
— Dave Thomas, Physicist and Skeptic

A Fascinating Book:
"The canvas of *Immortal Ideas* is the universe itself from the subatomic scale to the galactic and from the earliest times to the present. His brush is the scientific method and his pigments are rationality, humor and boundless enthusiasm. This well-written book is copiously and beautifully illustrated and sure to fascinate anyone who subscribes to Plato's dictum that the unexamined life is not worth living."
— Richard and Cory Wink, Teachers

The Power of Ideas:
"In *Immortal Ideas,* Harrison masterfully uses text and illustrations to elucidate how a few relatively simple ideas can have the most profound impact upon our understanding and appreciation of the physical world that surrounds us. From modest first principles, the author launches the reader into a

exhilarating exploration of the Cosmos that is both compelling and breathtaking. And, as the reader need not be an expert in either math or science, this book is accessible to virtually anyone who shares the belief that awesome power can reside in ideas."

— Jerry Larock, Elementary Teacher

A Joy to Read:
"The book is a pleasure on a number of levels — an overview of some important mathematical ideas, a recounting of history and an entertaining collection of insightful narratives. Having read the book once, I continue to pick it up to revisit topics of interest and each time I find myself intrigued by the subject matter. The book is well illustrated and specific areas of interest can be addressed without loss of continuity. A great read that makes accessible the beauty of mathematics. Well done!"

— Dr. Michael Schweigert, Occupational Medicine Specialist

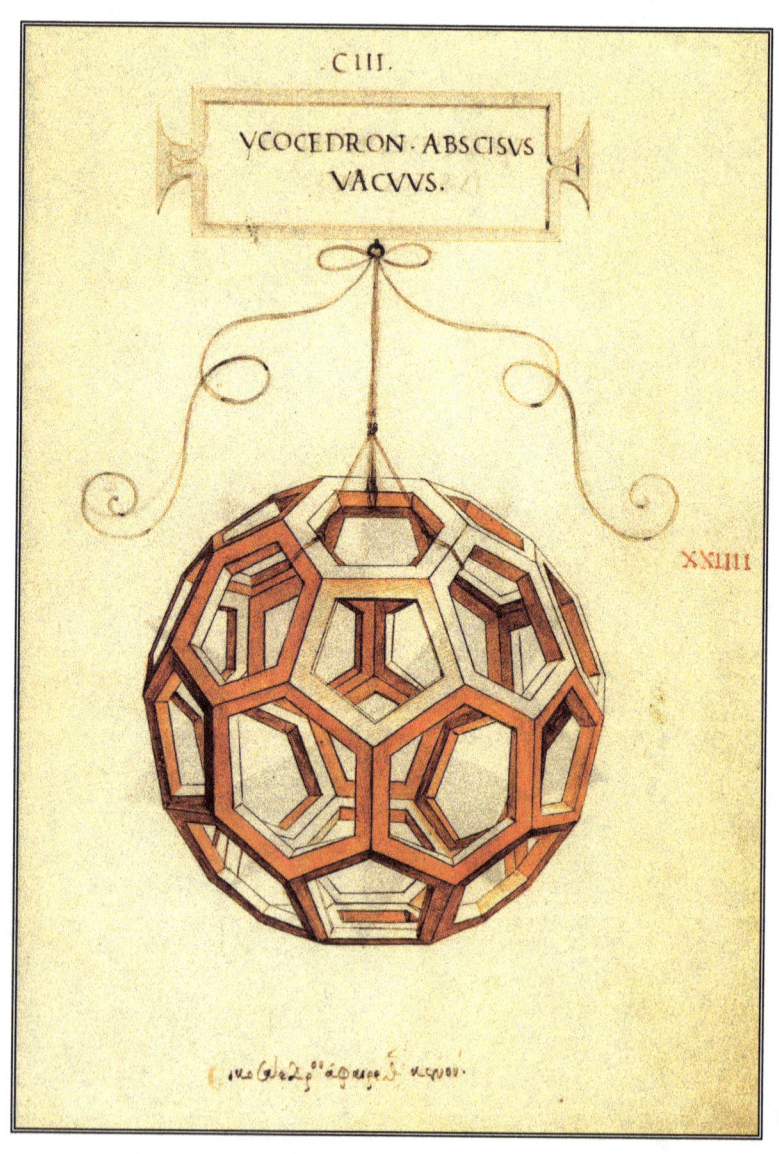

TRUNCATED ICOSAHEDRON
by
Leonardo da Vinci from *De Divina Proportione*

Immortal Ideas
Shared by Art, Science, and Nature

by

Gordon Harrison

 Prometheus Publications

Copyright © 2014 by Gordon Harrison

Prometheus Publications
Peterborough, Ontario
Canada

Immortal Ideas: Shared by Art, Science, and Nature. All rights reserved. No part of this book may be reproduced or transmitted in any form or by any means, electronic or mechanical without written permission from the publisher, except for inclusion of quotations in a review.

Library and Archives Canada Cataloguing in Publication

Harrison, Gordon, 1936-, author
    Immortal ideas : shared by art, science, and nature / by Gordon Harrison.

Includes bibliographical references and index.
ISBN 978-0-9879596-5-2 (pbk.)

    1. Ecology--Philosophy. 2. Symmetry (Biology). I. Title.

QH540.5.H37 2014   577.01   C2014-905489-0

Cover design by nature and the author

To Ruffy: Friend, Companion, and Clown

Who is forever chasing rabbits,
Birds, butterflies, and turtles
On the Happy Isles — home of all
Heroes and every brave being —
Nose down on the trail to happiness.

Ruffy

Animals

I think I could turn and live with animals,
  they are so placid and self-contained;
I stand and look at them long and long.
They do not sweat and whine about their condition;
They do not lie awake in the dark
  and weep for their sins;
They do not make me sick discussing
  their duty to God;
Not one is dissatisfied — not one is demented
  with the mania of owning things;
Not one kneels to another, nor to his kind
  that lived thousands of years ago;
Not one is respectable or industrious
  over the whole earth.
Walt Whitman, *Song of Myself*

ACKNOWLEDGMENTS

In writing this book I have been fortunate in having friends who could offer suggestions that have improved the text. Evelyn Andrews, my longtime companion and confidant, introduced me to David Timms, editor extraordinaire. David went through the entire book, word by word, offering innumerable improvements. I owe him a great debt. Evelyn herself has been an important sounding board and a source of ideas and unending support. My daughter, Jennifer, has listened to me rave about the book and has always been encouraging. After David's editing and my rewriting, Larry Keeley, a former teaching colleague and friend who shares a love of nature with me, proofread the final manuscript for lingering errors. And in his enthusiasm, Jerry Larock has led me to believe this book is better than I would have imagined. My friend Richard Wink read through Chapter 3 on chaos theory for any technical or theoretical errors. I am grateful to everyone.

I have to also thank Leo S. Bleicher for allowing me to use his computer-painted *Gold Fountain* shown on page 320. Similarly for Charles S. Lewallen's photograph of the water strider found on page 263. And, thanks to Ram Samudrala and Shriram Krishnamurthi for their montage *Pi à la Mode* on page 204. Lastly I must give credit to Nora Mickee for constructing Figure 5.4, the 153 Fishes in the Net.

Every reasonable effort has been made to contact holders of copyright for material or photographs used here. The author will gladly receive information that will enable him to rectify any inadvertent errors or omissions in subsequent editions. All the diagrams, however, and most of the photographs are my own.

An old proverb affirms, "Luck never gives; she only loans." But in the writing of this book, I've been Lady Luck's major beneficiary, and these gifts she can never take away.

Table of Contents

Acknowledgments . . . . **VI**

Preface . . . . **IX**

1. **The Measure of All Things** 1
   The Colossus of Rhodes * Epius and Phemius * The Delian Puzzle * Square-Cube Law * Size Does Matter * The Apple's Descent * Isaac Newton * The Heat Seekers * Form and Function * Life on Earth * The Tree Warriors

2. **The Perfect Five** 29
   Leonardo's Polyhedrons * Luca Pacioli and Johannes Kepler * The Ladder of Dimensions * Other Dimensional Cubes * Salvador Dali and the Fourth Dimension * A Natural Five * The Perfect Die * Pascal's Triangle * Cardano * William Blake's Dice

3. **Without Form and Void** 67
   Chaos Theory * The Pendulum * Chaos on a Calculator * Nature's Patterns * The Logistic Function * One, Two, Four, ... Chaos * Chaos on a Personal Computer * Bifurcation Diagrams * The Butterfly Effect * The Lorenz Attractor * The Rider in the Whirlwind * Contingent History * Ab-Surd Universe * Root Fractals * Long Night's Journey into Light

4. **Forms and Fractals** 125
   The Koch Snowflake * Benoit Mandelbrot * Fractal Dimensions * The Sierpinski Triangle * Self Similarity * The Chaos Game * The Odd and Even Triangle * The Garden of Simple Complexity * Julia Sets * The Mandelbrot Set * Seahorse Canyon * Departure

5. **Numbers: Supernatural** 163
   The Genesis of Numbers * The Sign of the Fish * 153 Fishes in the Net * Triangle Numbers * Fish Fry * Invention or Discovery * Numerical Miracles in the Koran * Michael Drosnin and Equidistant Letter Sequences * The Number Seven in Our Culture

6. **Numbers: Natural** 189
   Pi (π) and Curves of Constant Width * Meandering Streams and Pi * Biblical Value of Pi * Circle Squarers * Classification of Numbers

∗ Continued Fractions and Pi ∗ The Field of Dreams ∗ Fibonacci Numbers ∗ Phi (φ) — the Golden Number ∗ Sunflower Spirals ∗ Van Gogh's Sunflowers ∗ Numbers: Cultural ∗ The Pyramid of Giza ∗ Phi in the Parthenon ∗ Adam's Bellybutton ∗ Vitruvian Man ∗ Phi in Art ∗ Phi Foolishness ∗ The Truth about *The Da Vinci Codes* ∗ The Texture of Infinity ∗ Delta (δ) — the Feigenbaum Constant ∗ Paleolithic Cave Art

7. **LIFE ON THE EDGE** 243
   Animal Double Printing ∗ Archimedes' Tombstone ∗ Bouncing Balls and Photons ∗ Night Vision ∗ The Cowboy's Dilemmas ∗ *Starry Night* ∗ Queen Dido's Puzzle ∗ The Least-Action Principle ∗ The Skin on Water and Air ∗ Soap Bubbles ∗ Minimum Paths ∗ The Argument from Design ∗ Minimal Surfaces ∗ Cubical Soap Bubbles ∗ The Air Pressure Paradox ∗ Geometry of Soap Bubbles and Films ∗ The Dance of Life ∗ The Fox Knows Many Things

8. **SHATTERER OF WORLDS** .289
   Symmetry and Beauty ∗ Planes of Symmetry ∗ Handedness in Nature ∗ Reflections of Einstein's Face ∗ Inverted Drawings ∗ *Whom the Gods Love* ∗ Symmetry Groups ∗ Platonic Solids ∗ The Alhambra ∗ The Seven Frieze Patterns ∗ Broken Symmetry ∗ Instability — the Creator ∗ Art and Symmetry ∗ Nature and Symmetry Breaking ∗ Cosmic Fingerprint

9. **RIDERS IN THE WHIRLWIND** 329
   Join the Dots ∗ Spirals in Nature and Art ∗ Explosions in Nature ∗ The Geometry of Spirals, Meanders, and Explosions ∗ *The Lady and the Ermine* ∗ Truth in Nature, Art, and Mathematics ∗ False Patterns ∗ The Cluster Illusion ∗ The Birthday Puzzle ∗ The Hot Hand ∗ The Riders ∗ The Game of *Life* ∗ Emergent Properties ∗ How the Zebra Gets its Stripes

10. **THE SUM OF ALL THINGS** 367
    Epius and Phemius ∗ The Greeks and Deduction ∗ The Renaissance and Induction ∗ Art and Science ∗ Creativity in Science ∗ Is Seeing Believing? ∗ Illusions ∗ The "Lords of the Rings" ∗ The Deception Room ∗ Pascal's Triangle Upside Down ∗ Knowledge or Certainty ∗ Epius and Phemius ∗ The True Immortals

CHAPTER NOTES 389

INDEX 409

PREFACE

THE SPHINX ASKED OEDIPUS the famous riddle, "What goes on four feet in the morning, on two at noon, and on three in the evening?" "Man," he answered. "In childhood man creeps on hands and knees; in maturity he walks erect; in old age he uses a cane." It is not well known that the sphinx had a second riddle. "What does a man never think about in youth, and think of nothing but in old age?" "Death," I can hear many readers respond. Although the first riddle like the second is about aging, the latter has at its core the paradox of human existence: the ephemeral nature of life.

We fall in love, have children, amass wealth, build mansions, erect statues, become farmers, carpenters, presidents, teachers, nurses, soldiers. But there will come a time when all this will be forgotten. Our accomplishments will be like smooth stones in a stream, bearing no trace of their rough youth when they were giants. Eventually even these will be ground to dust and washed away.

Where is immortality? After just ten generations, your descendants have less than one part in a thousand of your original DNA. Of the Seven Wonders of the Ancient World, only the Pyramid at Giza remains, and it has lost much of its height due to erosion and vandalism. One day it will be no higher than an anthill. Where is immortality?

Gerolamo Cardano (1501–76) was one of those incredible Renaissance polymaths: doctor, mathematician, professor, scientist, writer, astrologer, occultist, and above all, gambler. Even in an age of great men and women, his life was extraordinary. When he was seventeen, his best friend Niccolò Cardano died. In a few months neither the boy's family nor his teachers spoke of him, his friends never mentioned him — he was only a fading memory. It disturbed Gerolamo that a life could be so insignificant. After intense reflection on this subject, he wrote his first essay about the ways a man could immortalize his name. Yet he himself left no material memorials, and of his two sons, the older was a murderer and the younger a thief. Where was his immortality? If our loins and our backs cannot create anything permanent, perhaps our minds can? This is where Cardano succeeded because five centuries after his birth many of his ideas are still recalled. Any intimation of immortality lies here with ideas.

Immortality through intellectual creation isn't a new thought. This book's theme comes from the poem *Heraclitus* by the ancient Greek Callimachus, which makes this point:

They told me, Heraclitus, they told me you were dead,
They brought me bitter news to hear and bitter tears to shed.
I wept as I remembered how often you and I
Had tired the sun with talking and sent him down the sky.

And now that thou art lying, my dear old Carian guest,
A handful of gray ashes, long, long ago at rest,
Still are thy pleasant voices, thy nightingales, awake;
For Death, he taketh all away, but them he cannot take.
Translation by William Johnson Cory (1823–92)

I choose to interpret *nightingales* as "creative ideas expressed through poems, paintings, sculpture, science, and mathematics."

We live in the greatest scientific age of all time; yet we live in the most innumerate of times. Our science writers need to inspire us with the beauty and power of *their art*, the same stirring splendor found in literature and music. We need writers like Shelley whose description of the mundane water cycle in the final stanza of "The Cloud" can rouse a class from somnolence. We need to touch people with the profound beauty common to art and science. This is my motivation!

Today we have journeyed to a distant place far removed from C. P. Snow's *The Two Cultures* where he described how art and science speak mutually incomprehensible languages. Now the world is even more fractured — we need to be healed.

Each chapter is built around a single idea that's common to art, science, and, nature. The last chapter is a summary of the previous nine. Although this book is nonfiction, sections of each chapter and the climax are developed by two semi-historical characters: one from the *Iliad,* the other from the *Odyssey* — the two blessed men of Odysseus. The older is Epius the engineer/scientist who with the sanction of Odysseus built the Trojan horse and ended the war. The younger is Phemius the poet/minstrel compelled to entertain the suitors in Odysseus' absence. On his return the hero slew the suitors but spared and blessed Phemius as a man inspired by god. Both scientist and poet have been alive for three millennia, interacting with history's great ideas through famous men and women.

If anyone tells you these two characters aren't friends, don't believe them. If anyone tells you they're imaginary, laugh. Down through the centuries they have been seen together in Athens, Florence, Paris, London, Berlin, Moscow and New York, producing magnificent art and astonishing science. It's unknown when they met after the Homeric age, but around 300 BC the citizens of Rhodes summoned Epius to complete the Colossus and Phemius to prepare the celebrations for its dedication.

The first chapter deals with the *square-cube law* and its unique consequences for all things living and dead. During the construction of the Colossus, Epius discovered this idea and realized some of its far-reaching consequences. Isaac Newton could easily have cited it rather than his law of universal gravitation to account for the apple's fall. The Colossus was the shortest lived of the Seven Wonders, standing for only 56 years, but the *square-cube law* will live a little longer. Immortality for the maker and the made lies with the creation of art and science. The themes in this book will live as long as sentient beings do.

"... *this place is sacred — thick-set with laurel, olive, vine; and in its heart a feathered choir of nightingales makes music. So sit thee here on this unhewn stone* ..."
Sophocles, *Oedipus at Colonus*

Gordon Harrison

Chapter — 1

The Measure of All Things

*To you, O Sun, the people of Dorian Rhodes
set up this bronze statue reaching to Olympus
when they had pacified the waves of war and crowned
their city with the spoils taken from the enemy.
Not only over the seas but also on land did
they kindle the lovely torch of freedom.*
Dedication attributed to Phemius, son of Terpias

AT THE ENTRANCE to New York City harbor on a small island stands a spectacular sight, a statue of a robed woman lifting a torch to the sky. This immense figure rises 152 ft (46 m) from foot to crown — higher if you include the pedestal. Originally called Liberty Enlightening the World, it's now universally referred to as the Statue of Liberty — the Colossus of New York.

People around the world immediately recognize this awe-inspiring figure, a gift from France to America. Yet few know that the architect, Frederic Auguste Bartholdi, found his inspiration in one of the Seven Wonders of the World — the Colossus of Rhodes. Over 2,200 years ago on a Greek island, the original stood at the entrance to another busy harbor. It was also built as a celebration to freedom. Reaching approximately the same height, this astonishing figure was one of the greatest engineering feats of the ancient world. On the base of Bartholdi's creation is a small tablet with a sonnet by Emma Lazarus acknowledging his archetype. It's titled "The New Colossus."

When Alexander the Great died in Babylon in 323 BC, his generals fought like jackals over the carcass of his empire. Alexander's boyhood friend Ptolemy won Egypt to which the Rhodians had pledged allegiance. One of Ptolemy's rivals, Antigonus, snatched what is now Turkey, and he straightway dispatched his son Demetrius to besiege and subdue nearby rebellious Rhodes.

After a protracted struggle involving immense siege engines, Demetrius sailed away in defeat. The Rhodians in exuberance at their good fortune and military skill decided to erect a statue to celebrate their victory. They commissioned Chares of Lindos, a sculptor, to draw up plans. He envisioned a statue of the sun god Helios standing nude on a pedestal with a cloak draped over his left arm, his right hand shading his eyes from the rising sun. This herculean labor took 12 years.

The fame of this structure spread so quickly through the ancient world that the statue came to be called the "Colossus." Each morning at the harbor entrance, the sun caught the polished bronze and made the god's figure shine. It stood for 56 years before breaking at the knees during an earthquake — the shortest lived of all the Seven Wonders. But for centuries after, huge pieces lay scattered along the harbor entrance.

Colossus of Rhodes
by Salvador Dali, 1954

A later visitor to this antique land, Pliny the Elder (AD 23–79), was deeply impressed by these remnants. He wrote:

> Even as it lies, it excites our wonder and admiration. Few men can clasp the thumb in their arms, and its fingers are larger than most statues. Where the limbs are broken asunder, vast caverns are seen yawning in its interior.

Chares did not live to see its completion, but legends and myths flourished about his work. After he was well into the planning and preparations, one story relates how the city fathers decided to double the height of the statue. In his scholarly book *The Ancient Engineers,* L. Sprague de Camp remarks on this tale:

> When the city decided to double the height, Chares asked for only twice the original fee, forgetting that the material would be increased eightfold. This error drove him to bankruptcy and suicide.

It seems, however, incredible that a man with the engineering skill that Chares must have had should not have known the *square-cube law* [my italics]. This law states that, if you increase the dimensions of an object while keeping its shape the same, the area increases as the square of the dimensions while the mass and volume increase as the cube. That is why no flying animal has ever exceeded about 30 pounds in weight, and Sindbad's roc, which bore off elephants in its talons, would be quite impossible.

Even with their fondness for monumental works, no Roman ever had the vision of Chares of Lindos, or the will of the Rhodians to carry it out. Until the specialized building materials of the 19th and 20th centuries, no statue remotely approached the enormity of the Colossus.

AFTER THE DEATH OF CHARES the city leaders began a search for an engineer able to complete the Colossus. Many names were proposed but only one seemed experienced enough for the task — Epius, son of Panopeus. A ship was dispatched to Athens to request his assistance. He readily agreed, subject to one curious condition. By the time he arrived on Rhodes, however, there was a strange rumor in the city about who he really was. Some said he was none other than the Epeios mentioned in the *Iliad* as a boxer and in the *Odyssey* as the builder of the Trojan horse. (The vase at the left shows Epeios presenting the finished Trojan horse to a Greek chieftain, almost certainly Odysseus.) But others pointed out that if he were the legendary Epeios, he must be over 800 years old! Whenever asked about this rumor, he would laugh uproariously.

TROJAN HORSE
AMPHORA

Unquestionably, his massive fists, great height, and bulk would have made him a formidable opponent in any boxing match. And his battered face and numerous scars spoke to the truth of his pugilistic past. If it weren't for his friendly manner and remarkable spirit, he would have been a frightening presence.

The curious condition Epius put on his employment was unrelated to any material reward. Instead he insisted his friend Phemius, son of Terpias, also be hired as the official poet and musician to commemorate the building of the Colossus. To this the city officials agreed because it soon became apparent that his companion had amazing artistic ability.

They were an unlikely pair. The disparity in their statures was as pronounced as the differences in their features and personalities. Often they were heard arguing at great length about topics no one quite understood. Yet it was soon clear these discussions were never personal, but always dealt with ideas. Furthermore, a reciprocal dependence between them showed itself in many ways. The larger protected the smaller from bullying work gangs and the like while the other made life easier by numerous small acts. Every evening he recited long sections from Homer and the Greek poets, accompanied by various musical instruments.

After Epius had reviewed the original building plans and the later size doubling demanded by the city fathers, he discovered the *square-cube law* and realized the problems it was causing. With patience and detailed models he explained all this so well to the officials that he got the extra men and materials needed to finish the construction. During this time Phemius composed songs, trained singers, and taught others to play a variety of wood and string instruments, and he wrote the world's first inscription *to freedom*. The completion of this world wonder by Epius took seven years, five more if you include the time Chares spent. The celebrations were to last an entire week.

On the morning of the dedication the whole city was out in boats or on rafts, promontories, and islands to see the first fingers of dawn touch the shining bronze on the sun god's face. Waiting in the pre-dawn glow, people pulled their tunics close about them to ward off the cold. When our great morning star rose out of the sea with his sundogs, the dazzling brilliance and enormity of the Colossus hushed the throng and held them spellbound by what they had accomplished.

The builder and the bard continued to reside on Rhodes where they were honored and respected. They bought a whitestucco home near Lindos that sat on the side of a cliff going down to the dazzling blue Aegean. Surprisingly, considering the shorter life expectancy in ancient times, they both lived to witness the collapse of the Colossus — more than half a century after its completion. Following this disaster, they left Rhodes. To add to the legend of their inconceivable longevity, reports say they appeared not to have aged after all those years.

THE DELIAN PUZZLE, also called doubling the cube, is one of the "three famous problems of antiquity." It leads directly to the *square-cube law*. The story of its origin went something as follows: the citizens of Delos were enduring a terrible plague (probably typhoid fever), many had died, and more were dying. Now the Greeks believed the gods — in this case Apollo — initiated all such disasters. Therefore the people sent a delegation to the oracle at Delphi to ask what could be done to appease his terrible anger. The Pythia, priestess of Apollo, told them to *double the size* of Apollo's cubical altar. So they built a new altar with each edge twice its former length, but the plague grew worse and death was everywhere. Those still alive from the first delegation returned to Delphi in confusion and anger. "We have done what Apollo demanded, but the plague worsens" they declared. The Pythia responded that they had increased the size of the cubical altar *eightfold,* not *twofold* as instructed and so, of course, the plague continued. The citizens turned to their mathematicians to find what the edge length should be to double the cube's size (volume). These learned men failed to find the exact length required (so did thousands of others in the following centuries). Fortunately the plague lessened its lethal grip and eventually life returned to normal. No thanks to the priestess, the mathematicians, or Apollo!

$$\text{area} = 6 \text{ units}^2 \quad \text{area} = 6(\sqrt[3]{2})^2 \text{ units}^2$$
$$\text{volume} = 1 \text{ unit}^3 \quad \text{volume} = (\sqrt[3]{2})^3 = 2 \text{ units}^3$$

DOUBLE THE VOLUME

Today we know the task the ancient mathematicians set for themselves was unattainable. Why? Classical Greek geometry allowed the use of straightedge and compass only, the so-called Euclidean tools. The length required to double the cube's volume, $\sqrt[3]{2}$, is what mathematicians call "a cube root irrational," and it can't be constructed with these tools. In 1837, more than two millennia after the Delian puzzle was first stated, the French mathematician Pierre Wantzel proved forever that it and the other two classical problems (trisecting the angle and squaring the circle) are unsolvable with such instruments.

We have seen two historical instances of misunderstanding over the *square-cube law* — the measure of all things. Because of this, and since it is this chapter's central idea, we'll take a detailed look.

SQUARE-CUBE LAW

Man is the measure of all things.
Protagoras (485–411 BC)

THE PYTHIAN PROPHECY by this mistress of ambiguity was conveniently open to two interpretations. The Delians' first response (below) is the law we're interested in while their second is one of the three classical problems mentioned above.

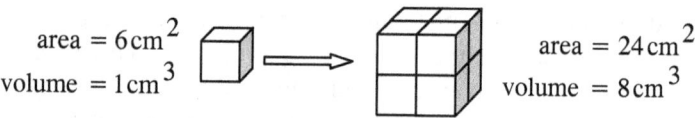

DOUBLE THE EDGE LENGTH

In the first case, doubling the length of each side of the cube increased the area *fourfold* (i.e. from 6 to 24) but the volume increased *eightfold* (i.e. from 1 to 8). In other words, the area increased as the square (i.e. $2^2=4$) while the volume increased as the cube of the edge length (i.e. $2^3=8$). When you double your height, your skin is 4 times while your weight is 8 times greater. The ratio of the surface area to the volume in the unit cube is 6:1, but in the 2x2x2 cube, it's 24:8 or 3:1. As a result, the *volume is growing more quickly than the surface area*. In a 6x6x6 cube, the terms of the ratio are equal (that is, 1:1). Increasing the size allows the volume to eventually *overwhelm* the covering area — this is evident from the following table's last row (Figure 1.1). See also the right column. Since these are ratios, the units may be any you wish.

Most remarkably, this law applies to *all shapes,* be they ever so simple or incredibly complex: double the height of anything and you increase the area fourfold and the volume eightfold. Whether living or dead, animal or vegetable, at your feet or in a distant galaxy, this law holds sway over all. Justly, it is *the measure of all things*. Stand down, Protagoras!

Edge Length	Surface Area	Volume	Ratio of Area : Volume
1	6	1	$6:1 = a:1v$
2	24	8	$6:2 = a:2v$
3	54	27	$6:3 = a:3v$
4	96	64	$6:4 = a:4v$
5	150	125	$6:5 = a:5v$
6	216	216	$6:6 = 1:1 = a:6v$
7	294	343	$6:7 = a:7v$
8	384	512	$6:8 = a:8v$
9	486	729	$6:9 = a:9v$
10	600	1000	$6:10 = a:10v$
⋮	⋮	⋮	⋮
100	60,000	1,000,000	$6:100 = a:100v$

The Square–Cube Law Table

Figure 1.1

MANY GOOD THINGS come from *applying* ideas to judge their depth and scope. We will listen to this nightingale's song (*see* the Preface) and see how high it can fly and witness the breadth of its dominion. Except for an intermission by Newton and Blake midway in the recital, the *square-cube law* program will include the following six songs under four headings:

SIZE DOES MATTER
- human size and its effect on strength
- giants and monsters, both human and animal

THE APPLE'S DESCENT
- apples of legend, literature, and science

Intermission by Newton and Blake

THE HEAT SEEKERS
- the sun, photons, and the planets
- the Ruandans and Inuit, mice and kinglets

THE TREE WARRIORS
- trees and leaves, and why they have so many.

SIZE DOES MATTER

IN GREEK MYTHOLOGY, Zeus charged two eagles with finding the exact center of the earth. He released one to the east and one to the west. They met at Delphi. A cone-shaped stone, called the omphalos, was placed in front of Delphi's Temple as a marker for the navel of the earth. And the Greeks further imagined they lived at the cynosure of the entire universe.

From the earliest times, *all* people have believed they were at the center of creation, perchance fashioned even in the image of god. The Eskimos still call themselves *Inuit* and the Cheyenne Indians of the Great Plains called themselves *Tsistsistas,* both meaning *The People.* The Hebrews have long referred to themselves as *The Chosen People.* Science has paved a broad path of withdrawal from this anthropocentrism, and the retreat has disturbed the equilibrium of the faithful. Copernicus and Darwin were only two — possibly the most important two — who laid the scientific bedrock for an objective view of man's true place in the universe.

In a related context, humans have long been thought of as intermediate in size, neither too large nor too small. Meaning the general number of creatures smaller than ourselves is believed to equal the number larger — we're at the center of the scale of life so to speak. Stephen Jay Gould dispels this myth-information in his popular book *Ever Since Darwin*:

> Most people think *Homo sapiens* is a creature of only modest dimensions. In fact, humans are among the largest animals on earth; more than 99 percent of animal species are smaller than we are. Of 190 species in our own order of primate mammals, only the gorilla regularly exceeds us in size.[1]

Let's consider both extremes of size and ferret out the consequences. First, the very small. The *Guinness Book of World Records* states that the shortest authenticated human to ever live was Pauline Musters of the Netherlands at a diminutive 23.2 in. (58.9 cm). Life was very different for her than for a woman of average stature — just consider the implications of this stature on her energy levels.

To ease the mathematics, we will investigate the kinetic energy a 3-foot person (E_3) can generate compared to a 6-foot individual — disregarding differences of gender. Now kinetic energy is energy of motion and equals one half times the mass (m) times the velocity (v) squared. The units need not concern us since we're interested only in the ratios. Consider the following general formula:

$$E = \frac{1}{2}mv^2.$$

Since the taller woman is twice the height of the shorter, she has eight times the volume or mass (8m). And *optimally,* the 6-foot woman can give a spear, axe, or her fist, twice the velocity (2v) of her diminutive companion. As a result, the formula shows that her kinetic energy (E_6) increases to the fifth power (i.e. $2^5 = 32$):

$$E_6 = \frac{1}{2}(8m)(2v)^2 = (32)\frac{1}{2}mv^2 = 32\,E_3.$$

Therefore, $E_6 = 32 E_3$.

This 32-fold increase holds only if the two participants are identical except for size. Nature is messy. Every outcome is the result of hidden causes that the scientist must uncover and then tease apart to learn what affects what. Here the foremost player is the *square-cube law;* this is its handiwork. The evolutionary implications of this huge energy difference are vast. Gould in the previously mentioned *Ever Since Darwin* clearly makes the case:

> At half our size, we could not wield a club with sufficient force to hunt large animals (for kinetic energy would decrease 16 to 32-fold); we could not impart sufficient momentum to spears and arrows; we could not cut or split wood with primitive tools or mine minerals with picks and chisels. Since these all were essential activities in our historical development, we must conclude that the path of our evolution could only have been followed by a creature very close to our size. I do not argue that we inhabit the best of all possible worlds, only that our size has limited our activities and, to a great extent, shaped our evolution.[2]

Evolution has no teleological principles — bodily changes promote survival right here, right now. We are the size we are because of who we are!

NOW SHIFT YOUR GAZE UPWARD to consider the very large — the giants, monsters, and behemoths of history and legend. Perhaps the most famous of these is the biblical Goliath of Gath. The assertion of I Samuel 17:4 that Goliath stood six cubits and a span (roughly 9 ft 8 in. or 295 cm) suggests confusion about the cubit's length, *or* more likely enthusiastic exaggeration by the Hebrew chroniclers. (Who among us hasn't stretched a story to increase its effect.) The 1st century Roman historian Flavius Josephus and some Septuagint manuscripts give Goliath's height at a reasonable four Greek cubits and a span (6 ft 10 in. or 208 cm). Historical literary references aside, there exists a second method proving extreme giants are a fantasy — at least as we imagine them. Galileo and the *square-cube law* will be our guides.

Given two exactly similar objects, one several times larger in linear dimension than the other, *the larger of these objects has less structural strength*. If allowed to grow unrestrainedly, the bigger object will ultimately collapse under its own weight like a whale on a beach. To accommodate a huge size, bodily proportions must change, in particular the leg bones. In his treatise on mechanics, *The Discorsi*, Galileo was the first to clearly establish this:

> From what has already been demonstrated, you can plainly see the impossibility of increasing the size of structures to vast dimensions either in art or in nature; likewise the impossibility of building ships, palaces, or temples of enormous size in such a way that their oars, yards, beams, iron-bolts, and in short, all their other parts will hold together; nor can nature produce trees of extraordinary size because the branches would break down under their own weight; so also it would be impossible to build up the bony structures of men, horses, or other animals so as to hold together and perform their normal functions if these animals were to be increased enormously in height; for this increase in height can be accomplished only by employing a material which is harder and stronger than usual, or by enlarging the size of the bones, thus changing their shape until the form and appearance of the animals suggest a monstrosity.

Imagine you're the unfortunate giant Goliath of Gath, and you continue to grow beyond the reasonable 6 ft 10 in. mentioned above. As your height keeps on increasing, your weight begins to crush your knees and ankles. To relieve this you must fall on all fours. As you further ascend the scale — even beyond biblical dimensions — your legs and arms noticeably shorten. Additional increases reshape and compress your neck to support your great head. Quite soon your knees seize, your feet become larger and more circular, your leg and arm bones grossly thicken. Ultimately you have the mass of an elephant; more specifically, you are one.

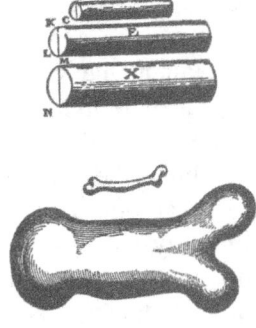

Figure 1.2

Figure 1.2 shows Galileo's original drawings illustrating the relationship between size and shape. To preserve the same strength, longer cylinders must be relatively much thicker than smaller ones. Likewise, large animals such as elephants and hippopotamuses must have disproportionately large leg bones. In the physical world, one can't increase the size or quantity of anything without changing its quality. The grotesquely enlarged insects, lizards, and gorillas of Hollywood horror movies — each just a photographic enlargement of the true creature — are a fiction.

THE APPLE'S DESCENT

APPLICATIONS OF THE SQUARE-CUBE LAW with its influence on strength and form abound in our world, existing under foot, overhead, and all around us. In springtime, the apple raises its five-petaled blossoms to the sun and the bees. Once fertilized, the heads begin to fall; the fruit forms, and gravity does its work. The volume of the apple grows as the cube of its linear dimension. Yet the strength of the stem grows *only* as the square of its linear dimension — the cross-sectional area. This is the *square-cube law* at work (*see* Figure 1.3). The stem would have to increase out of due proportion lest the apple fall, but in fact there is little distortion. For seed propagation, it's desirable that the mature apple falls. Only smaller varieties like crab apples hang on into the winter months to be eaten by squirrels and birds and so have their seeds widely broadcast in unique ways. Larger vine vegetables such as squash, melon, and pumpkin must rest their full weight on the ground.

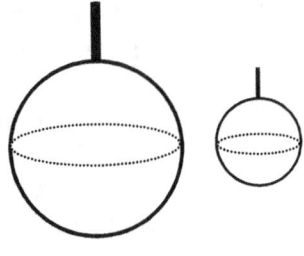

Figure 1.3

The apple runs as a minor theme all through legend and science. Linguists tell us that of the Proto-Indo-European language only 40 words survive in English; one of them is *apal* or *apple*. It's generally assumed that the fruit of the knowledge of good and evil from the Garden of Eden was an apple. Certainly it was an apple that Newton saw fall in the garden at Woolsthorpe Manor, starting the young man thinking about universal gravitation. And this force depends on the *inverse-square law*: the statement that gravity diminishes or varies inversely as the square of its distance from an object. That is, move an object twice as far away and gravity reduces to one-quarter of its previous strength. Newton's thoughts could just as well have centered on the *square-cube law*. Both laws are valid; each needs the other for the apple's descent.

CONTROVERSY HAS LONG SWIRLED around Isaac Newton, not about his science, but concerning his *scientific rationalism*. The debate began in earnest with the Romantic Movement. It was characterized by extreme dichotomies, each with an aspect of the truth: romanticism versus rationalism, heart versus head, faith versus skepticism, and holism versus reductionism. To confront Newton's public philosophical vision, the English poet and artist William Blake (1757–1827) unsheathed both of his creative swords — poetry and painting. His was a commanding voice:

> *Mock on, Mock on, Voltaire, Rousseau*
> *Mock on, Mock on; 'tis all in vain!*
> *You throw the sand against the wind,*
> *And the wind blows it back again.*
>
> *And every sand becomes a Gem*
> *Reflected in the beams divine;*
> *Blown back they blind the mocking eye*
> *But still in Israel's paths they shine.*
>
> *The Atoms of Democritus*
> *And Newton's Particles of light*
> *Are sands upon the Red sea shore,*
> *Where Israel's tents do shine so bright.*

In this powerful poem, Blake rejects the Ionian enlightenment of Democritus and the later mechanism of Newton for the rigid and unchanging revelation from Jerusalem. No journey, no search, no progress — just revealed, eternal truth!

Blake's second sword painted the richly colored background and strong form of *Newton* (Figure 1.4). It is, however, a portrait of a misguided hero or perhaps even a fallen angel like Milton's Lucifer in *Paradise Lost*. Newton's gaze is fixed on sterile geometrical diagrams, while the world of color and creativity swirl all around him. But strangest of all, Blake has placed Newton *under the sea,* implying he's from another realm — a creature alien to the rest of us. The compass in his left hand was a common symbol in art and literature. The year previous, the poet had used it in his *Ancient of Days* to depict god circumscribing the universe. Here the painter sees Newton as a demigod claiming his own barren realm for dispensing unequaled laws. In *Paradise Lost* (Book VII), Milton used the compass in a similar manner:

> *. . . in his hand*
> *He took the golden compasses, prepar'd*
> *In God's eternal store, to circumscribe*
> *This Universe, and all created things.*

Newton by William Blake,
1795, Tate Gallery, London

Figure 1.4

The poem and the painting tell us Blake didn't comprehend that reasoning itself can be a creative act. He confuses the *context of discovery* with the *context of proof.* The former is wild, unpredictable, multifaceted, nonlinear, fueled with endless cups of coffee, and by no means certain in its outcome. The latter is calm; some would say sterile, usually predictable, hiding the emotions and the human face of creativity in the logical, linear presentation of a discovery in a scientific paper or book. It struggles for the very objectivity that the act of creation doesn't allow, but truth demands.

Blake mistakenly imagines that Newton, like Milton's Lucifer, would rather "reign in hell than serve in heaven" — heroic but flawed. He fails to realize the renowned scientist is already in paradise. The state of rapture that knows the Law of Universal Gravitation unites *all things* in a web of mutual attraction, and appreciates that the three Laws of Motion explain the behavior of *everything* in this web. This profound understanding gives Newton a sense of union with the real world — the very state mystics dream of attaining.

Newton had a public and a private face. Neither Blake nor the people of Newton's day glimpsed his secret side. Even now, Newton's passion for the black arts is relatively unknown. Incredibly, he practiced both alchemy and gematria (*see* the Chapter Notes). And his obsession with the Holy Scriptures was such that he wrote over one million words on them. Unfortunately, most of it is still unpublished, and in sheer quantity, it amounts to more than all his scientific papers combined. This was the private face of Newton. Many of these writings, which he guarded in a black trunk in his private rooms, were anti-Trinitarian pamphlets. In 1888 the council of the University Library at Cambridge gave away the mathematical portion of these papers. Fifty years later, Sotheby's auctioned off the biblical writings, mostly on Daniel and Revelation. These are now dispersed around the world, never to be published.

Newton was an extraordinarily wild character in his private life. His often quoted, "I do not make hypotheses," meaning he did not deal in metaphysical speculation, was a public mask covering his private, intellectual passions. One foot was in the Enlightenment, the other mired in the muck of the Dark Ages. The real Sir Isaac Newton bears little resemblance to the stony creature William Blake penned or painted. Things are seldom as they seem.

The beauty of the rainbow is as open to the scientist as it is to the artist. On the other hand, the scientist knows how the rainbow is formed, the reasons for its secondary arc, and the basis for Alexander's Dark Band. From his studies in optics, Newton knew — and we can now say felt — a great deal about the rainbow's colors. By uncovering some of nature's layers of meaning, surely everyone more deeply appreciates her beauty. This is the real pot of gold buried at the end of every rainbow. Too few artists dig in this realm or have the tools to do so. Knowledge is no impediment, ignorance no boon to enjoyment.

Life is a sailing ship on the great ocean of time. We sail hither and yon, not knowing our destination but hoping, in Tennyson's phrase, that ". . . we shall touch the Happy Isles." Lashed by life's troubles, becalmed by dull routines, yet like Ulysses in old age, we long for grand adventures. These twin perils of danger and dullness, Scylla and Charybdis, challenge us. *To move* on the sea of time, we must have passions to fill our sails — passions like Blake's. *And to avoid* the sea's rocks and shoals, we must have a knowing hand on life's rudder — a hand like Newton's. We need both Blake and Newton!

THE HEAT SEEKERS

The sun is but a morning star.
Henry David Thoreau, *Walden*

A FAMOUS ASTRONOMER was lecturing on the birth and death of the sun when he noticed a hand waving frantically at the back of the auditorium. "Yes, young man, what is your question?"

"When did you say the sun would explode and become a red giant destroying the earth?"

"In *10 billion years.*"

"Thank heaven! At first, I thought you said one billion," declared the young man sitting down in relief.

Anything connected with astronomy inevitably involves large numbers — astronomically large numbers. A billion (1,000,000,000) is a huge quantity; so huge we have *no direct experience* of its true size. Consider our smallest unit of time, the "second." How old must a person be to have lived a billion seconds? I ask the reader to estimate the answer — use *your intuition* before reading further.

. . . .

Surprisingly, it takes almost 32 years of seconds to equal just one billion; an unusual human might live three billion seconds. Numbers like this transcend our ability to know through experience. They're just mathematical symbols on paper. Even so, allow me to use a few very large numbers.

Reflect on the sun, our morning star. To early hunter-gatherers, the night was so large and full of such unimaginable terrors that they must have rejoiced when the sun rose. Life on earth, then and now, depends on innumerable things; the first of these is nuclear fusion in the sun's core. (Possibly the community of life supported by deep ocean vents is an exception.) In this solar inferno of heat and darkness, light photons are created that immediately begin colliding with free electrons, protons, and alpha particles. Combine these impacts with a vast solar volume and you have a core essentially opaque to all electromagnetic radiation. These two factors, incessant collisions and massive volume, can result in photons taking a million years just to reach the sun's surface. This immense journey renders — to paraphrase Milton's famous oxymoron — "darkness made visible" with the explosive shower of photons we call our sun.

Consider the sun's area and volume — both astronomical numbers.

$$\text{sun's area} = 6{,}087{,}000{,}000{,}000 \text{ km}^2$$
$$\text{sun's volume} = 1{,}412{,}000{,}000{,}000{,}000{,}000 \text{ km}^3$$

By writing these gargantuan numbers as a ratio, we can reduce them to lowest terms and so more easily comprehend their relative sizes.

$$\text{sun's area} / \text{sun's volume} = 1/232{,}000$$

So, for every square kilometer (or square mile) of surface area there are 232,000 cubic kilometers (or cubic miles) of volume. This provides the newly created photons with an immense playground in which to bounce around before ever finding the surface's tiny exit gate. Their unbelievably long childhood is the major contributor to the 10 billion-year stability of our sun. This is the reason behind the professor's answer to the young man at the back of the auditorium. And all of this is just because volume grows more quickly than surface area — *the square-cube law at work.*

Planet or Moon	Ratio of Area : Volume
Earth	1/2126
Venus	1/2017
Mars	1/1132
Mercury	1/813
Moon	1/579
Io	1/605

Figure 1.5

To retain internal heat, the optimal shape is spherical no matter what the size. This form achieves maximum volume for minimum surface area. Let's look at some spherical members of the sun's family of planets and moons to consider their surface-area-to-volume ratios (*see* table to the left). The minor irregularities (mountains, valleys, and such) on planets and moons increase their total area. But, if the earth were shrunk to the size of a billiard ball, it would be smoother than a billiard ball!

As well as an external heat engine the earth has an internal heat source powered by radioactive decay. In some manner this molten core makes the continental plates drift and cause volcanism at their edges. Among the other inner planets, only Venus has some form of plate tectonics with active volcanism. This is to be expected since our sister planet closely mimics the earth's size and hence has almost the same surface to volume ratio. Mercury, our moon, and Mars have cooled considerably because of their smaller size and as a result have neither plate tectonics nor active volcanism to help cover past meteoric bombardment. For that reason, they remain cratered and rigid — a pockmarked legacy of a wilder past, a great longevity, but a senile present.

The moon and the minor planets do have some heat in their central core, but not enough to cause any surface activity. Also, due to their smaller size and hence smaller mass, their reduced gravity can't hold any substantial atmosphere to provide an environment for the sun to act as an external heat engine. None has surface erosion; thus, we have a second reason for their unchanging appearance for three billion years. Even so, one exception exists in the sun's family — nature always seems more creative than our theories. Jupiter's innermost Galilean satellite, Io, has the greatest volcanism in the entire solar system even though it has only two percent of the earth's volume. How is this possible? The answer lies with the internal heat generated by the incredible and uneven tidal tug-of-war "massaging" that Jupiter gives Io.

Of all the planets, Mars is the one of myths, legends, movies, and dreams. The astronomer and writer Carl Sagan claimed his imagination and interest in planetary travel were stimulated by science fiction stories about the Red Planet. One fine day his dream will be fulfilled, and a rocket will launch on humanity's most grand adventure — the first manned mission to Mars.

Again from his book *Ever Since Darwin,* Stephen Jay Gould presents an intriguing hypothesis about Mars. Since the Red Planet is intermediate in size between the earth and the moon, it should show evidence of past surface erosion and plate tectonics. Gould writes:

> About half the Martian surface is cratered; the rest reflects the activity of rather limited internal and external heat machines. Martian gravity is weak compared to that of the earth, but it is strong enough to hold a slight atmosphere (about 200 times thinner than ours). High winds course over the Martian surface and dune fields have been observed. The evidence for fluvial erosion is even more impressive. . . .
>
> Evidence for internal heat is also abundant (and rather spectacular), while some recent speculation plausibly links it with the processes that move the earth's plates. Mars has a volcanic province with giant mountains surpassing anything on earth. Olympus Mons has a base 500 km wide, a height of 8 km [should be 25 km] and a crater 70 km in diameter. The nearby Vallis Marineris dwarfs any canyon on earth: it is 120 km wide, 6 km deep and more than 5,000 km long.[3]

Now for Gould's speculation. With some evidence, geologists believe that rising plumes of molten material from the core-mantle boundary cause the earth's plates to move. Occasionally these plumes pierce the earth's surface at fixed points over which the plates ride. For example, the Hawaiian Island chain results from the Pacific plate moving over a fixed plume. Gould applies these thoughts to the Red Planet. Due to its intermediate size, Mars should be (or has been) more dynamic than the moon, but less so than the earth. Compared to Mars, the earth's crust is thin enough to break into plates and be moved by the core plumes. If these exist on Mars, then Mons Olympus may mark the site of a fixed plume, and the Vallis Marineris the edge of a monstrous plate — both fixed forever because of the thicker Martian crust. And this deep crust results from Mars' smaller size and relatively greater cooling surface area. Once more, we encounter the vast influence of our old friend the *square-cube law.*

SPHERICAL FORM is as good as it gets for heat retention. Any variation from this form increases the surface area. Flatten, elongate, pinch, wrinkle, branch, hollow out, grow appendages, sprout hairs, develop a tail — all of these and more have the same effect: heat loss increases. Of course, the deformation may be a trade-off, perhaps a need to lose heat. To cool our potatoes at mealtime, we mash or cut them thereby enlarging their surface area.

At the opposite extreme, consider an automobile radiator or an electric baseboard heater. They consist mostly of heat radiating surfaces; that's their function, and they do it very well. All creatures great and small have a position somewhere on this heat continuum, each according to its needs. *Homo sapiens* is also found here, more to the spherical end, but surprisingly in at least two locations.

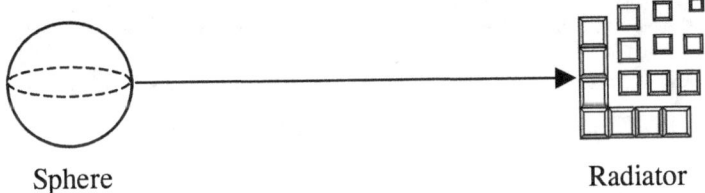

Sphere   Radiator

THE HEAT-LOSS CONTINUUM

Figure 1.6

The Ruandans of the Congo exemplify adaptation to hot climates. Their body type produces the largest skin area for humans relative to total body volume. Therefore, surface radiation is maximal and heat is most effectively dissipated. On the other hand, the short-limbed Inuit (Eskimos), with large, heavy torsos, exemplify bodily adaptation to arctic climates. Their shape produces the smallest skin area for humans relative to total volume. Surface radiation is minimal and heat is most effectively preserved. On the heat continuum above, the Inuit are to the left of the Ruandans. Both handsome racial types are products of natural selection (*see* Figure 1.7). The direction of Inuit evolution has happened before in hominid history. To exist during the last ice-age environment of Europe, skeletal remains of Neanderthals indicate they evolved a body type similar to, but millennia before, the Inuit — an example of convergent evolution.

20 / The Measure of All Things

LONG-LIMBED RUANDANS   SHORT-LIMBED INUIT

Figure 1.7

We have all heard that very small animals such as shrews have to eat more than their own body weight each day just to survive. A still smaller mammal is a practical and metabolic impossibility — after all, how much food could it catch or digest in any 24-hour period? Richard F. Burton in his thoughtful book *Biology by numbers*[4] presents clear evidence for varying metabolic rates among animals of different sizes.

Consider a mouse of 2 oz (60 g) and a cow of 1,300 lb (590 kg), each with a body temperature of approximately 98.6°F (37°C). Now postulate that *they have the same metabolic rate per unit body of mass*. By reductio ad absurdum, we will prove this statement false, and so its opposite must be true. Without sacrificing generality, we presume heat is lost to the environment at a rate that's proportional to the surface area. Furthermore, because the two animals in question have a similar shape, everything else follows by the *square-cube law*. (The reader may skip, without loss of continuity, the next seven lines and go right to the conclusion that follows it.):

The ratio of the masses is 1:10,000,
and so the ratio of their volumes is also 1:10,000.
Hence the ratio of their lengths is their cube roots,
namely, 1:21.5 or approximately 1:20 or $\frac{1}{20}:1$.
Reasoning from this, we see that for the mouse
the ratio of area to volume (mass) is $\left(\frac{1}{20}\right)^2:\left(\frac{1}{20}\right)^3$
which simplifies to 20:1.

Given this result and the postulate, the heat flow through a specific area of the rodent's surface should accordingly be only 1/20 of that same area of the cow's. It follows that the insulation would have to be 20 times more effective in the mouse to maintain its core temperature of 98.6°F (37°C). This suggests a fur thickness of at least 9 in. (23 cm), which would make walking rather comical. As a result, we have reasoned ourselves into a farce, so the *equal metabolic rates per unit body of mass* postulate must be false. The only logical conclusion is that mice — and more so for the smaller shrews — have significantly higher metabolic rates than cows.

If we don't live in the best of all possible worlds, we surely live in a fortunate one. Suppose large animals such as antelopes and moose had metabolic rates three or four times higher than they actually do. They would require three or four times as much pasture — a limited quantity — and so there would be considerably fewer of them. At the next higher nutritional level, the primary carnivores — leopards and wolves, say — also needing three or four times as much food, would therefore be exceedingly scarce. With the increasing rarity of wildlife at each higher level of the food pyramid, the entire natural world would be profoundly impoverished!

FOR HEAT RETENTION, insects have the worst possible shape and size. They're small and their bodies have multiple heat-radiating parts: three body segments, six long legs, two antennae, and often heat-dissipating wings. From this it's easy to infer that they must all exist at the temperature of their surroundings. But you would be mistaken! These creatures have been here since the Devonian Period of the Paleozoic Era — 350 million years ago. Moreover, for every human generation hundreds or thousands of insect generations pass away.

This bounty of time and progeny allowed for, as Darwin would say, "descent with modification." Natural selection's big gift to insects is the ability to *thermoregulate their bodies*.

Mammals range in size from shrew to elephant. Insects range from fairyfly to goliath beetle: a 500-thousandfold mass variation. The bumblebee is a colossus among midges. Yet this same bumblebee, flying from the sunshine into the shade, can cool at 1.8°F (1°C) per second, while a person cools at least 100 times more slowly. Heat loss is extremely rapid in small bodies because every portion of their mass lies near the surface — a consequence of the *square-cube law*. Conversely, heat gain is equally rapid.

Not all insects thermoregulate, but those that do use a variety of methods, chiefly the metabolic production of heat through thoracic muscle contraction. For example, a bumblebee at 41°F (5°C) must increase its metabolic rate 1,500 times in order to fly. But the higher metabolic rate is possible only after it first raises its thoracic temperature to almost 104°F (40°C). Every insect has its specific temperature requirements: some die from the heat of a human hand while others forage in the midday sun of Death Valley.

Since insects have so much difficulty retaining heat because of their small size and multiple parts, why have none evolved to human stature — the concocted giantism we have seen in so many Hollywood horror films? At various times, movie producers have given us massive ants, lumbering killer bees, and man-eating flies destroying people and devastating property. While enjoying these imaginative works, have you ever wondered why *real* insects are always so small? Why, thankfully, are there no enormous cockroaches?

This question has several answers, but a single powerful one will suffice. Julian Huxley in his *The Uniqueness of Man* reasoned that insects have been cut off from further "progress" by their breathing apparatus:

> The land arthropods have adopted the method of air-tubes or tracheae, branching to microscopic size and conveying gases directly to and from the tissues, instead of using a dual mechanism of lungs and a bloodstream. The laws of gaseous diffusion are such that respiration by tracheae is extremely efficient for very small animals, but becomes rapidly less efficient with increasing size [the *square-cube law*], until it ceases to be of use at a bulk below that of a mouse. It is for this reason that no insect has ever become, by vertebrate standards, even moderately large.

Insect "blood" doesn't carry oxygen, and, as Huxley wrote, insects don't have lungs. Instead, they breathe through spiracles, openings in their thorax and abdomen, which branch into tracheae allowing oxygen *in* and carbon dioxide *out*. And much of this is passively accomplished. However, as size increases, *the greatly reduced surface to volume ratio* complicates and eventually prohibits this method of breathing. With the previously mentioned bounty of time and progeny, insects have climbed too far up their individual evolutionary mountain to ever descend to the valley floor and then scale a different peak to be of human size.

LIFE ON EARTH requires heat regulation, either attaining or dispersing it. Humans use vast amounts of electricity and other forms of energy to heat their homes in winter and cool them in summer. Other life forms duplicate this in other ways. In the coldest place on earth — the Antarctic — at the coldest time of year, the emperor penguins breed. In an extraordinary feat of endurance and dedication the male cradles a single egg on his feet, covering it with a warm fold of abdominal skin. During this dark incubation vigil, he eats nothing for at least two months and loses half his body weight. Temperatures fall to an incredible –76°F (–60°C) and winds can rise to 112 mph (180 kph). We're left to wonder how he can survive in this unbelievably hostile environment and, most of all, incubate an egg. Truly, these penguins must have astonishing adaptations! What are they? The three most prominent are *size, shape,* and *behavior*. At 100 lb (45 kg) maximum, they're the largest of all penguins and the one with the most spherical shape — both adaptations lower the surface-to-mass ratio. But perhaps their most ingenious strategy is "huddling." The males gather in large rookeries of up to 1,000 birds forming a densely packed huddle of ten per square meter that slowly rotates so penguins on the outside can work their way into the middle and vice versa. This adaptation raises the "huddle temperature" 68°F (20°C) above the ambient and conserves precious body reserves. Huddling reduces the total surface area and the penguins form one, large, living organism.

Mice, squirrels, snakes, and numerous other creatures also huddle for warmth; humans do as well. In a cold bed, we draw up our legs into a fetal position and pull our arms close to our bodies in an effort to reduce our surface area by becoming more spherical, so to speak. Dogs curl up like a ball, geese, swans, and other birds tuck their heads under their wings — examples are all around us.

The following graph organizes many of the previous examples and generalizes Figure 1.6. The width of each shaded rectangle displays the variety of physical shapes within a class, for example, water buffalo to giraffe. And the height of each shows the range of actual masses, for example, shrews to whales. It's not drawn to scale.

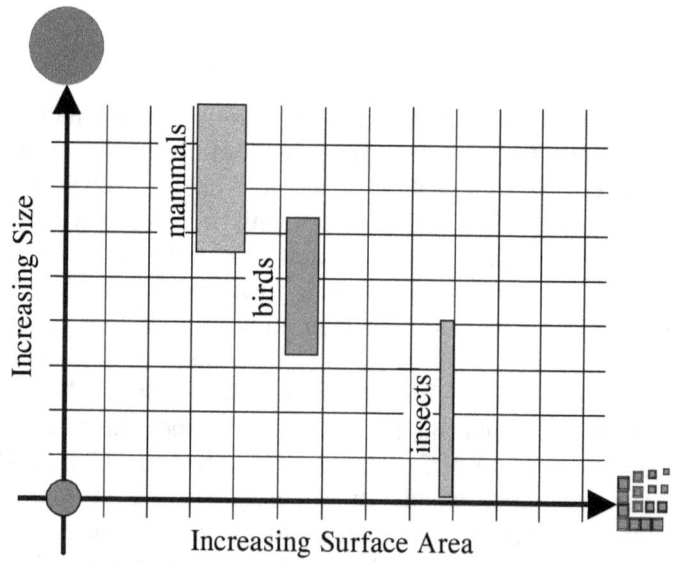

SHAPE, SIZE, AND HEAT RETENTION

Figure 1.8

We have seen that body *size* and *shape* are significant factors in how efficiently an animal responds to hot or cold climates. Two 19th century naturalists, Carl Bergmann and Joel Allen, formulated rules concerning these factors. In zoology, Bergmann's rule is a principle relating external temperature to the ratio of body surface to mass in warm-blooded animals. Birds and mammals in cold regions tend to be larger than individuals of the same genus or species in warmer regions. Bergmann proposed this principle to account for size as an adaptive mechanism to conserve or to radiate body heat, depending on climate. His principle is in fact an observation of the *square-cube law* applied to animals.

Allen's rule is a corollary to Bergmann's, stating that warm-blooded animals in cold climates have shorter protruding fingers, noses, tails, ears, and legs than another race of the same species in a warmer environment. Consider the short, stubby digits of the Inuit compared to the long, slender fingers of the Ruandans. Both these rules are combined in the graph on the previous page: Bergmann's on the vertical axis, Allen's on the horizontal. For example, as you move from right to left and simultaneously up, you *should find* bigger birds and mammals of colder climates.

In the hinterland of Ontario lives a tiny bird — a very tiny bird, the golden-crowned kinglet. The common black-capped chickadee of the same region weighs about 1/3 of an oz (9.4 gm), roughly double the mass of the kinglet at 1/5 of an oz (5 gm): the same as two pennies. When plucked, it's no bigger than the end of your finger. Heat flows from its body a hundred times greater than it does for humans. And kinglets are permanent residents of these forests where winter temperatures often fall to $-40°F$ ($-40°C$). For them to live in this ruthless cold is astounding!

Curiously, this reminds me of mathematician Ian Malcolm's comment in *Jurassic Park* after the park's creator John Hammond had declared his dinosaurs were genetically engineered to be utterly incapable of breeding. Malcolm countered with equal certainty that "Nature will find a way." And she often does. With the golden-crowned kinglet, she surely has — this bird somehow flourishes where it should perish. Its closest relative, the somewhat larger ruby-crowned kinglet, always leaves these regions for the winter months. The natural world is full of wonders and exceptions to rules like Bergmann's.

THE TREE WARRIORS

*Comparative anatomy is largely
the story of the struggle to increase
surface area in proportion to volume.*
J.B.S. Haldane (1892–1964)

URBAN LEGENDS are a fixture of modern life — perhaps they have always been part of human baggage. Often they're too nebulous and imprecise to positively discount, and they have a certain plausibility and charm. One such concerns the

18th century Canadian voyageurs, native and French explorers and fur traders. As the story goes, a band of these voyageurs westward bound from Montreal to Manitoba camped one evening on an island to escape annoying black flies and mosquitoes. The cook prepared a supper of warm beans and bannock, after which he scrubbed the pan with sand at the water's edge and hung it to dry by its large iron ring over a low tree branch. In his haste to depart the following morning, he forgot the pan, and the voyageurs paddled away into the tranquil, dark forest. Over 200 years passed before other canoeists happened to camp on the same island for the same reason. During their evening meal, one of them glanced upward and noticed the pan still hanging by its iron ring — now 50 ft (15 m) above the ground. The intervening centuries of tree growth had carried it skyward during its long vigil waiting for the cook to return. Or so the story goes.

This is an intriguing tale, quite believable on a first reading. But every rural resident knows this is an *urban* myth because trees don't grow from the bottom up. When a farmer nails his fence wire to a convenient tree, the passing years never move it upward. Rather the trunk grows horizontally outward increasing its girth and gradually enveloping the wire.

With humans and other animals, growth can occur in most body parts, and generally at different rates. As we mature, our bones, skin, and muscle all increase in size. On the other hand, trees don't develop like this. Instead, they grow by producing new cells, called meristems, but in only three places: the branch tips, the root tips, and the tree trunk. End of story!

Now that we understand *where* tree growth occurs, let's see *how* it happens. Consider the next two photographs (Figure 1.9): the one on the left is a 4 in. (10 cm) maple sapling, the one on the right, a 100 ft (30.5 m) mature maple (300 times taller). Assume trees grow by simple enlargement. That is, if you double their height, everything else is enlarged equally. Continued doubling would quickly make the individual leaves on the sapling larger than the ears on an elephant — so the assumption must be absurd.

Nonetheless, let's continue with the above assumption, and recall Figure 1.1, The Square-Cube Law Table. It applies to *all shapes,* including trees, be they very simple or vastly complex: double the height and you increase the area fourfold and the volume eightfold. If the height is multiplied by two, then the ratio of a : v (= area : volume) changes to a : 2v. If the height

SUGAR MAPLE SAPLING IN SPRING   MATURE SUGAR MAPLE IN AUTUMN

Figure 1.9

is multiplied by three, the ratio changes to a : 3v. The mature maple is 300 times taller than the sapling. Therefore if the ratio of area to volume in the sapling is a : v, in the mature maple this ratio becomes a : 300v — a phenomenal preponderance of volume over surface area. This extreme outcome is so bizarre that evolution would not allow it because trees could not feed themselves, nor could they "breathe."

From birth to death, each sapling's initial ratio has to be preserved. Every *living* tree, in every forest, on every continent, can be seen as a successful battle on the tree's part to maintain the high surface-area-to-volume ratio of the sapling. The vast system of branches on the mature maple is a testament to this truth. This section's opening statement by J.B.S. Haldane applies with equal power in the plant kingdom as in the animal. Trees are the true leaf warriors!

In the immense sprawling woodlands of North America, there exists a periodic parasite called the forest tent caterpillar. At seven to eleven year intervals, enormous population explosions of these leaf eaters defoliate entire districts. The caterpillars are everywhere, on homes, roads, railway tracks, cars, as well as trees. Black-billed cuckoos and other birds gorge themselves and their babies on them. Fortunately, these insects have only one life cycle per year. Later in the summer, stronger trees will grow a second leaf canopy; weaker trees die.

If during the following spring a second large infestation arises, many more trees will perish. Changing the ratio of surface area to volume in the growing season can be a fatal plague on trees.

In these same regions of North America, all the broadleaf trees annually lose their leaves. The winter's extreme cold not only threatens to freeze the liquid within the trees' cells, but also denies them one of their essential supplies — water. Although it lies all around them as snow and ice, in this solid state it's beyond their reach. So the trees of these great woodlands endure a drought as extreme as if they lived in Death Valley. All leaves, including pine needles, have small holes called stomata through which they exchange gases with the atmosphere in a process called transpiration (basically CO_2 in, O_2 out, plus some water vapor). In the desert-like conditions of a northern winter, losing water from the stomata can be lethal. Accordingly, trees dramatically reduce their surface area by shedding all their foliage. Even the larch drops its pine-like needles, and most conifers shed one third of theirs.

Now that the tree warriors have cast aside their shields, they can ease their assault, all their yearly battles to maintain the ratio are past, and *the square-cube law — the measure of all things* — can be forgotten. And so they rest. Weeks after October's leaves have blanketed the forest bed with a warm multi-colored quilt, the trees pull on a thick jacket of snow. And then they sleep.

Chapter — 2

THE PERFECT FIVE

*The chief reason for studying regular polyhedra
is still the same as in the time of the Pythagoreans,
namely, that their symmetrical shapes appeal
to one's artistic sense.*
H.S.M. Coxeter (1907–2003)

IN THE TWO-DIMENSIONAL WORLD of flatland, a figure may have three, four, or as many sides as you wish. When all the sides and angles are equal, we call it a *regular* polygon; there are an infinite number of such figures.

These are familiar objects in everyday life. From ancient times, peoples of many cultures have tessellated their floors and walkways with squares, triangles, and hexagons because they leave no gaps. The United States military built their headquarters in the shape of a regular pentagon, and the major American Baha'i temple is a nine-sided polygon. (This enneagon has also become the central icon in a New Age fad concerning personality types.) If you *fix* a regular (equilateral) triangle's perimeter, and then increase the number of sides to infinity — thereby forming a circle — you enlarge its area by 65 percent. Even deforming it to a square produces a 30 percent increase in area. So less complex peoples — who have traditionally built their huts, teepees, and igloos with a circular base — were practicing excellent mathematics: they maximized their floor area for a limited amount of wall material.

In our three-dimensional world, the analogue of the regular polygon is the regular polyhedron. These possess the following properties:

- Each face is a regular polygon.
- All polygons are the same size.
- All polygons around each vertex are identical.

We might assume that, like the polygons, these objects exist in infinite number. But in fact only five are possible — see the next two pages. As Lewis Carroll once remarked, they are "provokingly few in number."

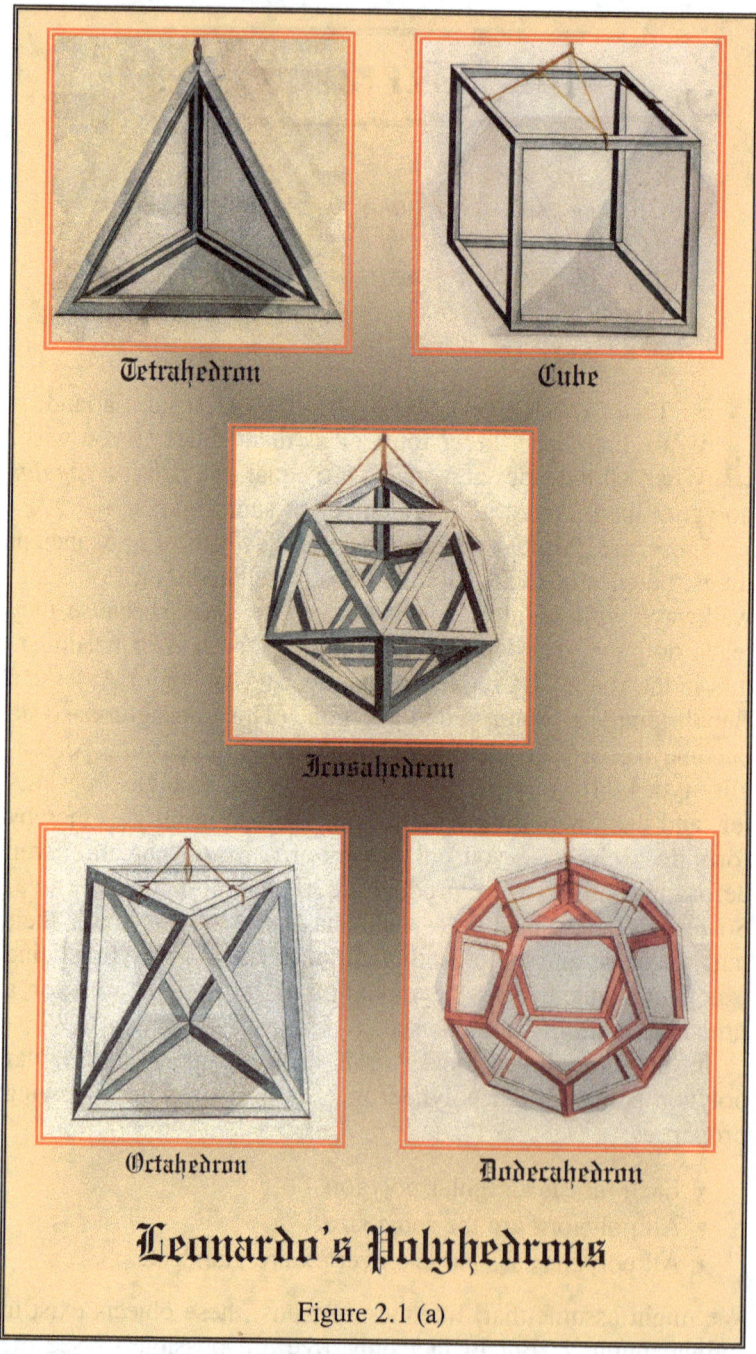

Figure 2.1 (a)

Da Vinci built exact models of the five perfect shapes, and from these he painted the impressive figures shown above for Luca Pacioli's book *De Divina Proportione* published in 1509.

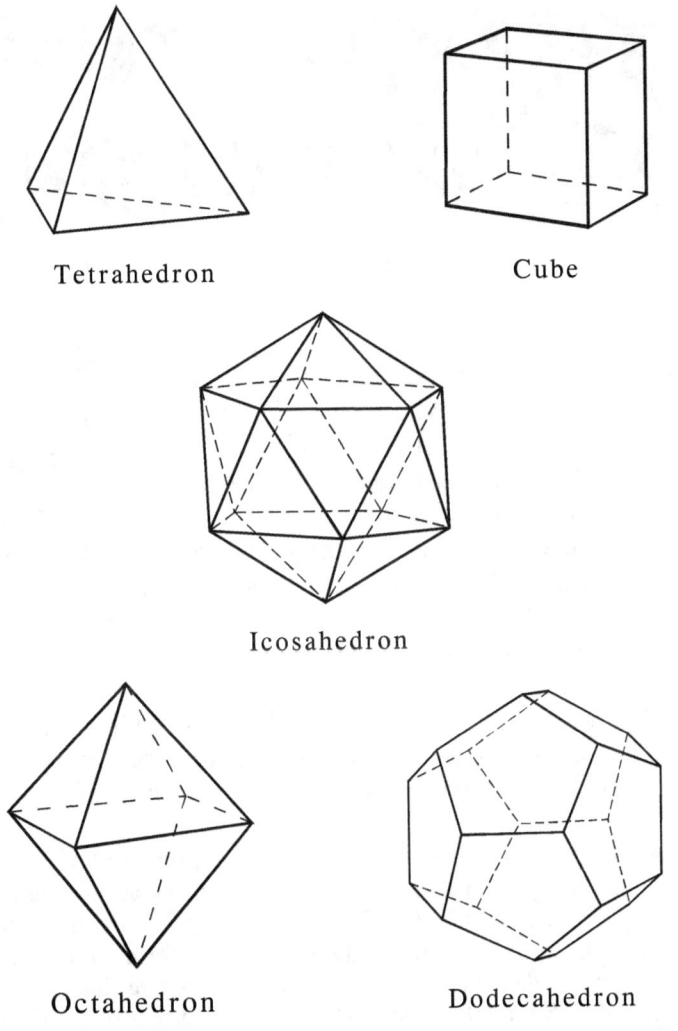

THE REGULAR POLYHEDRONS — TEXTBOOK STYLE

Figure 2.1 (b)

Who first discovered the regular solids languishes in the darkness of prehistory; you might as well ask who invented fire. The five ancient objects on the following page, from a collection of Scottish Neolithic stones in the Ashmolean Museum at Oxford, England, unmistakably show the basic patterns for a cube, tetrahedron, dodecahedron, icosahedron, and octahedron. These artifacts are more than 3,000 years old.

SCOTTISH NEOLITHIC POLYHEDRAL STONES

Excavators on Monte Loffa near Padua, Italy, found the Etruscan dodecahedron shown at the right. Apparently, this object was enjoyed as a toy at least 2,500 years ago. Gottfried Wilhelm Leibniz observed, "Men are never more ingenious than in inventing games."

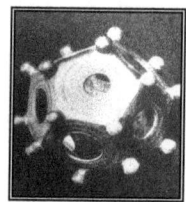

About the same time, the early Pythagoreans were investigating possibly three of the regular solids: tetrahedron, cube, and dodecahedron. And a remark in the first scholium of Book XIII of Euclid's *Elements* says that Theaetetus (417–369 BC) discovered the other two.

Theaetetus wrote a book, now lost, on the regular solids; historians think he was the first person to prove that, at most, only five can exist. We believe his proof went something like this:

> A corner of a polyhedron must have at least three faces — two would collapse on themselves. And the angles at any corner *must total less than 360°*. Consider the simplest face: an equilateral triangle with each angle 60°. We can form a corner by putting together three (180°), four (240°), or five (300°) such triangles. Six (360°) would lie flat and produce a tiling of the plane. So we have three possible ways to construct a regular solid with triangular faces: tetrahedron, octahedron, and icosahedron. Three and only three squares will similarly form a corner: four would again lie flat. This corner gives a possible cube. The same reasoning yields one possibility with three pentagons (interior angles are each 108°): the dodecahedron. It is impossible to go beyond the pentagon, because if we put three hexagons (interior angles each 120°) together, the shape again lies flat and we have a plane figure. This argument doesn't prove that the five regular solids can be constructed, but it clearly shows that no more than five are possible. And as the classical geometry textbooks would say, Q.E.D.

In his dialogue *Timaeus,* Plato (about 427–347 BC) gave the earliest description of all five regular polyhedrons with directions on how to construct them. This was the only Platonic dialogue studied or known in the West during the Middle Ages. Its main character, Timaeus of Locri, mystically associates each of the four primal elements of earlier Greek philosophers with the polyhedrons:

fire ⟶	tetrahedron
earth ⟶	cube
air ⟶	octahedron
water ⟶	icosahedron

And the 12-sided dodecahedron? Very conveniently, it related to the entire universe because the zodiac has 12 signs.

Consider this ancient association. Since these four elements were the fundamental building blocks of the material world, this relationship gave an exalted position to the regular solids. Curiously, this theory tells us much more about the ancient Greeks than about the physical universe.

One hundred years after Plato these figures continued to excite the best of the Greek intellectuals. The analysis of the Platonic solids, as they came to be called, forms the climatic final book of Euclid's *Elements.* And their beauty and symmetry haunted scholars from the days of Pythagoras through to the Renaissance.

THE CENTRAL INDIVIDUAL in the celebrated painting by Jacopo de Barbari shown on the following page is the Franciscan friar and mathematician Luca Pacioli. The robed friar stands behind a table filled with geometrical objects (slate, chalk, compass, and dodecahedron) while illustrating a theorem from Euclid, but his gaze is rigidly fixed on a striking glass rhombicuboctahedron half-filled with water. The identity of the other human, whose eyes are on the viewer, remains a mystery. Some art historians believe it's a self-portrait of a youthful Pacioli or his student, while others suggest it's the great German artist Albrecht Dürer.

Fra Luca Pacioli by Jacopo de Barbari,
1495, Museo di Capodimonte, Napoli

Figure 2.2

Every part of this painting, however small, has been brilliantly composed. The glass polyhedron is a stunning masterpiece of reflection, refraction, perspective, and a grand symbol of humanity's quest for perfection. This is the object of Pacioli's rapture. The entire painting epitomizes the deep Renaissance connection between art and mathematics — two paths to perfection.

An actual glass polyhedron, constructed by Pacioli, was used in the painting. The choice of this model was no accident: the friar himself discovered this form, so he was extremely proud of it. However, Archimedes almost certainly found it first, though this wasn't known during the Renaissance.

If perfect polygons of more than one type are joined, while keeping all corners identical, you can form the semi-regular solids. Exactly 13 exist, and by around 200 BC, Archimedes had discovered all of them. The figure on the next page is one of the most famous, and it's also the glass polyhedron of Barbari's painting.

Leonardo da Vinci sketched the first illustration of the rhombicuboctahedron (*see* figure below), also for Pacioli's book *De Divina Proportione*. As a matter of record, he drew all the magnificent pictures for this book from wooden models he or his students constructed. The friar was his special friend.

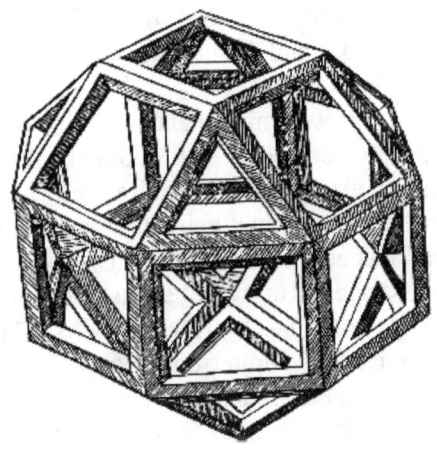

RHOMBICUBOCTAHEDRON BY LEONARDO

Figure 2.3

When the French army invaded Milan in 1499, Leonardo, fearing for his life, fled at night on horseback. After a decade of brilliant service, some of it military, to Ludovico Sforza, the prince of Milan, da Vinci took very few things with him, and only one person — Fra Luca Pacioli.

JOHANNES KEPLER

During the Renaissance, interest in regular and semi-regular figures was not confined exclusively to the world of art. After nearly 2,000 years, Timaeus' idea of element-polyhedral correspondence was picked up and expanded by the legendary Johannes Kepler (1571–1630), master astronomer/astrologer, and mathematician/numerologist.

His mother was accused of witchcraft; his aunt was burned at the stake. Kepler was a man from two worlds: one part sat securely in the superstitious dugouts of the Middle Ages while the other walked boldly onto the field of science and reason.

Kepler's accomplishments — considering his wretched youth and tortured life — are deeply puzzling. He defies all childhood development theories and all possible adult expectations. He transcended his social environment and showed that humans can be more than the sum of their parts. He came out of nowhere and stepped right up to take his place with Aristotle, Archimedes, and Galileo as one of the heroes of history. Eventually he would stand beside Newton himself in the pantheon of giants.

Only six of our present nine planets were known in Kepler's time. Fascinated by astronomy and geometry, he invoked both disciplines for an extraordinary inspiration. He reasoned that Saturn, the furthest planet then known, moved on the equator of a sphere — this has echoes of Pythagoras — with the sun at its center. He then supposed that a cube was constructed inside this gigantic sphere (*see* Figure 2.4 (b)). And inside this cube was another sphere on which Jupiter coursed the heavens. Jupiter's sphere in turn enclosed an inscribed tetrahedron, and in this Platonic solid, he inscribed a third sphere on which Mars moved. By using all the regular solids, he had room for six spheres — the exact number of the then known planets. Moreover, the ratios of these spheres' inscribed radii were the same as the ratios of the planets' distances from the sun. To Johannes Kepler, this correspondence and these ratios had a transcendent significance. His system follows:

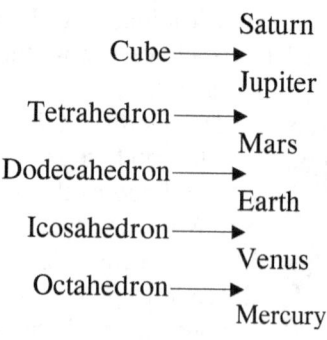

Figure 2.4 (a)

This idea must appear bizarre to any modern reader. But let's for the moment take it seriously and see what it was really all about. First, this theory is solidly Copernican. It makes no sense without the sun at the center. At a time when heliocentrism was almost universally rejected, Kepler made it the heart of his system. Second, it made clear why there were six planets and only six. Third, it replaced the ancient crystal spheres of the plenum with the regular solids. Fourth, it explained the ratio of the orbits of the six known planets. And last, this strange idea was incredibly fertile: it motivated Kepler to discover his three laws of planetary motion. Even more important, it laid a large part of the foundation on which Sir Isaac Newton built his celebrated world system.

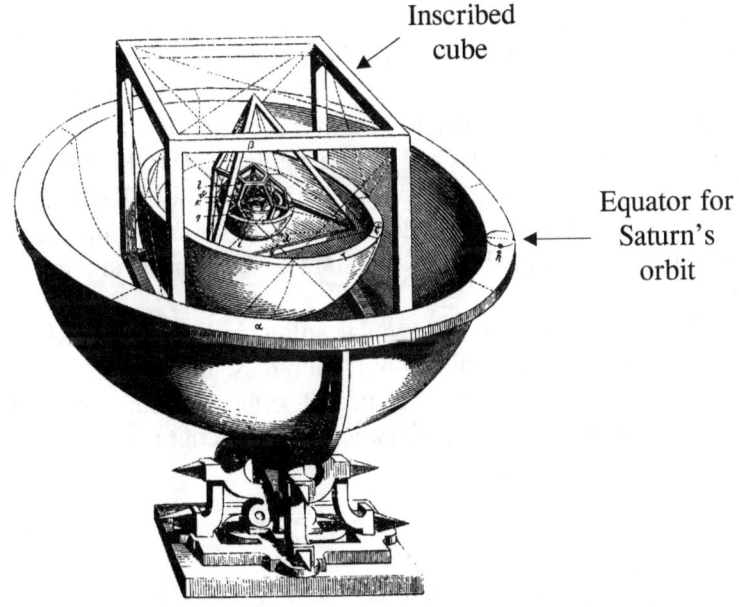

KEPLER'S SYSTEM FOR THE HEAVENS

Figure 2.4 (b)

Newton once said that if he saw further than others it was because he stood on the backs of giants. Kepler was one of those giants. By any reasonable standard, his bizarre idea has to rank as one of the most significant and fruitful scientific theories of all time.

THIS SECTION has been a shortened chronology of five remarkable objects. It began in Neolithic caves and tents — perhaps all across Eurasia — with small stone toys; Etruscan children played with them; Pythagoras analyzed them;

Theaetetus counted them. Greek philosophers wove them into their theories of the material world while Plato braided them into his famed dialogues. Euclid exalted them in the final book of his *Elements*. Throughout the Middle Ages, peasants, poets, and princes gambled with them. During the Renaissance, scholars and artists were haunted by their properties and perfection. Leonardo made superlative drawings and models of them. And in an amazing apotheosis — the very apex of their importance — Kepler placed them in the heavens among the stars.

In the 1990s German chemists discovered an astonishing molecule formed entirely from the metallic element molybdenum. Under proper conditions, 132 molybdenum atoms will form a sphere. And inscribed within this sphere is a perfect icosahedron of 12 corners and 20 triangular faces. In honor of Johannes Kepler, they called their molecular ball "Keplerate." The great astronomer would be pleased to know his grand vision had descended from the stars to be reborn among the atoms.

Does the history of the regular solids end here with Kepler and his spheres? Definitely not! But they lose their favored position as the pole star of science. The next section will show how post-Renaissance mathematicians have continued to study them, but in new directions.

THE LADDER OF DIMENSIONS

PONDER THE FOLLOWING PUZZLE. Take a sheet of white paper and a black felt pen; place your *left hand* flat on the paper and outline your fingers and thumb. Now for the question, is this outline a *left hand* or a *right hand?* Oh, I know I asked you to outline your *left hand*. But turn the paper over and notice that the black felt marker has bled through the paper to produce a *right hand*. So I repeat the question, is the outline on the paper a *left hand* or a *right hand?*

Someone might suggest we introduce a handless figure into flatland — with the heart on the left side. Doesn't the original outlined hand fit perfectly on the left arm? And so, we have the answer to our puzzle. Or do we?

The core of the conundrum lies in the confusion surrounding handedness in different dimensions. From three dimensions, the outlined hand is both left and right or neither if you wish. It all depends on your point of view. A left hand outlined on a frosty winter window appears as a right hand when viewed from inside the home. So, the handless figure can be introduced into flatland in *two ways*; in each way the hand will fit a different arm. We extra-dimensional beings determine which arm. Like gods, the decision is ours.

Take a step up the infinite ladder of dimensions, and consider a well-known thought experiment of Immanuel Kant[1]. Imagine the universe is completely empty except for a single naked human hand floating through space. Again the question, is it a left or right hand?

There are no *measurable* inherent differences between left and right hands, so a quantitative approach is useless in answering the question. And you can't visualize yourself looking at the hand and so seeing it as right or left. Why? That would be equivalent to putting yourself in three-dimensional space with your intrinsic sense of handedness. Recall, you were to imagine the hand in a completely empty universe. It seems apparent that we're unable to determine whether the isolated hand is right or left. Nor can we say whether the hand is large or small. These are all *relational* judgments, but no relationships exist here. Or do they?

Now repeat what we did to our outlined hand in two-space: introduce a handless human body near the solitary hand in three-space. The single hand must fit one of the arms, and hence its handedness will be determined. As Martin Gardner asks, in his celebrated book *The Ambidextrous Universe*[2], "Do you see the paradox confronting us?" If the solitary hand fits on the left arm, then it must have been a left hand before the handless body appeared. By this line of thought, Kant concluded there had to be some standard, some basis, or some way for calling it a left hand *before* the body materialized.

From this the philosopher falsely concluded he had solved this paradox by declaring that space itself has a structure. By this, Kant meant that this structure of space would allow us to determine if the solitary hand were left or right — all of this before the handless body was introduced. But the real solution lies elsewhere and is analogous to what we did in two-space.

Recall that in flatland we turned the paper over to change its relationship to the outlined hand. "Turning over" is equivalent to reflecting in a mirror. Suppose it's a snowy winter morning, and some mischievous child has written "clean me" on your car's back window. What do you see while sitting in the driver's seat looking in your rear-view mirror? Interestingly, you again see exactly "clean me." However, still viewing it from inside the car, but *only* through the back window, it's flipped to "ɘm nɒɘlɔ". Now, simply by looking at this flip in the rear-view mirror, you see its reflection. Any *even* combination of flips or reflections or both leave the original message unchanged. Here we had one horizontal flip plus one reflection; so, the writing is unaffected.

Can you have reflections in higher dimensions? That is, objects identical in all respects but their handedness. Yes, you can, and this duality exists in every n-space. Let's take a few steps from ground zero up the ladder of dimensions with analogy as our guide. One-dimensional line segments are mirrored by points; two-dimensional figures by lines; three-dimensional solids by planes; four-dimensional hyper-objects by solids, and on it goes. In a space of n dimensions, the "mirror" is an object of n−1 dimensions. Or alternately we can say any dual object of n dimensions will coincide with its reflection by "flipping" it through a space of n+1 dimensions.

Armed with this knowledge, let's return from our dimensional detour. Because of this reasoning, the solution to Kant's conundrum lies in which dual form the handless figure is introduced. Since we can view flat objects in both forms, analogously, a four-dimensional being would be able to view solid objects in both forms. On one of these, the solitary hand fits the arm closest to the heart; on the other, the arm furthest from the heart — just as with the outlined hand in two-space.

Perhaps we shouldn't be too hard on Kant for his faulty analysis. After all, n-dimensional geometry was unknown in his lifetime. Not until 1827 did August Ferdinand Möebius realize that rotating an object through the next higher dimension reverses its handedness. Martin Gardner[3] says an imaginary 20th-century Kant might put it this way: "Only the pure understanding of God Himself, who stands outside space and time, would see all pairs of enantiomorphic [reflected] structures, in all spaces, as identical and superposable."

EUCLIDEAN GEOMETRY can be generalized to every spatial dimension. This extension shouldn't be confused with Einstein's

4th dimension of time. With Euclid all dimensions are spatial and are found in the following fashion. Let a zero-dimensional point split into two points and move one foot apart. Move this one-dimensional line segment at *right angles* to itself forming a two-dimensional square. After this, move the square *perpendicular* to itself to produce a cube. The next step up the dimensional ladder cannot be visualized. Nonetheless, mathematicians ascended; so will we. By moving the cube at *right angles* to itself, we produce a four-dimensional cube (also called hypercube or tesseract). Long ago all the geometrical properties of hypercubes and beyond were worked out. Each rung of the ladder has a valid Euclidean geometry — just as valid as the plane geometry we learned in high school. All these points, lines, squares, . . . on the polytopes of the cube, are summarized in the following table.

	No. of Points	No. of Edges	No. of Faces	No. of Cubes	No. of Hyper-cubes	No. of 5-D Cubes	. . .
Point	1	0	0	0	0	0	0
Line	2	1	0	0	0	0	0
Square	4	4	1	0	0	0	0
Cube	8	12	6	1	0	0	0
Hyper-cube	16	32	24	**8**	1	0	0
5-D Cube	32	80	80	40	10	1	0
⋮	⋮	⋮	⋮	⋮	⋮	⋮	⋮

POLYTOPES OF THE CUBE

Figure 2.5

To find a specific table entry is surprisingly easy. Consider the number "8" (in bold) at the intersection of the hypercube row and the cubes column. How was it found? Simply take the "1" directly above, double it, and to this add the "6" to its left. Their sum is "8" — the correct entry. To verify the other elements use this recursive procedure. The out-of-sight entries to the left of the first column are all zeroes. Remarkably, you can generate the entire table in this manner except the "1" at the apex. Interested readers may wish to find the next row; the answer is in the Chapter Notes. A non-recursive formula for producing any table element will also be found there.

So, the hypercube is a four-dimensional object. Are there others? Yes, and sophisticated arguments prove exactly six "exist" — at least in a theoretical sense. Oddly, every dimension beyond four allows only three regular polytopes — analogs of the tetrahedron, cube, and octahedron. This is recapped below.

Number of Dimensions	Number of Regular Polytopes	Description of these Polytopes
1	1	line segment
2	infinite	regular polygons
3	5	Platonic solids
4	6	analogs of Platonic solids plus one (Chapter Notes)
5	3	analogs of tetrahedron, cube, and octahedron
⋮	3	analogs of tetrahedron, cube, and octahedron

Polytopes in All Dimensions

Figure 2.6

OF ALL THESE DIMENSIONS, the fourth has long been the playground for writers, artists, mystics, and frauds. In his curious short story, "And He Built a Crooked House" Robert Heinlein explores the hypercube's geometry. His protagonist, a Californian architect named Quintus Teal, designs and constructs a folded-down version of a tesseract that, during an earthquake, bends up into four-space. After a series of misadventures with its bizarre interior geometry, the prospective buyers and Teal exit the house through a window during an aftershock. Landing miles from the building site, they return to find the entire structure missing. Teal lamely tries to explain,

"It must have been that last shock, it simply fell through into another section of space."

Madeleine L'Engle wrote a highly praised children's book titled *A Wrinkle in Time* in which three young people travel through space and time via a tesseract. The children, Meg, Charles, and Calvin, encounter an evil entity that offers them complete security in exchange for their individuality and freedom — a modern version of the Garden of Eden.

Reasoning by analogy and other devices, it's possible to get some geometrical understanding of the fourth dimension postulated by these and other writers and artists. One such device is *unfolding*. Pretend you're a one-dimensional creature living in a line, like a railway car on a track. As a "line being," how can you "see" a square? The best that can be done is to take the square and by cutting through a corner unfold it into this "straight" world. As a result, you would see a square as four line segments.

Next, imagine that you live in the more interesting world of two dimensions, like the surface of a table. The problem now is to see a cube resting on this surface. Remember, you can't observe anything out of the table's top, so you perceive only the cube's bottom square (C, *look* below). By slicing seven of its edges, it unfolds into a cross of six squares. In flatland, this is how you would see a cube.

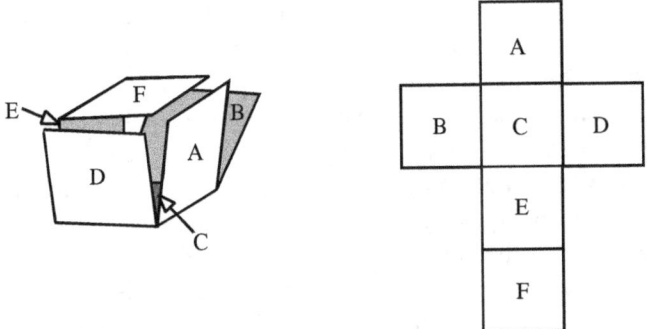

UNFOLDING A CUBE

Figure 2.7

Now we arrive at our three-dimensional world, wanting to understand something of the four-dimensional hypercube. What can we do? Exactly what was done before. Let's unfold it from four dimensions to three. This is accomplished by slicing off the cubes surrounding its single hypervolume. These are the eight cubes from Figure 2.5 (marked in bold) and Robert Heinlein's house before the first earthquake.

Conclusions from the previous three paragraphs:

- A straight-line creature sees a square as 4 line segments.
- In flatland, a cube is visualized as a cross of 6 squares.
- In three-space, we perceive a tesseract as 8 cubes.

Resting in the antechamber to a multidimensional universe, these eight cubes have fascinated writers, artists, and mystics alike. This enchantment and its grand opportunities for symbolism were cleverly exploited by the modern Spanish artist Salvador Dali.

Dali magnificently portrayed this imagery involving cubes and the fourth dimension in his masterpiece *Corpus Hypercubus* (*see* the next page). In front of Jesus' body are four small cubes implying a dimension beyond our own; they may also represent the nails driven through Jesus' hands and feet. Beneath the cross is a contrasting floor of two-dimensional squares with what appears to be an unfolded cube, like Figure 2.7. Now consider the large, ominous cubes forming the cross itself. How many can you find? Intriguingly, the dark cube at the back adds nothing to the painting's composition but for one seemingly unimportant point: it brings the total number of cubes to *exactly eight*. Dali knew his art; he understood his symbolism, and he also knew his mathematics.

The entire painting symbolizes Jesus as a being from the fourth dimension — Christianity's avatar sacrificed to cleanse a corrupt world. To dramatically emphasize this extra dimension, Dali has Jesus looking away from his mother Mary and the viewer as no other portrayal of the Crucifixion has ever done. This gesture depicts Christ's desire to leave this place of suffering and corruption to be with his Father in heaven's higher-dimensional universe.

The year after completing *Corpus Hypercubus* (1954), Dali painted *The Sacrament of the Last Supper*. As we have seen previously, Greek science and philosophy used the dodecahedron to symbolize the entire universe. In this later painting (*see* Figure 6.24), Dali once more applied mathematics to symbolize transcendent ideas and emotions. Behind Jesus and the Apostles rises a headless body, arms outstretched in the manner of the Crucifixion and surrounded by a gigantic dodecahedron. The implication is clear: Christ is a being from another time and dimension. And once again, his face is hidden.

Corpus Hypercubus by Salvador Dali,
1954, Metropolitan Museum of Art, New York

Figure 2.8

In addition to writers and artists, others have had an interest in the fourth dimension. The English philosopher Henry More (1614–87), one of the foremost representatives of the Cambridge Platonists, said "Spirits have four dimensions." But the most widespread fascination was among mystics and mountebanks who co-opted the idea to express their theories and enrich their pockets. In this extra dimension, they found a paradise of deception — both for themselves and others.

For instance, the American medium and magician Henry Slade became notorious when he was expelled from Victorian England for fraud connected with spirit-writing on slates. As a result, tremendous skepticism surrounded his reappearance as the primary source of evidence for the pseudo-scientific theories of the German astronomer Johann Carl Friedrich Zöllner (1834–82). Because of his association with the Slade, the astronomer became completely discredited. Nevertheless, Zöllner correctly realized that anyone with access to higher dimensions would be able to perform feats impossible for someone limited to only three. He suggested several test experiments for Slade that would demonstrate his hypothesis, for example, changing the handedness of seashells or removing objects from secured boxes. Although Slade never quite performed the stated tasks, he always managed to come up with something close enough to convince the incredibly gullible Zöllner. These experiences became the basis of the astronomer's inadvertently hilarious book *Transcendental Physics*. At the time, this opus and the claims of other spiritualists were wildly popular with the press and public.

For the mystics, what "evidence" was motivating their belief? Consider a two-dimensional being, Mr. T. Square, living entirely *in* a flat piece of paper on a desk. Also on the desk is a hexagonal-sided pencil with a worn eraser held by a ringed metal band. If you take this pencil and push it slowly through the flat paper, what would Mr. T. Square "see"? First, a black dot would appear, which almost immediately changes into a circle, that grows in size and suddenly transforms to a light brown color before morphing into a hexagon. For a while, the hexagon is constant but then changes into two alternating circles of different diameters before diminishing in size (as we come to the eraser) and abruptly disappearing. To Mr. T. Square, this must seem to be a supernatural event. By analogy, the mystics and spiritualists said that if a four-space object came into our three-space world, we would have feelings virtually identical to the flatlander. In a related context Arthur C. Clarke wrote, "Any sufficiently advanced technology is indistinguishable from magic."

Like the far side of the moon before the space age, the fourth dimension eludes direct examination. This protects the mystics from being proven wrong and the frauds from being shown devious. The 19th century was their great age of blossoming. Yet today they're in flower again, calling themselves "channelers" purporting to converse with spirits from other dimensions who are often said to be thousands of years old. These so-called ancient "entities" always seem to have names packed with vowels (Mafeo, Ramtha, Maitreya, and the like), and they forever talk like 60's flower children.

Time has shown their spiritualism/channeling produces only rotten fruit. Many who have eaten this decayed crop have never tasted the apples from the Tree of Knowledge. They still wander the valleys and search the caves of a perished Eden seeking the spoor of spooks and specters. The rest of us have marched eastward out of Eden.

A Natural Five

THE BEAUTY OF A LARGE, single crystal is arresting. The flatness of the faces and the sharpness of the angles are not what we normally expect from the disheveled natural world. Not all of the standard Platonic solids can be found in crystal form. However, the mineral galena forms cubes, magnetite commonly occurs as octahedrons, and the tetrahedrite's shape is obvious. Due to their *basic* pentagonal structure (*see* Figures 2.1 (a) or (b)), the regular icosahedron and dodecahedron never occur. This can be stated even more forcefully: a fundamental law of crystallography prohibits the latter pair from ever existing.

The reasons for this prohibition run something like this. Crystals are not enlargements of the mineral you see; instead, they're repetitions of *an original building block* (atom or molecule). As these building blocks are put together in three-dimensional space, some geometrical characteristics of the original block must be preserved. And if this block had the shape of a pound of butter, you could construct many forms — but not all. The edges don't necessarily have to be in alignment because the smoothness of the final aggregation results from a myriad of basic blocks. Every crystal is constructed in this way. Because of the particular angles in icosahedrons and dodecahedrons, no building block of the *correct proportions* exists. This limits the non-living world! In his marvelous book *Patterns in Nature,* Peter S. Stevens clarifies this point:

> Thus we find that crystals, which are assemblages of molecules, never have regular five-sided faces. In fact, no inanimate form exhibits pentagonal symmetry. No regularly pentagonal snowflake has ever fallen from the sky. Only animate forms, complicated forms, beyond the interminable stacking of identical molecules, have shapes with five equal sides.[4]

FOOL'S GOLD

Nevertheless, nature can almost do it. A common mineral called pyrite or fool's gold sometimes occurs as an imperfect dodecahedron whose pentagonal sides are not all equal (*see* picture to the left). This glistening yellow metal has launched more ships than the golden tresses of the legendary Helen of Troy.

Consider the truncated icosahedron, one of Archimedes' 13 semi-regular figures. The ancient mathematician's original writings on these polyhedrons have vanished, lost possibly during the looting and burning of the great Alexandrian Library. Or perhaps Christian monks of the Dark Ages scrubbed them from their precious vellum and parchment in an overly zealous desire to scribble their prayers and hymns — a common practice (*see* Chapter Notes). The encyclopedist Pappus in the 3rd century AD summarized Archimedes' work. From these synopses we can

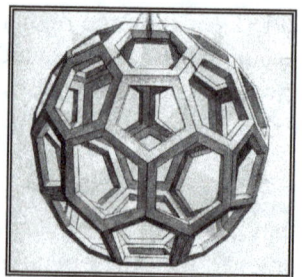

TRUNCATED ICOSAHEDRON

establish the ancient mathematician's priority and learn of the 13 semi-regular solids. On the other hand, the earliest known illustration of the truncated icosahedron is in *Libellus de quinque corpibus regularibus (On the Five Regular Bodies)* by the Renaissance painter and mathematician Piero della Francesca (1420–92). A few years afterward Leonardo da Vinci drew a magnificent "open" version (bottom of the previous page) for Pacioli's book *De Divina Proportione*. This open rendering allows us to see the polyhedron's inside as well as its outside.

Renaissance artists continued to be fascinated by these objects as this carving on the left from an Italian cathedral indicates. These polyhedrons are formed in the open style of Leonardo. At the top is an icosahedron bounded by 20 triangles with five of these meeting at each of the 12 vertices. Cutting off these vertices replaces each of them with a pentagonal face; it also converts each of the 20 former triangular faces into hexagons. The resulting truncated icosahedron is at the bottom of the carving. This derived form has 32 faces (20 hexagons and 12 pentagons) with 60 vertices.

The world is large and full of wonders! Some are slight and transient; others are major and paradigm shifting. Somewhere between these extremes resides the following 1985 scientific discovery. For decades, scientists believed carbon occurred in only two natural crystalline forms, graphite and diamond. Deeper study found a third crystal: a sphere of 60 carbon atoms (*see* below) arranged as a perfect truncated icosahedron like this book's frontispiece or a soccer ball.

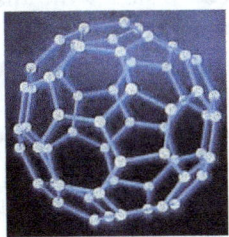
BUCKY BALL

SOCCER BALL

Figure 2.9

Like Kepler, whose celestial creations survive in molybdenum molecules, Archimedes would be pleased that his designs live on in the carbon crystal C_{60}. The very soot in his hearth, which he stirred on chilly winter evenings searching for bright embers, held one of his finest discoveries.

Plutarch was the biographer of the ancient world. In his *The Lives of the Noble Grecians and Romans,* he recounts the following famous story. During the Second Punic War the king of Sicily sided with Hannibal and the Carthaginians against the Romans. Almost immediately the Romans dispatched a general named Marcellus to subdue Sicily and besiege Syracuse, its capital and Archimedes' home. Due to the legendary war machines the great scientist had invented, this took much longer than expected. However, if nothing else, the Romans were persistent and ultimately they breached the walls of the city. Marcellus, in an uncommon show of generosity, ordered his soldiers to bring Archimedes to him alive. When they came upon the elderly mathematician, then in his 75th year, he was intently concentrating on some geometrical diagrams in the sand and he refused to come until he had finished the problem. Our ancient biographer writes that upon hearing this quixotic reply, ". . . the soldier, enraged, drew his sword and ran him through." In a few treasured pages, Plutarch records Archimedes' deeds and death in the otherwise forgotten life of "Marcellus," just another in an endless succession of Roman generals. Recalling this story, the philosopher Alfred North Whitehead said, "No Roman ever died in contemplation over a geometrical diagram."

Considering Archimedes' primacy in discovering the truncated icosahedron, it's natural to assume the new molecule, C_{60}, would have been named after him. Nevertheless, unlike Kepler, his memory wasn't similarly honored. Instead, the distinction went to the 20th century architect and inventor Buckminster Fuller. Hence, the molecule's formal name is Buckminsterfullerene, or colloquially, Bucky ball.

The Bucky ball is a geodesic sphere, though a very tiny one — it takes more than 25,000,000 of them aligned side by side to reach one inch (2.54 cm). It's also the roundest and most symmetrical molecule known. A survey of its geometric structure reveals an open hollow structure formed from 60 carbon atoms arranged into 20 hexagons and 12 pentagons.

The geodesic dome combines the sphere, the most efficient enclosure of volume, with the tetrahedron, the greatest strength for the least weight. It can withstand severe earthquakes and hurricane force winds. Paradoxically, it's the only structure that grows in strength, gets lighter in density, and cheaper to build as its size increases.

Fuller's geodesic domes are not as common now as in previous decades. Perhaps his convoluted writing style, sprinkled with neologisms and peppered with verbiage, turned away possible converts. Or perhaps the dome's ubiquity has robbed it of its novelty. Yet it's possible to consider Fuller's domes as the terminus of a polyhedral ancestry winding back through history to Neolithic playthings.

SPACESHIP EARTH PAVILION at Disney's Epcot Center is a full sphere.

Figure 2.10

The stability of the geodesic dome lies not in the strength of its individual parts, but in the way the entire structure distributes and balances the mechanical forces. Generally there are two such: *tension*, the pulling together of the individual struts, and *compression*, the effect of gravity. The struts that make up the framework are connected into triangles, pentagons, and hexagons, and each strut is oriented to constrain each joint to a fixed position. (Among the five Platonic solids, only the tetrahedron, octahedron, and icosahedron are structurally stable — a consequence of their triangulation.) The struts also map out the shortest paths between adjacent members — hence the name geodesic. All these different arrangements maximize the strength of the structure while minimizing the material required — as nature often does.

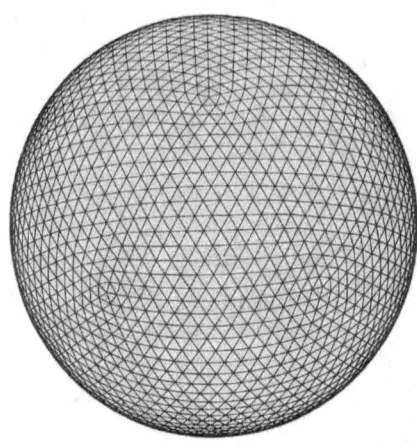

GEODESIC SPHERE

Figure 2.11

Consider a dome that has been enlarged into a sphere like the figure to the left. By definition, the disc of a great circle inside such a sphere passes right through the sphere's center, and any part of a great circle is a geodesic curve (*see* Chapter Notes). For economic reasons, these are the preferred flight paths of commercial aircraft.

What is truly remarkable here? What is worth further investigation? Study the above figure carefully; take the time to see where the triangles seem to "bunch." Detailed inspection reveals three evenly placed pentagons, one at each bunch point among a myriad of triangulated hexagons. Exactly how many pentagons are on the entire sphere? Well, there are 12. Always 12! Like the dozen pentagons on the dodecahedron, the Bucky ball, the soccer ball, the Spaceship Earth Pavilion, or the truncated icosahedron — count them. The intriguing question is, why always 12?

If you had only hexagons — whether equal or unequal, regular or irregular — any joining of them would *lie flat,* and a sphere would not, could not, form. To enclose just a hemisphere you need 6 pentagons; 12 for the whole sphere. This is not difficult to confirm. Interested readers will find a complete proof in the Chapter Notes. Surprisingly, the number of hexagons, when matched with the 12 pentagons, may be any number you fancy.

The *minimum* quantity of triangles required to enclose a solid is 4, as in the tetrahedron. The *minimum* number of squares required is 6, as in the cube. And the *minimum* number of pentagons required is 12, as in the dodecahedron. Since hexagons play no part in enclosing any polyhedron, they may be liberally joined with triangles, squares, or pentagons. As shown below, Leonardo da Vinci beautifully illustrated combining triangles with hexagons and squares with hexagons for *De Divina Proportione*.

 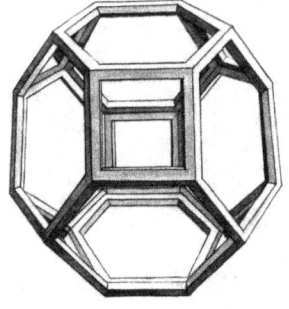

TRUNCATED TETRAHEDRON
with 4 triangles
and 4 hexagons

TRUNCATED OCTAHEDRON
with 6 squares
and 8 hexagons.

The living world is as restricted as the non-living by the nature of three dimensions. Even so, life has transcended the lifeless by assuming forms more varied and more wonderful than any mere stacking of molecules can ever accomplish.

Reflect on the radiolaria, microscopic single-cell marine organisms that have been wandering through all the world's oceans for more than 500 million years. Their varieties are innumerable, their shapes fantastic, and their appeal instant. In his celebrated book *On Growth and Form,* Sir Darcy Thompson says of them:

> Moreover, the radiolarian skeletons are of quite extraordinary delicacy and complexity, in spite of their minuteness and the comparative simplicity of the 'unicellular' organism within which they grow; and these complex conformations have a wonderful and unusual appearance of geometrical regularity.[5]

The naturalist Ernst Haeckel (1834–1919) drew hundreds of radiolaria like the exquisite structure below. However, despite its specific name *hexagona,* the sheer abundance of its surface figures allows a variety of polygons. As we have seen, no network of hexagons alone can enclose a sphere. Closer inspection of the figure — as Haeckel himself said — also reveals squares and pentagons. On occasion even a heptagon may be found, but these like all polygons of six or more sides play no part in closure.

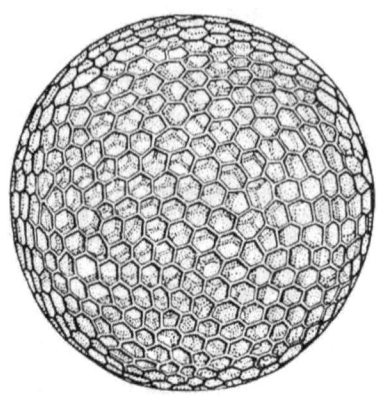

Aulonia hexagona

More apparent, but paradoxically less noticed, is a second type of regularity found in both the living and non-living worlds. That is the perfect *three-way corners* with angles of 120°. Because of the forces attempting to minimize the tensions among polygons, these are everywhere on the exoskeleton of *Aulonia hexagona*. This lessening of tensions is accomplished by minimizing the distances between adjoining corners over the whole surface, and angles of 120° do just that. Chapter 7, "Life on the Edge," has a section on soap bubbles and surface tension that further explores and explains this idea.

Imagine your height were shrunk to the size of the radiolaria — that's about as large as the dot on this letter "i." How do these creatures appear now? In a phrase, they're similar to the greatest rococo masterpieces. In a sentence, they're watery pockets of protoplasm hung on glass-like structures of incredible intricacy decorating an oceanic Louvre of innumerable rooms each awash with its own fantastic procession of specimens. One extraordinary room (*see* Figure 2.12) holds an entire gallery of Platonic solids. Granted, they're generally spherical, but all the vertices are perfect in number and exact in position. Since not every corner is three-way, surface tension can't be the only force at work here. Nevertheless the exteriors are mostly a lacework of hexagons with the expected corners and angles.

In what must be considered a phenomenal achievement of talent and temperament, Ernst Haeckel drew all these wonderful forms, and he published them in his renowned *Monograph of the Challenger Radiolaria*. D'Arcy Thompson writes of Haeckel's Platonic radiolaria:

> It is at first sight all the more remarkable that we should here meet with the whole five regular polyhedra, when we remember that, among the vast variety of crystalline forms known among the minerals, the regular dodecahedron and icosahedron, simple as they are from the mathematical point of view, never occur.[6]

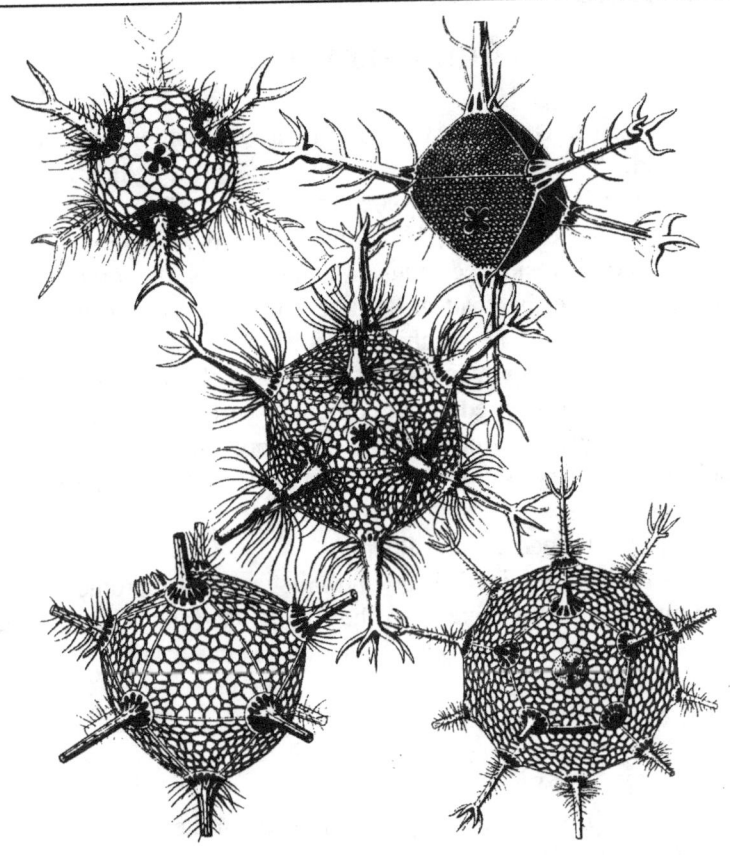

HAECKEL'S PLATONIC RADIOLARIA

Figure 2.12

The British research ship H.M.S. Challenger on her voyage of discovery dredged these life forms up from the mud and muck of the ocean depths. What are we to learn from these exquisite objects? Does all this unseen beauty require an explanation, or is nature simply prodigal with her productions? Where does this splendor originate? As with all nature, there are forces at work here: surface tension, the near absence of gravity, water pressure, chemical actions, and so on. The abundant radiolaria, extravagant in their fortuitous beauty, are purely a byproduct of these mindless forces — blind watchmakers all. Stephen Jay Gould asks, "Shall we appreciate any less the beauty of nature because its harmony is unplanned?" We know the answer!

THE PERFECT DIE

TWO GREEK GENTLEMEN out for a walk came to a fork in the road. Since the road was unfamiliar, they didn't know which branch to take. The older man, Epius, thought the road to the right was correct, believing it was the shortest way to their seniors' residence. Out of contrariness the other man, Phemius, chose the road to the left. They argued spiritedly for some time but without reaching a decision.

Finally the older man said, "Let's each flip a coin, and the one who correctly predicts the outcome will choose the road." The other agreed. Accordingly, they both took a coin from their pockets and flipped it so that they could compare their results.

"I call two heads," said Phemius.

"I say they're mixed, head and tail," replied the other. Sure enough, they were mixed, so they took the road to the right. For some time they strolled along in silence, while Phemius pondered his lost.

After a while, as they were nearing a bend in the road, he spoke out, "You tricked me!"

"How?" said Epius.

"You only won because your coin had two heads."

"No, no, definitely not. My coin was perfectly normal, as you can see." To prove this, he showed it to him. "Nonetheless, I did have *twice your chance* of winning, but not by cheating. And oddly, if my coin had had two heads, the game would have been fair and square." For a long long time Phemius was silent — some say for weeks, even months — as he tried to determine the nature of this deception.

PHEMIUS' CONFUSION is understandable. No less a person than the French mathematician Jean le Rond D'Alembert (1717-83) had the identical problem. Like many probability difficulties, the solution lies in correctly counting the number of all possible outcomes. Let H = head and T = tail; now let's count. Each coin could have a head (HH). Or one could have a head and the other a tail (HT), or the reverse (TH) — these two outcomes are different, and each must be counted separately. Lastly, both could have tails (TT). As a result, there aren't just three cases, but four: HH, HT, TH, TT. Therefore, Phemius, who wanted two heads (HH), had only one chance in four of winning. On the other hand, Epius, who wanted a mixed pair (HT and TH), had two chances in four, or double his friend's probability.

If you flip three coins and try to list all the outcomes, frequencies, and probabilities, then the situation is more complex. Let H^3 = HHH, H^2T^1 = HHT, and so on. The following table summarizes all possible results for three coins:

Outcomes	H^3	H^2T^1	H^1T^2	T^3	Totals
Frequencies	1	3	3	1	8
Probabilities	1/8	3/8	3/8	1/8	1

The outcome H^2T^1 means two heads and one tail are obtained; the frequency 3 tells us the number of ways this may be done (HHT, HTH, THH). Order has no importance when finding the frequency of the outcomes. As expected, the complexity with four coins is greater, and so it goes with each additional coin.

Good fortune favors fools, children, and mathematicians. In this case, an unexpectedly easy method exists to find these frequencies for *any number* of coins: specifically, Pascal's triangle (*see* Figure 2.13). The leftmost column, labeled "n" gives the number of coins; the topmost row labeled "r" gives the number of tails; their intersection is the frequency for the associated outcomes. Rows and columns are counted *from* zero. By employing the triangle's fourth row, consider the case of four coins. You have 1 way to get H^4; 4 ways to get H^3T^1; 6 ways to achieve H^2T^2; 4 ways to flip H^1T^3, and finally 1 way to get T^4. At each step, progressing left to right, the heads decrease and the tails increase by one each. The least likely chances occur at the ends of the triangle (probability = 1/16); the highest in the middle (probability = 6/16). This table of numbers may appear of little importance except for winning trivial coin tosses, but, as we shall see, that naive assessment would be wrong — wildly wrong.

You can appreciate the simplicity with which rows are formed: any number outside the triangular array is 0. Each entry is the result of adding the number directly above to its left neighbor. As an example, observe that for the bold-shaded trio in the table below, **6 + 4 = 10**. The interested reader may wish to develop row seven — the answer is in the Chapter Notes.

Blaise Pascal (1623–62), the eponym for the triangle, was a French scientist, mathematician, and religious writer. In his sickly, wretched life he produced a wealth of inventions and discoveries, including the hydraulic pump and the first mechanical computer. Although he didn't discover the triangular pattern named after him — poets and writers knew it six centuries earlier — he was the first to prove many of its intriguing properties.

$n \backslash r$	0	1	2	3	4	5	6	...
0	1	0	0	0	0	0	0	0
1	1	1	0	0	0	0	0	0
2	1	2	1	0	0	0	0	0
3	1	3	3	1	0	0	0	0
4	1	4	**6**	**4**	1	0	0	0
5	1	5	10	**10**	5	1	0	0
6	1	6	15	20	15	6	1	0
⋮	⋮	⋮	⋮	⋮	⋮	⋮	⋮	⋮

PASCAL'S TRIANGLE

Figure 2.13

To solve certain problems, the renowned Persian poet/mathematician Omar Khayyám (d. 1123) used this array of numbers, so to this day in modern Persia (Iran), it's still referred to as Khayyám's triangle. Bertrand Russell in his *History of Western Philosophy* writes, "Omar Khayyám, the only man known to me who was both a poet and a mathematician, reformed the calendar in 1079." In the Edward Fitzgerald translation of the famous poem *Rubáiyát*, Khayyám refers to his mathematics and the way he hoped to be remembered in the following two quatrains:

> Ah, but my Computations, People say,
> Reduced the Year to better reckoning? — Nay,
>   'Twas only striking from the Calendar
> Unborn Tomorrow, and dead Yesterday.
> (quatrain 57)

> And when, oh Sákí, you shall pass
> Among the Guests Star-scatter'd on the Grass,
>   And in your joyous errand reach the Spot
> Where I made One — turn down an empty Glass!
> (quatrain 101 — the last)

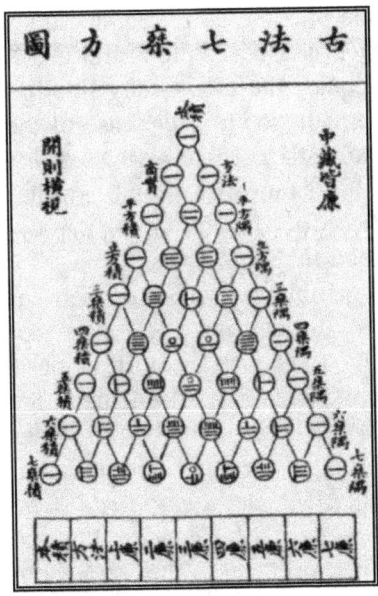

CHINESE TRIANGLE

Figure 2.14

In 1303 the Chinese mathematician Chu Shih-Chieh printed the *Precious Mirror of the Four Elements* containing the drawing shown to the left. This form of the triangle is centered rather than left-justified so that every entry is the sum of the two numbers directly above — left and right. No doubt the Chinese knew of the triangle long before this book was printed.

Between the Persian and the Chinese triangles is an Arabic text titled *The Luminous Book on Arithmetic* by As-Samaw'al ibn-Yahya al-Maghribi (d. 1180), a Jew who converted to Islam. This manuscript includes an illustration of the first eleven lines of the triangle.[7] The author doesn't claim the table is his invention, but credits an earlier mathmatician from the late 10th century. And we can be certain this is not the start of the trail but only the absence of still earlier written records. Everything has a past before it has a history. People in China, Persia, Arabia, Europe, and elsewhere have found this triangle intriguing and useful, but who first discovered it is lost in the mists of history.

THE READER HAS A RIGHT to ask what Pascal's triangle has to do with the Platonic solids. Quite a lot actually! One of the marvels of this array is its ability to morph — we'll see more of this in later chapters. Like Proteus, the ancient Greek god of the sea, it changes form but its character remains constant. Here it may be a pyramid, there an inverted array, and over here an expanded table of frequencies. Or it could have its odd and even numbers replaced by black and white dots to reveal a hidden fractal pattern, and so on. Changeable in form; constant in function.

Although we generally picture dice as being six-sided, Pascal's triangle can be viewed as a probability table for two-sided dice such as coins. The cube plus the other regular solids are the only perfect polyhedral dice; perfect to the extent that if you roll one — all things being equal, literally — any side has the same probability of turning up (or down in the case of the tetrahedron). Shortly we will see a Pascal-like array for *any number of cubical dice*. And even this new array can be further expanded to include probabilities for all the regular polyhedrons. First, let's establish a need for such tables.

GEROLAMO CARDANO

Like gambling itself, the men involved in discovering the "laws" of probability have a colorful past. Gerolamo Cardano (1501–76) gave the first and still largely unrecognized account of this — a century before Pascal — in his *Liber de Ludo Aleae (The Book of Games of Chance)*. Oystein Ore in the preface to his thoughtful book *Cardano: the Gambling Scholar* characterizes him as follows:

> The Renaissance bred many a man of genius and wrought many a fantastic life story, but even in such distinguished company does the life of Gerolamo Cardano, physician from Milan, readily catch the imagination. A brilliant start as a scholar, the rectorship of his university while he was a student, many years in attempting to build a professional career, all brought him only to the humiliating state where, with wife and child, he had to seek shelter in the poorhouse. But from here he began a one-man frontal attack upon the practices of the medical profession in his day with such success that some years later he ranked with his friend Vesalius in the forefront

of physicians of the world, and with flattering offers the crowned heads of Europe vied for his services.

. . . .

As a mathematician Cardano was probably the leader in his century; his work on the "Great Art" [algebra] has been characterized as the first that goes decisively beyond the attainments of the giants of classical Greek mathematics.[8]

Compared to Galileo, Cardano was mathematically sophisticated. Consider their different approaches to solving dice problems. For two dice, 36 (6x6=6^2) possible permutations exist, and by direct enumeration these can be found. The slang terms "snake eyes" (1,1) and "boxcars" (6,6) refer to the hardest rolls to get. Dicing with three cubes was enormously popular in ancient Greece and Rome, chiefly among the upper classes. In his dialogue the *Laws* (Book 12), Plato correctly states that "the dog" (1,1,1) and "an Aphrodite" (6,6,6) are the most difficult throws with three dice since they each occur in only one way. For three dice, 216 (6x6x6=6^3) permutations arise, and with sufficient diligence, these too can be found. The great Galileo Galilei did exactly this after an Italian nobleman asked him why 10 happens more frequently than 9 when three dice are tossed. He correctly showed that of the 216 possible permutations, twenty-seven summed to 10 and twenty-five to 9. Yet decades earlier Cardano had worked out all these frequencies and much more, and he did it by *indirect methods* rather than a brutal frontal attack. For four dice, 1296 (6x6x6x6=6^4) permutations are possible, but to list all these by Galileo's method would induce number dementia.

As Oystein Ore has shown, Cardano was the first person to fully work out the basic probability theorems. Unfortunately, after a life spent in accomplishment and controversy, strange tales clouded his last years in Rome. Some said he walked the streets and piazzas with the undirected gait of a lunatic. One story persists — almost without doubt apocryphal, but with Cardano you can never be certain — that by astrology he predicted the day of his own death. When that day came and he was still alive, he committed suicide so as not to ruin his reputation for casting perfect horoscopes. Yet this aside, as his biography demonstrates, the addition of his life is a very large sum: here was a man not without greatness in an age of great and cruel men.

TABLE OF FREQUENCIES FOR ANY NUMBER OF CUBICAL DICE

r \ n	0	1	2	3	4	5	6	7	8	9	10	11	12	13	14	15	16	17	18	19	20	...	frequency [sum]
0	1	0	0	0	0	0	0	0	0	0	0	0	0	0	0	0	0	0	0	0	0	...	$1 = 6^0$
1	1	1	1	1	1	1	0	0	0	0	0	0	0	0	0	0	0	0	0	0	0	...	$6 = 6^1$
2	1	2	3	4	5	6	5	4	3	2	1	0	0	0	0	0	0	0	0	0	0	...	$36 = 6^2$
3	1 [3]	3 [4]	6 [5]	10 [6]	15 [7]	21 [8]	25 [9]	27 [10]	27 [11]	25 [12]	21 [13]	15 [14]	10 [15]	6 [16]	3 [17]	1 [18]	0	0	0	0	0	...	$216 = 6^3$
4	1	4	10	20	35	56	80	104	125	140	145	140	125	104	80	56	35	20	10	4	1	...	$1296 = 6^4$
...	...	...	...	...	...	...	...	...	...	...	...	...	...	...	...	...	...	...	...	...	...	...	

Figure 2.15

In spite of the extreme complexity in counting all the cases with three or more dice, good fortune again favors fools and mathematicians. By generalizing Pascal's triangle to an array where every entry is the sum of six numbers — the one above and the five to its left — all frequencies for any number of dice can be readily found (*see* Figure 2.15). For clarity, under the frequencies for *three dice* their sums have been printed (i.e. [3], [4], [5], . . . [18]). These are found by adding the matching "n" in the left column and the "r" in the top row.

Use the preceding table in the following manner to find the probability of getting [10], when rolling 3 dice. Look down the leftmost column to find 3, the number of dice, and then across the top to locate 7. Their sum, [10], is the total you want, and their intersection, 27, the frequency you seek. The sums necessarily begin at 3, the lowest total achievable with three dice, and the aggregate of the frequencies is 216. Therefore, the probability is 27/216 (=1/8) — Galileo's hard-won result.

Dice are simple random-number generators, making them ideal for games and gambling. This has been so since the days of the early Egyptians. The four-sided tetrahedron is the least popular because it barely rolls and only randomizes four numbers. Being easy to make and quick to roll, the cube is the most widely used. As for the others, specimens of octahedral dice have been found in the tombs of the ancient pharaohs and are still used today. Dodecahedrons and icosahedrons have been employed in fortune telling and games such as dungeons and dragons. But whatever their shape, the frequency table for every one of them is a Pascal-like triangle akin to Figure 2.15. These arrays have all the properties of the original triangle, just generalized.

Analogous frequency tables can be constructed for figures with any number of sides regardless of whether they're regular or not. This gives a complete, simple solution for what is called the "generalized dice problem" — history's greatest gambling question.

A MINOR MYSTERY occurs in William Blake's painting *Soldiers casting Lots for Christ's Garments* concerning the dice: what is their sum? The soldier at the bottom left has just rolled, and judging by his generous grin, it's a winning roll. From any

64 / **The Perfect Five**

Soldiers casting Lots for Christ's Garments
by William Blake, 1800

Figure 2.16

viewpoint, only three faces of a cube or die are visible. Under magnification, *each* of Blake's dice reveals *two* dots on one face and *one* dot on another. Furthermore, the top on the left die probably has *three* dots, while the top on the right is indiscernible. See the painting above and diagram on the next page:

In the Western world, two conventions are always observed in the manufacturing of dice. First, opposite faces total seven — the ancient Egyptians started this; second, all dice have identical handedness. If you hold a die so that you see its 1-2-3 faces, then the numbers go counterclockwise — always. Nonetheless, Blake broke the second convention in *Soldiers casting Lots for Christ's Garments*.

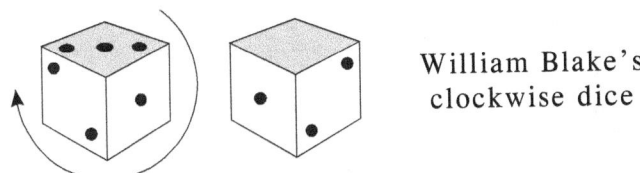

William Blake's clockwise dice

If we reasonably assume that both his dice have the same handedness, then the sum of their tops can be found. To be clockwise *à la Blake,* the die above right must have a 3 facing down, and so by the first convention, its top must be a 4. As a result, *the tops* are 3 and 4 for a total of 7 — a common winning number for millennia. And so the soldier smiles!

A PREVAILING MYTH of our time regards extraterrestrials as humanoid in appearance. This is abundantly evident from our movies, science fiction stories, and television shows, for it would be difficult to have empathy with a creature too exotic in shape. And non-humanoids are characteristically malevolent. The most enduring TV series of the last quarter of the 20th century is unquestionably *Star Trek* and its spin-offs; they follow this legend faithfully. So, friends and sometime foes are Ferengis, Vulcans, Klingons, Romulans, Bajorans, and so on — all humanoids.

This chapter opened with one of history's first proofs of anything: Theaetetus' on the possible number of regular solids. The proof implies that the nature of three dimensions limits the kinds of structures that can exist. The gods themselves cannot construct a sixth Platonic solid! For that reason, when the Ferengi and *humaaaan* gamblers of *Star Trek* fame meet at Quark's Bar on Deep Space Nine to play the game tongo, they roll the regular solids — the ones we know. And everywhere else — even at unimaginable distances in galaxies far far away — beings of whatever shape, humanoid or not, toss only *the perfect five*.

CHAPTER — 3

WITHOUT FORM AND VOID

And the earth was without form, and void . . .
Genesis 1:2 (KJV)

First there was Chaos, the vast immeasurable abyss,
Outrageous as a sea, dark, wasteful, wild.
John Milton, *Paradise Lost*

BY COMMON CONSENT, the greatest works of English literature are the King James Version of the Bible, the plays of William Shakespeare, and the works of John Milton. The opening quotations declare *chaos* was without form, misshapen, and empty, like the surface of the sea or the turbulence of a cascading river. Peoples everywhere had similar beliefs about chaos as featureless and undifferentiated. Ancient Greek and Chinese creation myths state that in the beginning the heavens and earth were one, and all was chaos. And out of the chaos, the gods forged a cosmos for humanity to inhabit.

As the gods began to fade, scientists found previously undiscovered order in the natural world, and they did this brilliantly! Well-behaved equations reigned everywhere, from the rules of Kepler and Galileo to the laws of Newton and Einstein. Scientists aspired to explain all of nature with them; the French mathematician Pierre Simon marquis de Laplace said as much. *Chaos* and *cosmos* were not seen as opposite ends on a continuum but rather as absolute opposites. If any phenomenon appeared partly explained and partly chaotic, everyone assumed further detailed study would expose its deterministic structure. This was as self-evident as Euclid's fifth postulate, as clear as geocentricism, as true as heavier objects falling faster than lighter ones. And, of course, it was false, enduringly false!

Scientists do not recognize a time when the universe was entirely chaotic, not even "In the beginning." The first sentence of Carl Sagan's book *Cosmos* sums up the modern attitude: "The Cosmos is all that is or ever was or ever will be."

In addition, contemporary scientists do not see "chaos" in the historical context explained in the first paragraph. Compare these definitions, old first, and new second:

> *Chaos* (kā´äs) n. 1. the disorder of formless matter and infinite space, supposed to have existed before the ordered universe 2. extreme confusion or disorder 3. [Archaic] an abyss; chasm
> *(Webster's New World Dictionary of the American Language)*

> *Chaos*: stochastic [probabilistic] behavior in a deterministic system (modern mathematical definition by the Royal Society in London, 1986).

The old definition is easily understood, but the new appears self-contradictory. After all, how can you have random behavior in a deterministic situation? Isn't the adjective *deterministic* the antonym of *probabilistic*? In this chapter, I'll attempt to explain this paradox.

Together with relativity and quantum mechanics, chaos theory was the third in a triumvirate of 20th century paradigm-shifting ideas. The reason for its later arrival is easily explained — unlike the idea itself. It all begins with differential equations, which come in abundant varieties. Those of the linear kind are well-behaved and *comparatively easy* to work with. For example, the addition of any two solutions is a third solution. In his excellent book *Does God Play Dice?*, Ian Stewart describes differential equations:

> Equations that involve rates of change are referred to as differential equations. The rate of change of a quantity is determined by the difference between its values at two nearby times, and the word 'differential' consequently permeates the mathematics: differential calculus, differential coefficient, differential equation, and just plain differential. Solving algebraic equations, not involving rates of change, is not always easy, as most of us know to our cost: solving differential equations is an order of magnitude more difficult. Looking back from the end of the 20th century the big surprise is that so many important equations can be solved, given enough ingenuity. Entire branches of mathematics have sprouted from the need to understand a single, crucial, differential equation.[1]

To a great extent classical mathematics dealt with *linear* equations because they're usually far easier to solve than the *nonlinear* variety. If necessary, nonlinear terms were even dropped out to make the equations simpler to solve (e.g. replace $\sin(\theta)$ with θ). You get your answers where and however you can. This was the classical credo — and it worked spectacularly for mathematics. But for physics, it really impaired progress by solving the wrong equations. Scientists of the late 20th century discovered, as they should have long suspected, that nature is uncompromisingly nonlinear! Draw a curve, any curve. Is it a straight line, i.e. linear? Certainly not! And so it is with the entire world. The Polish-American mathematician Stanislaw Ulam pointed out that to call the study of chaos "nonlinear science" was similar to calling zoology "the study of nonelephant animals." The phrase nonlinear science puts the emphasis in the wrong place! (It's probably no coincidence that the words *rule* and *straightedge* are synonyms.) Why does all this matter? Well you see, it was mainly in the behavior of nonlinear equations that chaos lurked, waiting to be discovered.

Henri Poincaré (1854–1912), the celebrated French mathematician, first stumbled on chaos in celestial mechanics while working on the unsolvable three-body problem (*see* Chapter Notes). He was horrified! But it could have been found earlier in the behavior of simple quadratic equations, rightly viewed.

Beyond scientists and mathematicians ignoring nonlinear equations, a second reason for the late development of chaos theory was the absence of high-speed computers. Computers define the present age; life without them is inconceivable. Differential equations are not easy to solve; nonlinear differential equations are generally impossible. Faced with this, what is a scientist to do? One answer is not to attempt the impossible; don't solve them! Do something completely different! What would that be? Graph their phase portraits using a computer. A phase portrait is a god's-eye view of a dynamical system for *all times*, rather than a particular solution for a particular time and for fixed initial conditions. These portraits give the entire range of possibilities for a dynamical system because computers draw marvelously detailed graphs of intractable nonlinear differential equations.

As a detour, consider what an historically curious thing graphing phase portraits was — it went against the tide of scientific thinking of the last four centuries. Let me explain. The late Jacob Bronowski described the original, but opposite, transition from god's-eye view to instantaneous view in "The Music of the Spheres," the fifth chapter of his inspired book *The Ascent of Man*. This transition laid the cultural groundwork for Newton and Leibniz to discover the differential calculus, which is about nothing if not instants in time. Bronowski writes:

> The perspective painter has a different intention. He deliberately makes us step away from any absolute and abstract view. Not so much a place as a moment is fixed for us, and a fleeting moment: a point of view in time more than in space.[2]

He also points out that earlier paintings without perspective illustrated scenes and people *for all time*. Their landscapes were static and fixed; their portraits, secure and serene like never-changing, never-ending cycles in Cathay.

With the development of perspective, Renaissance masters strove to catch an *instant in time!* Consider da Vinci's celebrated *The Last Supper* shown below. The vanishing point to which all eye movement flows (emphasized by superimposed white lines) is precisely in Jesus' brain. The entire picture is not a god's-eye view of time and space as were most of the earlier versions of this final meal. Instead, Leonardo captures the dramatic moment when Christ says, "I tell the truth, one of you is going to betray me."

The Last Supper by Leonardo da Vinci, 1498,
Church of Santa Maria delle Grazie, Milan

Figure 3.1

Now that our detour is over, let's return to the highway. So there you have it: the two major reasons for the late discovery of chaos theory. The first was the abhorrence of nonlinear equations because of their intractable nature; the second was the absence of computers to graph them until the end of the 20th century. Of course, these are related — the second eventually became handmaiden to solving the first.

ALL THE GREAT SCIENTISTS have at least one story demonstrating their mental prowess or indicating some eccentricity. A few of these tales may be apocryphal, but it seems out of place to question their genuineness. We recall Archimedes running naked through the streets of Syracuse shouting, "Eureka! Eureka!" after discovering the principle of buoyancy. Or Newton in the garden at Woolsthorpe seeing the apple fall and realizing that gravity also reaches into space, binding the moon to the earth in an eternal celestial waltz. Other stories persist. One concerns the mathematician Karl Friedrich Gauss as a child prodigy. At the age of ten, his teacher, to punish the entire class of boys for rowdy behavior, assigned them to add all the numbers from one to a hundred. Gauss instantly wrote the answer on his slate; he went to the teacher and placed it upside down on his desk saying *"Ligget se,"* Low German for "There it is." Eric Temple Bell in his *Men of Mathematics* describes the scene further:

> Then, for the ensuing hour, while the other boys toiled, he sat with his hands folded, favored now and then by a sarcastic glance from Büttner [his teacher], who imagined the youngest pupil in the class was just another blockhead. At the end of the period Büttner looked over the slates. On Gauss' slate there appeared but a single number. To the end of his days Gauss loved to tell how the one number he had written was the correct answer and how the others were wrong. Gauss had not been shown the trick for doing such problems rapidly. It is very ordinary once it is known, but for a boy of ten to find it instantaneously by himself is not so ordinary.[3]

The reader may enjoy trying to duplicate Gauss' feat. His method and answer are in the Chapter Notes.

At this point, we're concerned with a story about Galileo and his seminal studies of the pendulum. As a young man, he attended the University of Pisa to study medicine among other things. One day while attending services at the duomo, his attention began to wander either because of the heat or the dullness of the homily. The church windows were open and a *varying* breeze caught a chandelier, causing it to swing. With his quantitative mind, Galileo noticed that the time of any swing through an arc — whether large or small — was always the same. In other words, the period remained independent of the amplitude. To time each swing he used his pulse beat. He reasoned that the extra height of a larger arc gave it the additional velocity to complete the swing in the same time as a smaller arc. He later proved that the bob's mass also played no part in the pendulum's period. However, the length of its rod does — just as the beat of a hummingbird's wing has a significantly shorter period than the pelican's. With these observations, Galileo started a river of study on the pendulum that flows vigorously to this day.

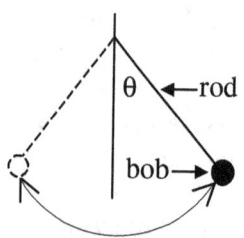

GALILEO'S PENDULUM

The pendulum's position may be graphed on the usual (x,y) plane, but with some difficulty because its motion is basically circular. Then again, you could more easily use the *angle* between the rod and the vertical, and the *velocity* of the bob itself. Mathematicians have an arsenal of measuring units and devices; they normally choose the easiest one for the job at hand. Here it's the angle (θ), and the velocity (v). Graphing these on the ($\sin(\theta)$,v) plane produces one variety of *phase portrait for the pendulum* — see below.

Consider an idealized pendulum's phase portrait where the horizontal coordinate is the sine of its angle and the vertical coordinate its velocity. Each circle represents an arc of a different length — the larger the circle the larger the arc. The arrows stand for the direction of the implicit time flow. Movement counter-clockwise (ccw) is positive, clockwise (cw) negative. Come join me as we ride the bob of the pendulum to explore the full

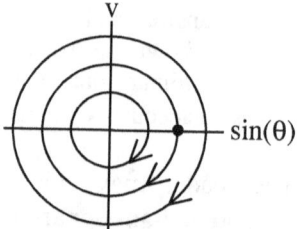

PHASE PORTRAIT OF THE PENDULUM

meaning of its phase portrait. Think only of graphing the pendulum's angle and velocity; not its left-right movement. At time zero, say, we mount the bob at its extreme rightmost position (the black dot on the middle circle). The velocity is zero; the angle is at its maximum; we're on the positive θ axis. At the instant the bob is released, it accelerates negatively in the cw direction. As we cross the vertical axis at the bottom of the arc, the bob achieves maximum negative velocity and the angle is zero. Quickly we rise and decelerate to a halt in the leftmost position; the velocity is now zero and θ is a negative minimum. Instantly the bob moves in the other direction; our velocity is now ccw or positive, but the angle is still negative. *In our phase portrait, we're above the horizontal axis proceeding upward and to the right.* Remember, this graph is on the $(\sin(\theta), v)$ not (x, y) plane. When we cross the positive vertical axis the angle is zero, and the bob attains maximum velocity. In the last quarter of the full period, we decelerate to zero and return to our original starting position on the positive θ axis. We dismount and recognize we have traveled but a single unique trajectory in an infinite family of circles.

All the above-described motion of the pendulum is highly idealized; we see no chaos here. Galileo consciously disregarded the nonlinearities of friction and ever-present air resistance. Only heavenly bodies are free of both. But there is a much deeper problem with his model, although most of it's still taught as the gospel truth in high schools across America. Concisely put, this model is false! Or more kindly, we could say it's just a fair first approximation of a real world pendulum. The difficulty lies with the changing angle: it creates a tiny nonlinearity in the equations. At small arcs (amplitudes), the error is almost nonexistent, but nonetheless it's always there.

Long ago scientists discovered that great truths often grow from probing minute irregularities — these can be portals to new fields of understanding. Considering this, we will try an equation that closely models the behavior of the same simple pendulum being watchful for any nonlinear terms. These will ultimately lead us into greener pastures. Recall the Law of Conservation of Energy: the total energy of a system plus its surroundings is constant. Here the system is a frictionless (fictional) pendulum in motion. The energy of this system has two forms: kinetic (K.E.) and potential (P.E.). As the bob smoothly swings, the K.E. and P.E. flow seamlessly from one to the other like water from glass to glass all the while maintaining a constant amount of fluid. So for the pendulum this law states:

$$K.E. + P.E. = \text{constant} (= c)$$

By choosing the units so that the bob's mass and the rod's length are 1, "K.E. + P.E. = c" can be modeled by:

$$\tfrac{1}{2}v^2 + \sin(\theta) = c$$

Here θ is the angle as before, and $\sin(\theta)$ is the bob's *horizontal distance from the origin* (usually just called x). Now we can easily solve this equation for v and graph it on the $(\sin(\theta), v)$ plane again:

$$v = \pm \sqrt{2c - 2\sin(\theta)}$$

As my old mathematics professors would say, "without loss of generality" we may replace 2c with C to get:

$$v = \pm \sqrt{C - 2\sin(\theta)}$$

Pick a value for the constant "C," say 1.5, and work out $\sqrt{1.5 - 2\sin(\theta)}$ for θ from $-180°$ to $+180°$ in increments of $10°$. If the term under the root is negative, discard it. If not, plot two points at the present value of θ on the $(\sin(\theta), v)$ plane: i.e. $(\sin(\theta), \sqrt{1.5 - 2\sin(\theta)})$ and $(\sin(\theta), -\sqrt{1.5 - 2\sin(\theta)})$. For this value of the constant, you get one of the ovals in Figure 3.2.

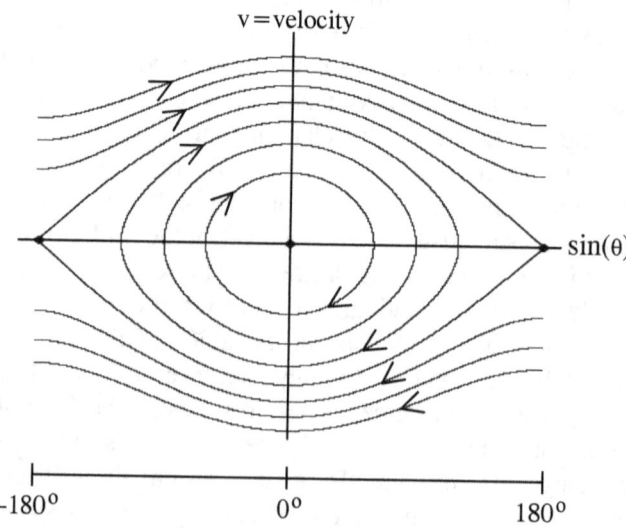

PHASE PORTRAIT OF A PENDULUM

Figure 3.2

The arrows, as previously, indicate the implicit time flow in the phase portrait. The Chapter Notes include a QBasic program to generate this graph. Some readers may enjoy playing with it to see the effect of changing the constant's value. For others the following description and table should suffice.

Each curved line of the graph represents a different value of the constant "C." Or stated alternately, each curve represents a different energy level for the system. The dot at (0,0) symbolizes the pendulum at rest, hanging motionless with K.E. and P.E. both zero. On the other hand, the regular ovals stand for the arcs of different amplitudes. And the transition trajectory, the "oval" with pointed corners, is on the verge of going over the top. And yes, the pendulum, when given sufficient energy (a large push, say), can go over the top and become a propeller. When this propeller/pendulum spins ccw (positively), it's one of the eyebrows of Figure 3.2. When it spins cw (negatively), it's one of the wrinkles below the ovals. As you progress outward from rest at (0,0), the energy levels regularly increase. The effects of all these different values for the constant "C" are summarized in the following table:

Values for the constant "C"	Description of the resulting curve
C < -2	No points whatsoever
C = -2	A single point
-2 < C < 2	Regular ovals
C = 2	The transition "oval"
C > 2	Eyebrows or wrinkles

Constant Change

The figure on the previous page shows a full period. However, the entire graph in our frictionless model goes to infinity in both directions as seen below.

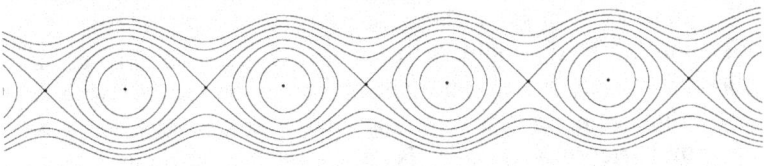

Figure 3.3

Figures 3.2 and 3.3 are the very essence of regularity. Again, we see no chaos here! So perfect is this regularity that the Dutch scientist Christiaan Huygens — just ten years after Galileo's death — used the pendulum as the mechanism in the first "grandfather's clock." Huygens employed a system of falling weights to transfer the proper amount of energy to the pendulum to prevent friction and air resistance from bringing it to rest. The chronometer became the emblem of classical science — the symbol of the clockwork universe.

Neither Figure 3.2 nor 3.3 represents the pendulum in the most realistic fashion. Ian Stewart in his book *Does God Play Dice?* explains why:

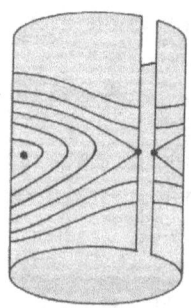

Figure 3.4

How does a geometric circle 'know' that −180° = +180° [Figure 3.3 above]? It knows because it wraps round and joins itself. This gives the circle a very different topology from the line, and explains why we're having problems: we're trying to use numbers, which live on a line, to represent an object that has the wrong topology. No wonder we have to wriggle on the hook a bit!

To get a more faithful picture of the pendulum motion — one whose geometry accurately reflects reality — we do the same thing. We *wrap the whole picture up* in the horizontal direction, to bring the left- and right-hand edges together, and physically force −180° and +180° to coincide. In other words, we roll the sheet of paper up into a cylinder [*see* Figure 3.4] [4].

Where's the chaos? To find it, our model must ever more closely mirror the real world, and the real world contains friction and air resistance — technically called dampening forces. A dampened pendulum will rather quickly spiral or wobble to rest, but still shows no chaos. We need one concluding element: a periodic driving force. It can be applied to the bob or the rod. Happily, the reader is probably familiar with this, for who hasn't given a periodic push to a swing, forcing it to maintain a fixed amplitude or perhaps go ever higher and higher? It all depends on how much force you use. As with any pendulum, it accelerates coming down and decelerates going up, amid friction and air resistance exerting continuous dampening effects. Assume you apply a *regular periodic push*.

At this point, if you have rarely done this, your intuition will *spectacularly* fail you. Inexperience says, no matter what the initial height, the motion will soon settle down to a regular back and forth pattern with the swing reaching the same height each time. Often this happens. Bizarrely, however, the motion can turn wildly erratic, going high, then low, but never finding a steady state and never exactly repeating any pattern of swings that came before. (Experienced swing pushers, like moms and dads, know this; for that reason, they limit the strength of their regular pushes.) So at last, we have found chaos!

What does chaos look like? The short answer is not much. At least the actual swinging of the pendulum itself reveals little but a jumble of erratic movements, but phase portraits are entirely different creatures. Previously our portraits were graphed by using the horizontal distance and the bob's velocity on the $(\sin(\theta),v)$ plane. Now we will use the angle, and the pendulum's angular velocity, ω, (omega) in degrees per second on the (θ,ω) plane. See Figure 3.5 below — it's worthwhile to thoroughly study this graph.

This figure is an accumulation of hundreds of pendulum swings — graphing always the angle, θ, and the angular velocity, ω. First, we notice that chaos has a distinct outline: a tight pattern of closely packed trajectories.

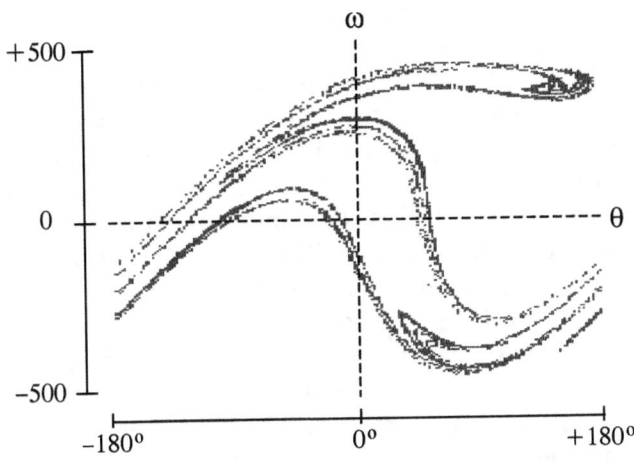

PHASE PORTRAIT OF A CHAOTIC PENDULUM

Figure 3.5

This design, properly called a *strange attractor*, has a deep geometry all its own: the incredible fractal geometry made famous by Benoit Mandelbrot — but more of this in the next chapter. Since θ runs from -180° to +180°, roll the graph into a cylinder in the horizontal direction as we did previously in Figure 3.4. Now observe that the sides exactly match, making the entire phase portrait continuous.

At this point lots of readers may wish they had a physics laboratory of their own to conduct pendulum experiments. Well, maybe they do. The modern teacher is the Internet, and numerous sites exist to familiarize us with chaotic pendulum behavior. Many of these have excellent moving graphics that mimic real behavior when different values are inserted for the constant "C."

The mathematician/educator Alfred North Whitehead deplored what he called "inert ideas." By this he meant ideas merely received into the mind without being tested, thrown into fresh combinations, or applied to the real world. And Whitehead was a man renowned for his unworldliness and unbelievable powers of abstraction. He observed that some ages sparkled with thought and enthusiasm while others lay crippled under the heavy burden of inert ideas. Can his damning charge be held against the preceding pages on pendulum chaos?

Absolutely not! The pendulum was a kind of lab rat for the emerging science of chaos even though it had been a standard example of classical mechanics. What a paradox! What a shock this would be to scientists from the beginning of the 20th century. Decades ago I recall my secondary school teacher telling the class that all physics had been discovered. It's the same song about the end of science we hear many singing today. Yet even apparently well-understood things can harbor surprises. It turns out that *every* regular oscillator is a relative of the pendulum; sometimes the equations are identical. Give nonlinearity its due, for chaos lurks in dynamical systems everywhere and at all times. Chaos has reversed the overspecialization and departmentalization prevalent in learning institutions around the world, which Whitehead so deplored. Why? Because this theory — starting with the pendulum — cuts across different fields of study and combines disparate disciplines with a stunning display of unifying power. Even your heart shows regular pendulum-like behavior; regrettably, it can also be chaotic.

SOME SCIENTISTS DECRY the word "chaos" as a title for this brilliant new theory. But science, and mathematics in particular, has always had poor names for rich ideas. "Real numbers" aren't real. "Complex numbers" aren't complex, they merely have two parts. And "imaginary numbers" are as real as any number. "Improper fractions" have never been observed doing anything indecent, and "irrational numbers" aren't in need of psychotherapy. Historically, these labels just reflect our mental struggles when confronted with novel ideas. On the other hand, many researchers think the title "chaos" is well suited to this new science. We just need to get used to it.

Chaos has the form of a fractal — a self-similar graph to be fully examined in the next chapter. But its meaning will be explored in the following section. Despite Genesis, chaos is neither without form nor void!

IDYLL IN THE SUN

Chaos was first, but next appeared broad-bosomed Earth . . .
Hesiod, *Theogony*

EPIUS TURNED TO HIS OLD FRIEND and said, "Phemius, come see my new calculator."

"What do you need that for? You can always use the computer in the common room of the residence."

"I know, but this is quicker and easier to study chaos theory with. I'm reading Ian Stewart's book *Does God Play Dice?*" replied Epius.

"Chaos! That's the story of our lives."

"You're thinking about chaos in the original sense — the one Hesiod used in his creation story. Today scientists have given it a new and precise definition. You take the words and make them mean what you want."

The poet nodded, "Explain this new meaning to me."

"I will with the help of this calculator. You know, Phemius, if I could take one thing back to ancient Athens to amaze and confound the sages, this would be it."

"Why not a laptop?"

"That would work too. But this you can put in your pocket and pull it out like a magic talisman. Aristotle would faint —

thinking I'm one of the Olympians. Remember what Arthur C. Clarke said, 'Any sufficiently advanced technology is indistinguishable from magic.' The philosopher would be impressed."

"Impressed, hell! He'd be speechless."

"Phemius, a moment ago you said our lives have been chaotic, but I think that's a good thing. Utopias, if there are any, must be dull, boring places — too much order, no wilderness, no danger. Something like this place on bad days."

"Enough of that. Show me the new chaos idea."

"Sit with me on this grassy knoll with our backs to the sun. Take the calculator. Here let me turn it on first. Punch in any number, and press the square root button $(\sqrt{})$. Press it again and again until you see something happening. After about 30 iterations, what answer do you get?"

"I get 1. Is that what always happens?"

"Yes, the system has reached a steady state. No chaos here," concludes Epius. "Now enter 0.7, say, and press x^2 repeatedly. Where does it go?"

"I get 0, another steady state. What about using a number larger than 1?"

"Try it."

He does, and after a few iterations, the calculator had overloaded and says *error* because the answers were on their long road to infinity. Epius puts the calculator in radian mode and gives it to his friend, asking him to enter a number and press the cosine button about 40 times. He does and it goes to the curious steady state of 0.739085133 and just sits there — still no chaos.

"Can one of these function buttons go to two numbers?" Phemius asks.

"Enter 153, or whatever you please, and press the reciprocal button $\frac{1}{x}$ and you'll get 0.0065359477 — another press and 153 reappears. We say the iteration has a period of two, meaning if you hit the button twice you get back what you started with. So the answer to your question is Yes."

"I need to see an example that produces chaotic results, not steady states or periods. But how can you possibly get chaos from a calculator unless it's broken?" asked Phemius.

"We'll get there. Mathematics, like music, has a learning curve. Let's invent a new button, x^2-1, say. Start with a value between -1 and 1, hit the x^2 key, and then $-1 =$. Repeat for a while. Soon you'll be jumping back and forth from 0 to -1 — another period-2 function. Nothing novel here either."

"Epius, you're torturing me."

"Have patience. Here's another case: $2x^2 - 1$."

"But that's almost identical to the previous one."

"Humor me, Phemius, I'm older than you. As before, enter any number from -1 to 1, square, multiply by 2, subtract 1, and press the equal sign. Repeat repeatedly. I'll wait. . . ."

"OK, I've iterated about 50 times . . . no pattern yet . . . it's very slow . . . will this thing ever settle down? . . . how can this be? All these digits appear random, yet this function is almost identical to the previous one."

"Phemius, you've just entered the realm of chaos — modern chaos. Have a seat; be comfortable here. This is a large room. From something as simple as $2x^2 - 1$ you get an infinite, random sequence of numbers. Chaos out of order."

"That's bizarre!"

Epius unfolded a piece of paper from his pocket. "See, I've made a graph of some of my iterations of this function. Look, there's no pattern, nothing repeats, just randomness."

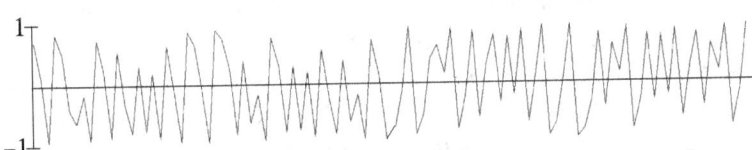

THE NUMBER OF ITERATIONS OF $2x^2 - 1$

"Epius, I feel like a walk in the sun. This idea of chaos has impressed me — more than anything I can recall for a long time. It's so anti-intuitive — directly opposed to what we learned at our mother's breast. I need a while to reflect on it." With surprising agility the two old friends walked across the park, over a hill, and were gone. An hour or so later they sauntered back to the residence where the staff, right on time, offered them soup and sandwiches on the outdoor patio.

After lunch the poet turned to the scientist and said, "Chaos theory is fantastic, but I believe I can duplicate its results. Get a pen and paper because I'm going to give you a list of random numbers — as many as you wish, 100,000 or twice that."

"Phemius, our lunch wasn't that good. Be careful what you promise."

"Are you ready?" The poet began rattling off apparently random numbers. After Epius had filled two pages, both sides, he begged for a stop.

"What's the point?" the scientist asked.

"As I understand it, if I put the identical number into $2x^2 - 1$ that I did the first time I'll get the same random, or should I say pseudo-random, sequence again. Is that right?"

"Yes," replied Epius. "I always said you were a quick study."

"Well, I can repeat those four pages of numbers again from beginning to end, maybe even from end to beginning if I take my time — just like $2x^2 - 1$."

"That's impossible. Show me."

Phemius began repeating the numbers, slowly at first, but faster as he went along. He was an unstoppable force with a mission to complete. Now it was Epius' turn to be amazed. For some time they sat silently, each with his own mystery to contemplate.

"OK, Phemius, explain to me how you did your magic."

"You should be able to figure it out. I memorized *The Iliad* and *The Odyssey* millennia ago — a common practice in the days before Gutenberg. When we were boys in ancient Greece, each character of our alpha-beta did double service as both letter and numeral because $\alpha = 1$, $\beta = 2$, ... $\iota = 10$, ... $\rho = 100$, ... $\omega = 800$. Sometimes I've noticed you still scribbling Greek letters for numbers — old habits die hard. So my feat of memory was easy. The first sentence of Homer's second epic puts the emphasis on Odysseus' ability to reason:

This is the story of a man, one who was never at a loss.

I just translated *The Odyssey* from Greek letters into numbers, and voilà!" [*See* the Chapter Notes for page 14.]

"Ingenious! You're right, I should have known. But since in *all languages* letters occur with different frequencies, an analysis of your digits would show something was strange. I once heard that when reading English, you're 130 times more likely to find an 'e' than a 'z'. That's quite a spread."

"But at a first look, Epius, none of that is obvious."

"I agree, but on the other hand, any analysis of $2x^2 - 1$ would show that its digits occur with equal frequencies."

"So your numbers were pseudo-random and mine were only pseudo-pseudo-random."

"Yes, that's true. . . .

"You know, Phemius, speaking of Odysseus, not only was it his idea (some say with the aid of Athena) to build the Horse, but he also drew up the plans that I used to construct it. He was a fountain of ideas. Once I heard his men complaining that there was nothing more they could possibly do to win a certain battle. Odysseus reprimanded them saying, 'There are always options.' Think of all the heroes who died at Troy: Patroclus, Memnon, Antilochus, Hector, and the great Achilles. All these and more perished at that terrible place, but Odysseus ended the war, not by brawn but by brain. He lived to sail away; sadly, none of his brave companions from Ithaca had his ability to survive the years of wandering. We're fortunate to have been blessed by such a man."

"You and I are living his tradition of working with ideas — maybe that's what he intended."

CHAOS IN THE NATURAL WORLD

Nature and Nature's laws lay hid in night:
God said, Let Newton be! And all was light.
Alexander Pope (1688–1744)

IN A MOMENT OF RARE HUMILITY Sir Isaac Newton spoke of himself as merely a child gathering seashells and occasionally finding a smoother pebble or prettier shell than ordinary. While, as he put it, the great ocean of truth lay all undiscovered before him. His metaphor was more accurate than he could have imagined. Newton's light was dazzling; it illuminated the heavens and vast areas of the natural world. But nature is complex, very complex! And the entire ocean of turbulence — the vast nonlinear world — lay all undiscovered before him.

For survival and reproduction, every living creature must recognize certain patterns. Life on earth depends on this. Many birds time their egg laying to match the seasonal peak abundance of specific insects. Bears arrive at berry fields only when the fruit is ripe. In the northern hemisphere turtles lay their eggs on the south-facing slope of hills to increase their chances of hatching, and the list goes on. The continued existence of early humans depended on seasonal pattern recognition of wild fruit ripening and animal migration. Infirm and elderly group members were valued for their deep knowledge of such things. For example, Shanidar man, from a cave in northern Iraq, was a 40-year-old Neanderthal killed accidentally in a rock fall some 46,000 years ago. His list of ailments and illnesses was impressive: one side of his body was withered, his right arm had been amputated above the elbow, he had terrible arthritis, and he was blind in his left eye — the result of horrific facial and skull wounds. All his teeth were abnormally worn down from chewing hides to soften them and possibly as a means of manipulation in lieu of his missing right arm. Quite clearly he could not have survived without the care of his fellow hunter-gatherers. Maybe they had altruistic feelings toward him, but more likely he, at least in later years, had a thorough knowledge of vital animal movements and wild harvest times. On many levels, pattern recognition equals survival!

All peoples, in every land, at all times, have searched for nature's designs. Many claim to see patterns where none exist. Today scientists hear the call as clearly as did Paleolithic man. It cries through the corridors of history, and the words are always the same: "find nature's patterns"!

THE LOGISTIC FUNCTION

IN THE 1960S ROBERT MAY, one of the United Kingdom's top scientists, became fascinated with the way living populations of birds, animals, insects, and bacteria changed over discrete time intervals. Each time interval represented a reproduction period, which could be hours for bacteria or years for birds and mammals. Unlike humans, who are always sexually active, wild creatures have a single breeding season so generations don't normally overlap. Discrete rules of the type May sought are termed "difference equations," to be clearly distinguished from "differential equations," which run over continuous time intervals. The distinction between these two types of equations is similar to the distinction between an analogue and a digital watch. Nonetheless, their behaviors can be practically identical.

Let's see how we might model population growth with a simple difference equation that feeds on itself. Mathematicians call feedback loops an iteration process, and they mirror the real world. After all, next year's population (ignoring dampening factors) must depend on this year's numbers. A standard example of a feedback loop is a loudspeaker picking up its own sound and rapidly building to a screech.

The simplest difference equation occurs when the output is identical to the input, implying immediate stability, for example, (0.3, 0.3) on the graph below. Consider its equation where the subscript "t" means the present time, and "t+1" is the next time. The factor "1" is the growth rate:

$$x_{t+1} = 1x_t \quad (a).$$

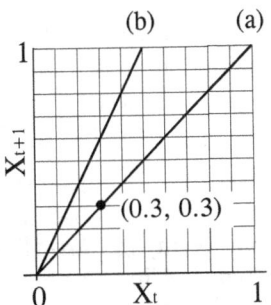

This is a straight line through (0,0) at a 45° angle to the horizontal axis. Entertain a higher growth rate, say 2:

$$x_{t+1} = 2x_t \quad (b).$$

In this case, the population doubles after each breeding interval: 3,000 this season (i.e. x_t) produces 6,000 the next (i.e. x_{t+1}). In general we have $x_{t+1} = mx_t$ where m is the growth rate, either high or low. When $m > 1$, we have a classic Malthusian system that will eventually — in days or decades — exceed any upper bound if unrestricted by considerations of food or morality. This is reminiscent of the human population boom. And political or religious institutions supporting unfettered human procreation unknowingly encourage a single outcome: eventual extinction through starvation.

Any realistic model must, however, include restrictions of food and space if not morality. Scale the setup so the resulting values fall between 0 and 1: extinction and maximum capacity. That is, the greatest population the environment will support is x=1 (*see* graph). If there are x insects, say, then 1-x is a measure of the space nature permits for further population growth. This implies that m(1-x) is a more accurate growth factor, considering space restrictions. Consequently, replacing m by m(1-x) in the above general equation produces our final model:

$$x_{t+1} = m(1 - x_t)x_t$$
or rearranging we get
$$x_{t+1} = mx_t(1 - x_t).$$

This is the famed *logistic equation*, which exhibits all the behavior found in complicated chaotic systems. Such a simple formula appearing so innocent! Who would have suspected that this function could produce anything intricate or even mildly disordered?

One, Two, Four, ... Chaos

THIS RULE, THE LOGISTIC EQUATION, is our archway into a comprehensive understanding of chaos and fractals:

$$x_{t+1} = mx_t(1 - x_t)$$

where x_t goes from 0 to 1 and m from 0 to 4. The initial value of x_t is of no consequence as long as it's between 0 to 1. In the long term, the growth rate "m" controls all terminal behavior when it's in the range 0 to 4. Later we will look at x_t and m outside their normal values.

More particularly, if m lies in the interval $1 < m \le 3$, the logistic equation terminates at *one number* regardless of the starting x_0. For example, choose m=2 and $x_0 = 0.3$, say, and by applying the rule repeatedly with t = 0, 1, 2, ... we get the following sequence:

$x_0 = 0.3$	initial value
$x_1 = 0.42$	$= 2(0.3)(1-0.3)$
$x_2 = 0.4872$	$= 2(0.42)(1-0.42)$
$x_3 = 0.4996$	
$x_4 = 0.4999$	.
$x_5 = 0.4999...$	.
$x_6 = 0.5$	.
$x_7 = 0.5$	
$x_8 = 0.5$	

And there it stops. This is a point attractor sometimes called a pole or steady state of the dynamical system. To prove the sequence stops at 0.5 substitute this number into the logistic equation: $2(0.5)(1- 0.5) = 0.5$, as expected. In general, the point attractor in the range $1 < m \le 3$ is always $(1-1/m)$. By direct substitution, this too may be verified.

What about m in the interval $0 \le m \le 1$? Here x, (1 - x), and m are all between 0 and 1, therefore their product quickly converges on the point attractor "0." We say the system goes to extinction — a common fate with many species today.

Both these fix-point outcomes, 0 and [1− 1/m], have their counterparts in the natural environment — that's a small part of the reason the logistic rule is a good real-world model. Sometimes animals can become so reduced in numbers — through over-hunting, habitat loss, and pollution — that they can't find another to mate with. Consider blue whales in the vastness of the dark oceans. Below a certain number, these behemoths will go extinct, and eternities could pass away and such incredible creatures will never be seen again. As with the natural world so with our function — both can exit.

Alternatively, natural dynamic systems may attain a steady state like our rule sitting on and repeating 0.5. Recall this is a scaled system, scaled to fit the interval 0 to 1 so that 0.5 ($m=2$) could represent a certain quantity of individuals per acre. Much of the natural world settled on an equilibrium point early in its history. This is something of what is meant by the timeworn phrase *the balance of nature*.

FINDING STATIONARY POINTS, whether by hand calculation or computer, can be bothersome. There is a better way! Historically, mathematicians have swung between two fashion extremes: those who relished using geometrical diagrams and those who assiduously avoided them. Pierre Simon Laplace prided himself on shunning all drawings in his already difficult to understand *Celestial Mechanics*. To compound matters, he often wrote, "Thus it plainly appears" and cavalierly omitted the entire proof. Nicolas Bourbaki, the Russian émigré who now lives in France, avoids geometry and excludes pictures and drawings from his writings on the works of a fictional group of eminent mathematicians.

In the opposing camp, but still in France, Henri Poincaré at the turn of the century did some of his greatest work with and about geometry — including the first tentative steps in chaos theory and fractals. At the end of the century, Benoit Mandelbrot, with his ubiquitous fractals permeating all the sciences, brought geometry home to stay.

So, this better way to find stationary points is, of course, geometry! Specifically, *cobweb diagrams*.

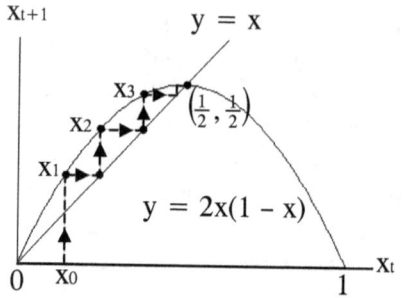

Steady State Cobweb Diagram
of the Logistic Function, m=2

Figure 3.6

The outcome from repeated iterations of the logistic function can be found geometrically with cobweb diagrams — see Figure 3.6. First graph $y=2x(1-x)$ to get a smooth downward opening parabola cutting the horizontal axis at 0 and 1, our desired interval. Next, draw the diagonal line $y=x$ on the same graph. Pick x_0, an initial value from the interval, and draw a vertical line (cobweb), and note where it touches the parabola at x_1. From there draw a horizontal line to hit the diagonal $y=x$ at (x_1, x_1). Now the magic! From this point draw a new vertical line to again touch the parabola, but now at x_2. And again from here draw a horizontal line to hit the diagonal at (x_2, x_2). Repeat, to form "steps" or "cobwebs" between the line $y=x$ and the parabola. The coordinates of successive "risers" of the staircase are the successive terms of x_t.

In the case above where $m=2$, the cobwebs move up the diagonal and then home in toward the point where the parabola and the diagonal cross, *namely* (0.5, 0.5). This is a fixed point, so stability is the natural outcome because the cobweb goes inward. When $0 \le m \le 1$, the cobweb steps still move inward but descend to 0 rather than ascend to $[1-1/m]$.

PERIOD-DOUBLING BEHAVIOR occurs when $m > 3$. When $m = 3$, the fixed point is only marginally stable; convergence is exceedingly slow. This indicates the function is undergoing a dynamic change. Indeed any value slightly larger than 3, say 3.1, becomes unstable and the cobwebs spiral outward. Ultimately these settle down and alternate between two points. As before, any number chosen from the interval 0 to 1 has the same terminal behavior. The controlling factor is always m, never x_0.

Mathematicians say the function has bifurcated and this can happen any number of times leading, as we shall see, to chaos — deterministic chaos! First let's iterate the logistic function for $m = 3.1$ and $x_0 = 0.3$ although the seed value doesn't matter as long as it's from the interval 0 to 1. We get the following sequence:

$x_0 = 0.3$	initial value	
$x_1 = 0.651$	$= 3.1(0.3)(1-0.3)$	
$x_2 = 0.7043$	$= 3.1(0.651)(1-0.651)$	
$x_3 = 0.6455$		
$x_4 = 0.7092$		
$x_5 = 0.6392$		
$x_6 = 0.7149$		
. . . .		
$x_{42} = 0.7645$		
$x_{43} = 0.5580$		
$x_{44} = 0.7645$	(high)	
$x_{45} = 0.5580$	(low)	

By 45 iterations, the function has settled down on *two numbers* each accurate to four decimal places: a high of 0.7645 and a low of 0.5580. For m in the interval $3 < m < 3.4495$ ($= 1 + \sqrt{6}$), the function always jumps between two distinct numbers — *see* cobweb diagram, Figure 3.7 below.

These alternating points model numerous animal cyclical fluctuations. Perhaps the most well known are the regular variations in the lemming populations and the subsequent dependent cycles of the arctic foxes and owls. Spectacular swings of snowshoe hare populations are also common in the Canadian hinterlands and boreal regions. This phenomenon can be traced back over 200 years in the fur-trapping records of the Hudson Bay Company and has been aptly termed the "10 year cycle." The actual interval varies from 8 to 11 years, averaging out at 9.6. Following a year or more later are the dependent lynx populations (snowshoe hares are their main prey). This trailing cycle is also well documented over 200 years by the same company.

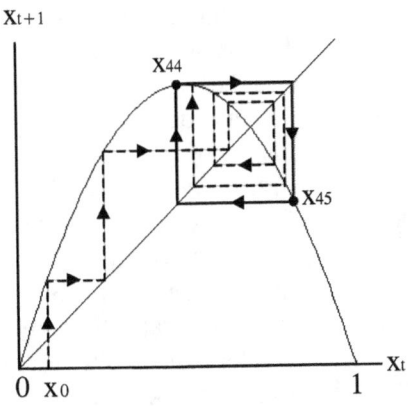

Period–Two Cobweb Diagram of the Logistic Function, m=3.1

Figure 3.7

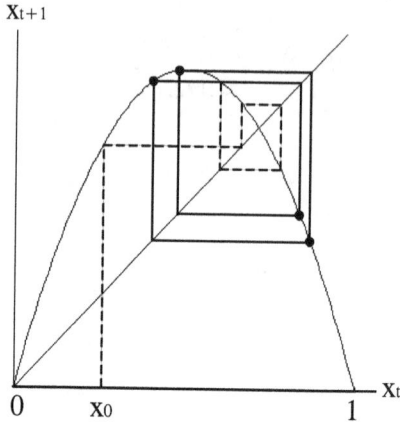

Period–Four Cobweb Diagram of the Logistic Function, m=3.4

Figure 3.8

By increasing m beyond 3.4495, say to 3.5, the period-two attractors become unstable, and bifurcate into a period-four cycle (*see* Figure 3.8). These four distinct states may be more representative of wildlife fluctuations between highs and lows if these swings take more than a single breeding season.

Above 3.5440 the four cycle bifurcates again to an eight cycle. Between 3.5644 and 3.5687, there are 16 distinct terminal numbers. Beyond this, the logistic function quickly cascades to 32, 64, 128,.... An emergent property materializes by 3.5699, something unexpected, even unacceptable — *chaos!* The period has doubled an infinite number of times; now it's over, finished, finis. Don't think of infinity as an immensely large number. *It's not!* Infinity is an entirely different creature, not just a quantity larger than any other. Even its arithmetic is unique. We're not in Kansas anymore.

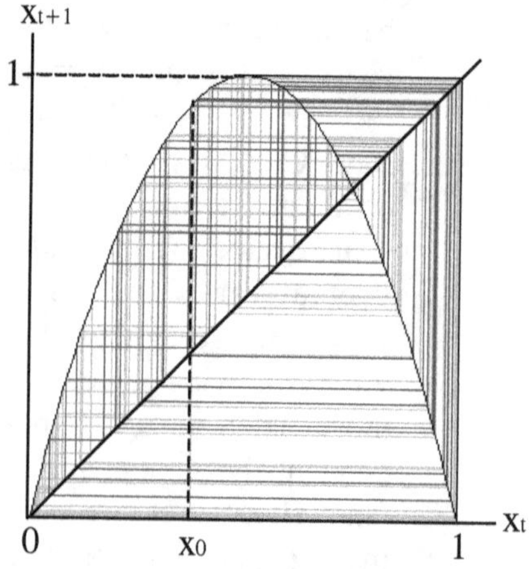

The figure to the left is an attempt to depict chaos by a cobweb diagram (*see* Chapter Notes for an excellent website to interactively explore these diagrams). But a more accurate illustration would render the entire figure black with every point on the parabola touched twice. And how many different points would that be? Quite a few actually.

Infinite–Cobweb Diagram of the Logistic Function, m=4

Figure 3.9

The parabola reaches a maximum height of 1. So the question can be rephrased as, "In how many ways can numbers occur or be placed on the line between 0 and 1?"

Let's count the ways. First are all the fractions, rightly called *rational numbers* — these are infinite in number. Are there others? Yes, the *irrational numbers* — these are also infinite. All these rationals plus irrationals constitute every number/point in this interval. Collectively these are the *real numbers*. The logistic function, when it goes chaotic, touches all the real numbers from 0 to 1; in fact, it touches them exactly twice: once on the way up the parabola and again on the way down.

These rational and irrational numbers *don't* alternate along the interval with any pattern whatsoever. Here the landscape is unfamiliar, unexpected, bizarre:

- Between any two rational numbers is another rational number.
- After zero, there is no *next* real number.
- Both rational and irrational numbers are infinite.
- The set of irrational numbers is greater than the set of rationals numbers. This means infinities come in different sizes — incredible.
- An infinite number of infinite sets of different "sizes" exists.

If this isn't Kansas, then where is it? In the late 19th century, the German-Jewish mathematician Georg Cantor solved all the paradoxes surrounding infinity, some dating back to ancient Greece. Bertrand Russell thought Cantor to be the greatest genius of his time; there were other opinions. But today his ideas about this strange land are universally applauded for their consistency and power. David Hilbert, another eminent German mathematician, declared, "From the paradise created for us by Cantor, no one will drive us out."

Return to the logistic function when $m > 3.5699$, the domain of chaos, and step back to broadly view what has happened here. We observe that this function is the essence of simplicity and determinism. Yet, weirdly — this word is not strong enough — it cloaks chaos in its core: deterministic chaos. This seems to be an oxymoron, like Milton's "No light; but rather darkness visible" from *Paradise Lost*. Oxymoron or not, that's the way the function behaves; that's also the way much of nature works.

Chaotic behavior also occurs in probability dynamics such as coin flipping, dice rolling, autumn leaves falling to the forest floor face up or down, gene shuffling at conception, and so on. Is there a difference between deterministic chaos and probabilistic chaos? Imagine you get some data on each. Could you tell which was which? They're both varieties of randomness. By just inspecting the final data, however, and not the process that led to it, you cannot distinguish one from the other. If it walks like a duck, quacks like a duck, look likes a duck, then it must be a duck! And a mighty duck it is!

MOST PERSONAL COMPUTERS using a Windows interface have a *Microsoft QBasic* programming language under the applications icon. You may enjoy typing in the simple program listed below — it only has 10 active lines — and changing the values of m (*see* Figure 3.10) to acquire a good understanding of the logistic rule's behavior. Since "REM" means "remark" in QBasic, it's not part of the program, and may be omitted.

```
REM  Play with the following two initial values to see what happens.
m = 3.5
x = 0.3

REM  The loop below gives the function time to settle down.
FOR  k = 1 TO 50
     x = m * x * (1 - x) : REM This is the feedback loop.
NEXT k

REM  This loops prints the final 10 iterations.
FOR  k = 51 TO 60
     x = m * x * (1 - x)
     PRINT x
NEXT k

STOP
```

REM Under the "Run" menu press Start. Under the "View" menu press Output Screen. Use the "Esc" key to return to the program.

Remember to keep the growth factor "m" in the interval 0 to 4, and the starting value "x_0" in the interval 0 to 1, although the particular seed used is unrelated to the outcome. When m=3.5, as in the program above and the table below, the final values are always this period-four cycle:

$$0.8749 \rightarrow 0.3828 \rightarrow 0.8269 \rightarrow 0.5008$$

The following table summarizes the effect of the different growth factors on the logistic function.

INTERVAL	ATTRACTOR	EXAMPLE
$0 \leq m \leq 1$	Goes to 0 — extinction	m = 0.2
$1 < m \leq 3$	Goes to $1-1/m$	m = 2.3
$3 < m < 3.4495$	Period-2 attractor	m = 3.1
$3.4495 < m < 3.5440$	Period-4 attractor	m = 3.5
$3.5440 < m < 3.5644$	Period-8 attractor	m = 3.548
$3.5644 < m < 3.5687$	Period-16 attractor	m = 3.568
⋮	⋮	⋮
$3.569945 < m < 4$	Chaos	m = 3.8

TABLE OF GROWTH FACTORS — m

Figure 3.10

The logistic function is an excellent model to explain something of nature's chaotic dynamics. That chaos occurs in nature is unquestionable. We find it in weather systems, measles epidemics, animal and insect population fluctuations, water turbulence, and so much more. But *reality* should never be confused with its *mathematical model*. Reality — the world out there — is a great deal richer than anything ever captured in the butterfly net of our poor equations. Nonetheless, the logistic function is a finely tailored suit on the solid body of reality.

Many scientists find it inexplicable how well mathematics explains the natural world. It's also mysterious how much mathematics nature knows and how long it has known it. These are deep philosophical problems best left to other authors.

IN A LAND FAR AWAY AND LONG AGO lived four very wise men who were advisors to a young Indian rajah. What was unusual about these men, other than their great wisdom, was their blindness. Blindness they alleged was an asset for it allowed them to deeply contemplate life without being distracted by the things of this world.

The rajah was not convinced that this was so. He was young and had newly inherited his kingdom from his father, and these wise men had been his father's advisors. The young rajah decided to test their renowned wisdom. Now in this region of India elephants were unknown, but the rajah during his education had traveled widely and seen them often. The wise men had never traveled since they had no interest in the outside world. So the young lord had an elephant brought to his land and led into his court. Then the four wise men were commanded to advise the rajah on what manner of object/creature this was. One of the blind sages found the beast's tail and avowed it a kind of broom. Another grasped a hind leg and loudly affirmed it a beech tree. Still a third mistook the elephant's trunk for a giant snake and quickly moved away. The last of the advisors, rumored to be the wisest of all, smoothly ran his hand along a huge tusk while the entire court was hushed in anticipation of his answer. At last he declared this a mighty instrument of war, a great warrior's spear, to the amused rajah and the smiling courtiers.

After witnessing this piecemeal picture of the elephant — apparent to all present but the advisors — the rajah retired them, each to contemplate his own navel where they could do no further harm. And the young rajah became his own advisor.

At this moment, we're in the position very like the "wise advisors." We see only the pieces of the puzzle that compose the logistic function. Our elephant is Figure 3.10, the table of the growth factors m. We need a global view for *all m, from 0 to 4*, that's every value, rational and irrational, in the interval, not just pieces and points. So let's construct a new graph — a remedy for our blindness — with $0 \leq m \leq 4$, on the horizontal axis, and $0 \leq x \leq 1$ on the vertical. Every x on this curve will be an *attractor* of the feedback function after at least 50 iterations. In other words, the function has had time to settle down to its terminal value(s). For any m outside the interval 0 to 4, the logistic function spirals to negative infinity.

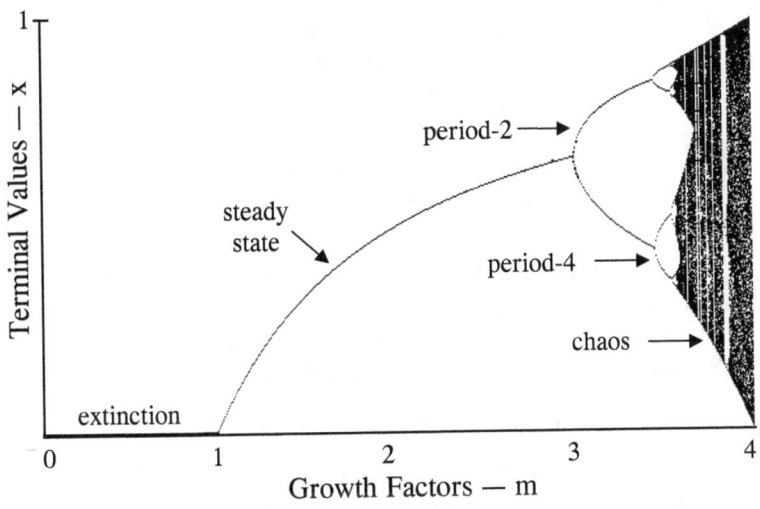

BIFURCATION DIAGRAM ONE

Figure 3.11

The full title for the above figure is the *Bifurcation Diagram of the Logistic Function;* it shows the complete effect of the growth factor m — the whole elephant, so to speak. As m increases from 0 through to 4, we can visually follow its influence on the function. While you trace along the curve, recall a few of the corresponding cobweb diagrams. Each vertical slice, for any m, gives a quick snapshot of the matching cobweb graph with all the steps/webs removed leaving just the attractor(s). For example, when m is less than 3, the attractors are points, the points of the extinction and steady-state curve.

Ah, but the interesting part of the graph begins when m equals 3: this point is just marginally stable; here convergence is exceedingly slow. At any value immediately beyond 3, the associated cobweb diagram (*see* Figure 3.7) spirals outward to two points, giving the bifurcation graph two branches. At about 3.4495 a period-4 cycle emerges — *see* cobweb Figure 3.8 and the above figure. After this, these bifurcations come thick and fast — 8, 16, 32, ... — and then suddenly stop. Beyond a particular point, a limiting value generally termed the "point of accumulation," periodicity transforms into chaos. This period-doubling cascade effect is reminiscent of some lines from Lewis Carroll's famous nonsense poem *The Walrus and the Carpenter:*

> *Four other Oysters followed them,*
> *And yet another four;*
> *And thick and fast they came at last,*
> *And more, and more, and more —*
> *All hopping through the frothy waves,*
> *And scrambling to the shore.*

So, the bifurcation diagram sums up the full behavior of the logistic function. As the growth factor m increases, we appear to see an increasing degree of disorder. Figure 3.11 seemingly offers a road map of the function's behavior:

$$\text{steady state} \longrightarrow \text{periodic} \longrightarrow \text{chaotic}$$

However, the situation is not quite that neat. There is an enigma here, a riddle worthy of Lewis Carroll himself. Experienced artists, scientists, and mathematicians have learned to pay close attention to odd or anomalous objects and outcomes. A little bump in the ground, a small unevenness in the results, an unusual reading; this is where the gold will lie buried. Consider the history of such: Alexander Fleming's discovery of penicillin in a spoiled petri dish, Karl Jansky's recognition of the extraterrestrial origin of radio background noise, or Charles Darwin's realization that the different beak configurations of Galapagos' finches implied evolution from a common ancestor. These discoveries were not wholly serendipitous. All these breakthroughs, and thousands of others, came about because perceptive men and women refused to gloss over or disregard something that hundreds of others had missed.

Where is the puzzle hiding in the bifurcation diagram? Well, it's right before our eyes. As you can see in Figure 3.11, entire regions of the graph are black. Why the chaotic region is not exclusively black is our mystery. More particularly, a few vertical strips are white, islands of order in a sea of chaos. So, our neat summary — running from steady state to periodic to chaotic — is not correct, or at least not the whole story.

Consider Figure 3.12 on the next page. Here the previous bifurcation diagram has been enlarged in the interval from 3 to 4. Notice that the wide vertical white strip (indicated by a question mark) in the right section has a period of 3. As researchers like to say, "Period 3 means chaos."

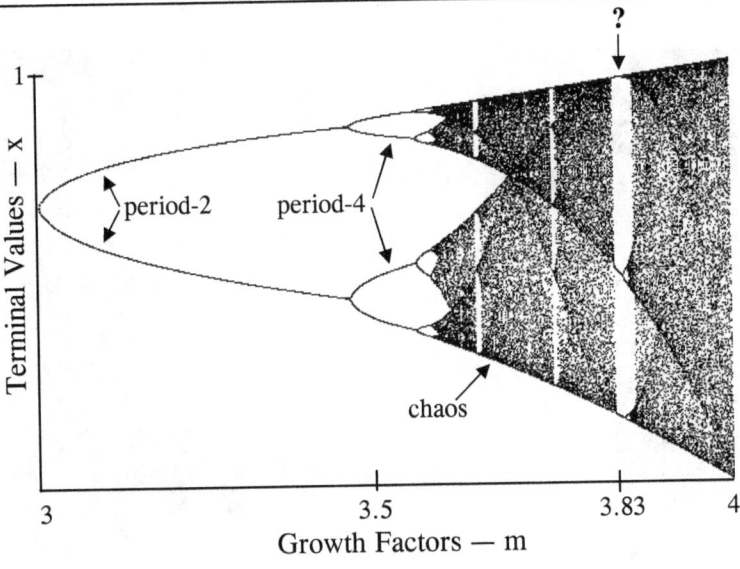

BIFURCATION DIAGRAM TWO

Figure 3.12

According to the table in Figure 3.10 — the pieces of the elephant — chaos starts when the growth factor m is greater than approximately 3.5699. Let m be 3.83, well inside the chaotic zone, and using any seed find the terminal values for $x_{t+1} = 3.83x_t(1 - x_t)$, the logistic function. Surprisingly, after 50 or so iterations it settles down on a cycle of just three numbers. Generating something simple from something surrounded by extreme complexity is unexpected:

$$0.1561 \rightarrow 0.5047 \rightarrow 0.9574$$

If m is increased slightly beyond 3.83, new periods emerge in multiples of 3, that is 6, 12, 24, 48, . . . quickly cascading into chaos. So, as we unfold the layers of this graph, we come face to face with an incredible complexity arising out of the deceptively simple logistic function. Generating extreme complexity by simple means is unexpected and liberating. Ecclesiastes 1:9 was mistaken:

> *The thing that hath been, it is that which shall be;*
> *and that which is done is that which shall be done:*
> *and there is no new thing under the sun.*

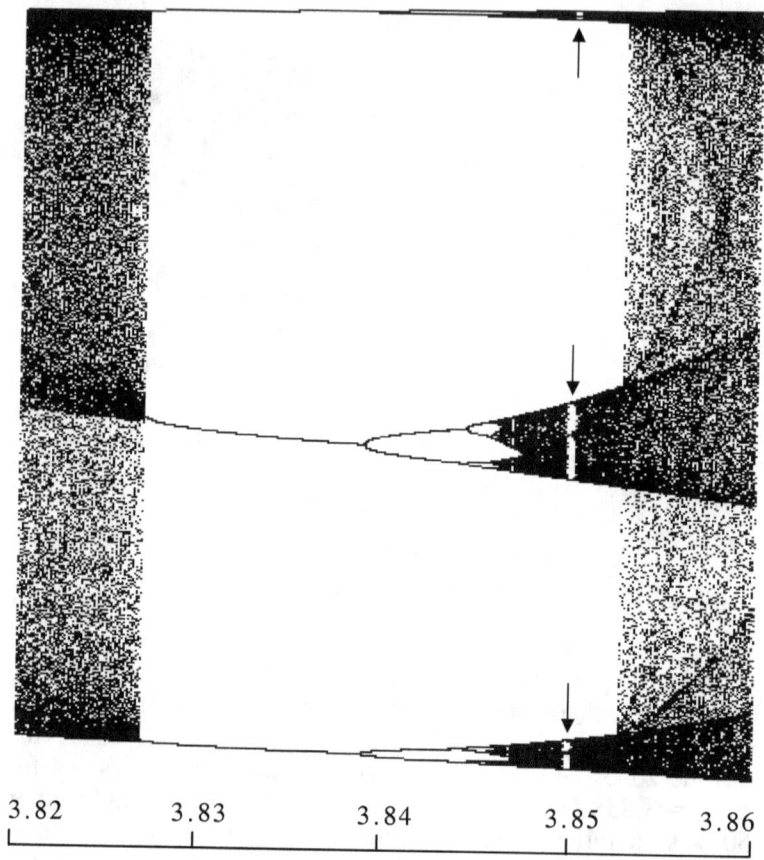

BIFURCATION DIAGRAM THREE

Figure 3.13

To comprehend a little of this vast complexity, see the above graph. It's an expanded view of the white strip on either side of 3.83. Note the abrupt transition from chaos to period 3 on the left; note also the reverse transformation from period doubling to chaos on the right. Strikingly, we detect three new images, homunculi so to speak, reminiscent of the entire bifurcation graph. We say this figure is self-similar, a wonderful topic to be explored in the next chapter. Three additional white strips (indicated by the arrows) are glimpsed with still more homunculi — graphs within graphs, worlds within worlds, ad infinitum.

The American physicist Mitchell Feigenbaum uncovered much of the magic in these bifurcation diagrams. Incidentally, in German the name "Feigenbaum" means "fig tree." Hence, these figures are commonly referred to as such — a scientific pun.

Inspired by Feigenbaum's discoveries, other researchers pushed his studies into new territory. Incredibly an utterly different variety of function, $x_{t+1} = m\sin(x_t)$, has the same fig-tree diagram as the logistic function. And the mystery deepened further when researchers uncovered that *any polynomial function* that rises initially and falls later has the identical graph. This is more than astonishing; it's bizarre! It's as if you had found that one shoe size fits everyone or one hat size fits all. A profound universality lay buried deep in the process of iteration and with the fig tree — Feigenbaum had found it.

The Butterfly Effect

IN THE EARLY '60S EDWARD LORENZ, a meteorologist with mathematical inclinations, made a landmark discovery. He found that no matter how much data he collected, his weather predictions — and everyone else's — would *never* be accurate in the long run. And the cliché *in the long run* could mean as little as a few days. This is not a matter of refining our models or discovering new ideas; it's part of the nature of things. Neither the awe-inspiring power of the Cray supercomputers, nor the mythic reliability of the *Farmers' Almanac,* nor the alleged absolute verisimilitude of your aunt Mildred's corns will prevail! In the end, the unpredictability of the weather will always triumph.

To model weather systems, Lorenz programmed three and sometimes as many as twelve nonlinear equations and calculated their *long-term behavior* — a classic deterministic prediction setup. He did this on what would now be considered a primitive and noisy computer, a Royal McBee. Of course, the result of each calculation was fed back in for the next number crunching: tomorrow's major conditions depend on today's major trends, or so everyone thought.

James Gleick in his definitive book on this subject, *Chaos: Making a New Science,* described the context of discovery:

> One day in the winter of 1961, wanting to examine one sequence [printout of the Royal McBee's calculations] at greater length, Lorenz took a shortcut. Instead of starting the whole run over, he started midway through. To give the machine the initial conditions, he typed the numbers straight from the earlier printout. Then he walked down the hall to get away from the noise and drink a cup of coffee. When he returned an hour later, he saw something unexpected, something that planted a seed for a new science.[5]

Now, this second run should have duplicated the last half of the first run, but it did no such thing. The numbers diverged significantly; all similarity vanished. Serendipitous discovery requires a great deal from the discoverer. Lorenz could easily have dismissed this divergence as just another computer glitch such as a loose wire or a broken vacuum tube. Fortunately, on further reflection he saw the difficulty was in the numbers he had typed in. The Royal McBee printed out three decimal places, but its memory stored six; Lorenz had entered the shorter numbers. The assumption, whether conscious or not, was that one part in a thousand was of no consequence. At this moment he clearly realized this supposition was false, surprisingly false! With this, the *Butterfly Effect* — sensitive dependence on initial conditions — was born in the brain of one perceptive individual.

The Butterfly Effect can readily be demonstrated with the assistance of our old friend the logistic function. Let $m = 4$, the most extreme growth factor; it's at the far end of the chaotic region. Also let the initial condition of the first series be $x_0 = 0.4000$, and $x_0 = 0.40001$ for the second series — a minute difference of only one part in ten thousand. After just 19 iterations (*see* Figure 3.14 on the following page), the first seed of the first series iterates to 0.0608 and the second seed to 0.3950. This divergence, or sensitivity to initial conditions, is more evident in the graph below the table. At x_{12} a slight separation appears while at x_{17} the series is nearly at a maximum distance apart of one. Paraphrasing his most memorable and colorful expression, Lorenz declared, "The flap of a butterfly's wing in Brazil can set off tornadoes in Kansas."

$X_{t+1} = 4X_t(1 - X_t)$		
Iteration	$X_0 = 0.4000$	$X_0 = 0.40001$
Series	—◆—	—■—
X_0	0.4000	0.400010
X_1	0.9600	0.960008
X_2	0.1536	0.15357
X_3	0.5200	0.51994
X_4	0.9984	0.99840
X_5	0.0064	0.00635
⋮	⋮	⋮
X_{14}	0.0351	0.0005
X_{15}	0.1356	0.0018
X_{16}	0.4689	0.0072
X_{17}	0.9961	0.0286
X_{18}	0.0154	0.1111
X_{19}	0.0608	0.3950

A BUTTERFLY EFFECT TABLE

Figure 3.14

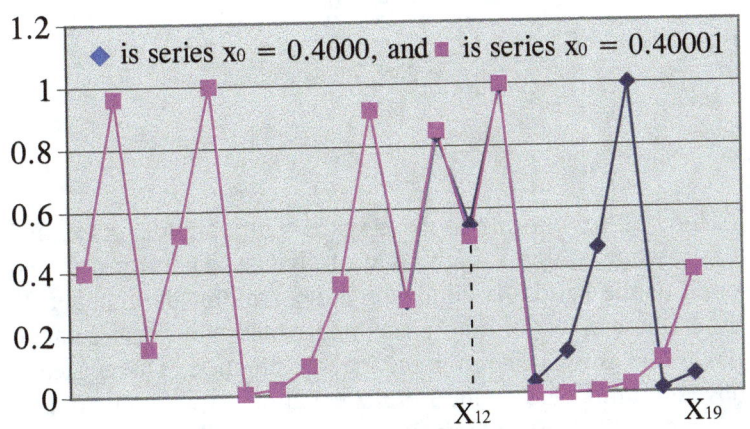

A BUTTERFLY EFFECT GRAPH

Figure 3.15

Consider the nature of the problem that produced the Butterfly Effect in the Royal McBee. Internally computers and calculators crunch numbers to a greater accuracy than they display. It's the difference between the memory register and the display register. You can test this for yourself on any calculator, not just the McBee. Here's what to do: take one and divide by seven to get 0.142 857 142 857. . ., a rational and hence by definition a period-repeating decimal — six in this case. Of course, other fractions will do. Typically, a small calculator will display ten digits of accuracy (i.e. 0.142 857 142). From this subtract 0.142 857 to get a difference of 0.000 000 142. In the next and last step multiply this difference by 1,000,000 and press the equal (=) button. Magically more decimals, previously hidden, will be dragged from the memory register into the display register. Mine showed three more (i.e. 857). Another application of this simple algorithm will produce no further hidden decimals. My calculator displays 10 digits but works internally with 13.

It's a common scientific gambit to simplify a problem, at least initially, in an attempt to comprehend its essential elements. Lorenz did this when he discarded most of the equations of a previous weather researcher to a bare minimum of three. He was left with a stripped down model of heat convection: the rising of hot air. His trio of equations has now become a classic:

$$\frac{dx}{dt} = 10y - 10x$$

$$\frac{dy}{dt} = 28x - y - xz$$

$$\frac{dz}{dt} = xy - \frac{8}{3}z$$

Here we're working in three-space with variables x, y, and z; "t" is time; "d/dt" is the rate of change. The terms xz and xy tell us the system is nonlinear. Furthermore, the number 28 pushes the convection into chaos and hence plays a roll analogous to the growth rate m in the logistic function. This unsteady convection modeled by the Lorenz equations is without wobbles and wheel-like. Paradoxically, it actually does reverse its direction repeatedly unlike the wagon wheels in old western movies, which only appeared to. And the convection with reversals is non-periodic, without pattern, and hence chaotic.

To see this behavior in phase space, the three equations must be solved. Like the biblical Jacob toiling for Leah and then again for Rachel, young mathematics students labor seven years to become proficient at solving linear differential equations and seven more trying in vain to do the same for the nonlinear variety. Lorenz' three can be integrated and solved using a fourth order Runge Kutta method. Happily, we won't do this. Trickery, guile, and cunning are also good. As well, brute force numerical computation works, and this was what Lorenz used.

The result is the now famous Lorenz' "strange attractor" which displays some rather remarkable behavior and represents a landmark in the field of chaos — *look* down. Despite its flat appearance, this attractor is three-dimensional; the trajectories in the center of the figure don't truly intersect. Intersection would imply the paths pass through identical values, and this would make the convection periodic and hence non-chaotic, which it is not. These are the figurative Brazilian wings causing chaos over Kansas.

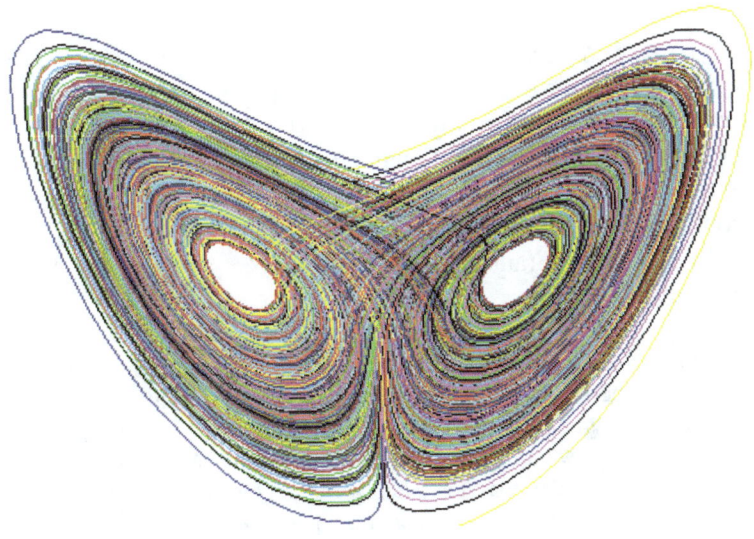

THE LORENZ ATTRACTOR

Figure 3.16

We're in the domain of chaos, but a strange confined chaos with an explicit form and always inside the finite space of a box. Repeated magnification reveals an infinite regress of restricted paths none ever duplicating its neighbor however close — a deep self-similarity. There is form here, the form of the fractal. The flow of time is indicated by color. And each wing represents convection in either a clockwise or a counter-clockwise direction. As the trajectories wander about randomly, a striking bilateral symmetry unfolds like the wings of a butterfly. There is meaning here. This strange attractor is neither without form nor void.

THE RIDER IN THE WHIRLWIND

Into this universe, and Why not knowing,
Nor Whence, like Water willy-nilly flowing;
And out of it, as Wind along the Waste,
I know not Whither, willy-nilly blowing.
Rubáiyát of Omar Khayyám

A FOLK WISDOM RESONATES to the beat of the butterfly's wing, an echo we all hear. Had I not gone to that party, I would never have met my wife. Had I left the house a few seconds earlier/later I would have been involved in the car accident. Had I not taken my daughter to the dentist that morning, I would have been at the World Trade Center.

In Benjamin Franklin's *Poor Richard's Almanac* for 1758 you can read:

> And again, he, Richard, adviseth to circumspection and care, even in the smallest matters, because sometimes a little neglect may breed great mischief, adding, for want of a nail, the shoe was lost; for want of a shoe the horse was lost; and for want of a horse the rider was lost, being overtaken and slain by the enemy, all for want of care about a horseshoe nail.

In this homily, *For Want of a Nail*, we hear a clear reverberation of the Butterfly Effect. Earlier versions predate Franklin by at least 150 years; some are probably centuries older.

Ideas gain power when they're clearly articulated and used: Edward Lorenz did this for the Butterfly Effect. This power is a broad indication of an idea's worth and longevity. New visions are then created; former structures are rebuilt; others demolished. History is one such house.

Human history is an excellent example of non-periodic behavior. Civilizations may rise and fall, but events *never* happen in the same way twice. Small actions can change the world — such as the birth of a new virus! Some historians thought that the heroes and despots of previous ages controlled the unfolding of events. Others considered that these defenders and destroyers were really carried along by the flux like wood chips in a great whirlwind of waves and water. Sensitivity to initial conditions — and there is a near infinite regress of these — speaks more positively of this second view. But truly, neither alternative commands history: minuscule actions and/or broad events often rule for a while until shoved aside by new usurpers.

By finding patterns in time's passage, men and women thought to understand the events of the past. They longed to give existence meaning by discovering cycles and consistency in their lives, in the life of their country, and in the lives of previous civilizations. From Daniel in the distant past to Arnold Toynbee in our time, historians have known this to be their major task: find history's design. Down through the centuries they have pursued this chimera only to have it vanish in the flap of a butterfly's wing.

Oh, like the weather we may know the seasons, different climates, and the average of this and that — enough to map out a strange attractor for the system dynamics, but never any of the intriguing details and definitely not the long-term outcome.

This constantly changing cascade of events was immortalized by the 5th century Greek philosopher Heraclitus (not the person in the poem *Heraclitus*) in his maxim, "Man never bathes in the same stream twice." Or as some later wit remarked, "Not even once." So complete was Heraclitus' belief in the flux that he declared the sun to be created anew each day like the morning's cooking fire. For him the only unchanging thing was change itself.

Plutarch's *Parallel Lives* sponsors the hero as history's herald and hammer. (The sheer scope of his *Lives* is immense — some 800,000 words in 1,300 pages of fine print in the unabridged Dryden translation.) But Plutarch makes no claim to being an historian. At the beginning of his "Life of Alexander the Great" — paralleled with Julius Caesar — he writes, "It must be borne in mind that my design is not to write histories, but lives."[6] Accordingly, it's difficult to classify the *Lives* as history, biography, or philosophy. Even his few surviving assessments of the parallel lives, for example, "The Comparison of Demosthenes and Cicero" dwell more on differences than similarities. Plutarch's timeless studies are in a class of their own construction, so neither chaos theory nor the Butterfly Effect can find much fault with them. Modern historians have been less fortunate.

Just after the First World War Oswald Spengler, a German historian, wrote a scholarly tome that caused quite a sensation among intellectuals around the globe. In *The Decline of the West*, Spengler made two central points. First, histories of various cultures can be shown to follow a similar pattern; second, all aspects of culture — art, politics, mathematics, and science — have related underlying principles that vary only slightly.

Underpinning these two central points is Spengler's basic thesis: that cultures and civilizations are living organisms in their own right, just like plants, animals, and humans. For Spengler this was no metaphor, but an external reality. Civilizations are like flowers he would say; they start as seeds; wiggle with religious fervor into small green sprouts; grow to maturity; blossom with culture and wealth; turn to seed, corrupt, and die.

Without reference to either chaos theory or the Butterfly Effect, Bertrand Russell in his *History of Western Philosophy* evaluates Spengler's cyclical conjectures while discussing Hegel's theory of history. Russell with his famous irony writes:

> Like other historical theories, it required, if it was to be made plausible, some distortion of facts and considerable ignorance. Hegel, like Marx and Spengler after him, possessed both these qualifications.[7]

In England 20 years or so after Spengler, Arnold Toynbee published his monumental 12-volume opus *A Study of History*. At the time, this was generally recognized as one of the greatest achievements of modern scholarship. Toynbee's thesis is not unlike his German compatriot's: by analyzing the rise and fall of 26 civilizations, he concluded that history is cyclical.

Modern detractors view Toynbee as essentially a classical and Christian scholar who personally imposed order by seeing reflections of classical history wherever he looked.

However valid the reasoning of the history critics, Toynbee's and Spengler's executioner was modern chaos theory with its Butterfly effect. History is an immense, complex dynamical system, incredibly sensitive to initial (prior) conditions. History lives in the same house as weather. Nothing ever truly repeats. Every path is unique; every event, distinct; every life, original!

Finding patterns where none exist is part of our biological heritage. We're not passive observers of the landscape, but active participants in it. We dance, but we dance together. The power of the scientific method is its ability to disentangle the dancers: the subjective from the objective. Toynbee brought his culture and early education — he was reading the Greek and Latin classics at eight — to his study of history. He publicly lamented his lack of a science background. Through these dual lenses of Christianity and Classicism, he viewed all humanity and saw more than was there. We all do!

ALL RECORDED HUMAN EVENTS, what we call history, is such a minuscule span of time it could almost be overlooked. But prehistory, and particularly pre-prehistory, is where the action happened. Here the panoply of life on earth unfolded. Here life originated and started on its immense evolutionary journey.

The latest research indicates that life began at least 3.6 billion years ago in the Precambrian Era. Now ages or numbers of that magnitude are just black marks on white paper. Human beings, still living on average the biblical lifespan of three score and ten years, have no psychological or emotional understanding of them. Let's attempt to remedy this in two ways. First, if the height of a six-foot (183 cm) man represents all geological time from the inception of life to the present, then *written history* is less than the thickness of the epidermal layer on the dome of his skull.

There is another way, a second path to a deeper emotional comprehension of the immensity of evolutionary time. An obscure American journalist, Langdon Smith, wrote a poem of 108 lines titled "Evolution." Martin Gardner in his book *Order and Surprise* has written the only existing account of Smith and his one poem. What follows is a quotation from Gardner and the poem's first stanza and final four lines:

> The poem conveys what Darwin had in mind when he wrote, at the close of his *Origin of Species*, "There is grandeur in this view of life." The epic surge of evolution, from its humble beginning in the dark sea to the mellow light of Delmonico's [restaurant], is caught in this poem as it has not been caught in any other poem before or since.

> *When you were a tadpole and I was a fish,*
> *In the Paleozoic time,*
> *And side by side on the ebbing tide*
> *We sprawled through the ooze and slime,*
> *Or skittered with many a caudal flip*
> *Through the depths of the Cambrian fen,*
> *My heart was rife with the joy of life,*
> *For I loved you even then.*
>
> . . . .
>
> *Then as we linger at luncheon here,*
> *O'er many a dainty dish,*
> *Let us drink anew to the time when you*
> *Were a Tadpole and I was a Fish.*[8]

It has been a long long time getting from there to here, and the road has been anything but straight. We never climbed an evolutionary ladder or ascended an evolutionary tree. The best metaphor we have is a bush or multiple bushes. No life form evolves in a predetermined fashion or direction: this is the false concept of orthogenesis. We're on our way to nowhere not knowing whence or whither — as the *Rubáiyát* says, "*Into this universe, and Why not knowing, Nor Whence, like Water willy-nilly flowing.*" We neither descended from angels nor ascended from apes.

The very word "evolution" itself is misleading implying a direction, specifically an upward direction. Many dictionaries give an incorrect definition for it — even as their primary meaning. Somewhere they will say we're evolving *upward from the simple to the complex*. Yet occasionally organisms, especially parasites, adapt by becoming less complex: the classic example is the tapeworm, just a head with a segmented stomach.

The correct definition implies no road map, and the preferred phrase is the one Darwin often used, "descent with modification." The notion that evolution has a direction — and worse an overriding purpose — is outmoded and false. We have existed in a vast whirlwind of events over immense eras, forever *adapting to the here and now* or perishing. This is the condition of all organisms; it's the human condition, and we must deal with it.

Opposed to the soft yearning for orthogenesis is the tough concept of contingent history: the idea that no particular path of evolution is inevitable. All outcomes are contingent on a multitude of quirks and accidents. It's possible, for example, that one early vertebrate worm was responsible for the evolution of all later vertebrates. Had some accident such as climatic change eliminated that worm, human beings would never have existed.

In his splendid book *Wonderful Life*, Stephen Jay Gould fully develops the concept of contingent history with numerous detailed examples from the Burgess Shale. The book's title comes from Frank Capra's famous Christmas movie *It's a Wonderful Life* — Hollywood's unsurpassed example of contingency. The reader may sense a whispering familiarity about this idea, and rightly so. Previously we spoke of it as the Butterfly Effect.

What has been said about the Butterfly Effect on human history over the last few thousand years applies with equal force from then back to the period when life first began — about 3.6 billion years ago. In the final chapter of *Wonderful Life* Gould imagines seven possible worlds as life might have been. Yet the combinations stirred by the Butterfly Effect are such that he could as easily have imagined seven million. Consider three:

FIRST: For over two billion of these years — more than half the time for all life on earth — only primitive prokaryotic cells existed. These simple structures are without organelles: no nucleus, no mitochondria, no chloroplasts, and no paired chromosomes. Bacteria are prokaryotes; they were the dominant life form then, some say they still are. Their lonely two billion year reign provides no sense of direction or urgency in evolution.

Whatever Butterfly Effect or other event ultimately initiated the development of eukaryotes (cells with organelles — a necessary precursor to multicellular life), it could as easily have been delayed by another two to ten billion years or perhaps for eternity.

If we ever find extraterrestrial life, it might be bacteria in a slime mould. No wonder they haven't sent or responded to our radio signals.

SECOND: It has been proposed that roughly every 26 million years an extraterrestrial object impacts the earth causing mass extinctions of entire species. Whatever the merits of this theory, we do know that 65 million years ago the reign of the dinosaurs ended because an asteroid impacted off Mexico's Yucatan peninsula. This chance event made possible, but not necessary, the blooming of new species of mammals and later *Homo sapiens*.

But the winner in the extinction sweepstakes occurred 250 million years ago marking the Permian-Triassic boundary. At that time, 96% of all living species became extinct. By the skin of our teeth, one of our distant ancestors made it through this tiny corridor. There is some evidence that smaller rather than larger species manage better during these extinctions, but no safe passage guarantees are ever issued.

The maelstrom of events created by these impacts could easily have extinguished all life on earth. By chance, by luck, by beating all the probabilities, a few creatures survived, and one of them evolved to self-consciousness — that's a case of matter reflecting on itself, the hallmark of *Homo sapiens*.

THIRD: For our final scenario we will let Gould speak for himself of the species closest to his heart and ours. The metaphor is a movie theater running the tape called "Life":

> Run the tape again, and let the tiny twig [on the bush] of *Homo sapiens* expire in Africa. Other hominids may have stood on the threshold of what we know as human possibilities, but many sensible scenarios would never generate our level of mentality. Run the tape again, and this time Neanderthal perishes in Europe and *Homo erectus* in Asia (as they did in our world). The sole surviving stock, *Homo erectus* in Africa, stumbles along for a while, even prospers, but does not speciate and therefore remains stable. A mutated virus then

wipes *Homo erectus* out, or a change in climate reconverts Africa into inhospitable forest. One little twig on the mammalian branch, a lineage with interesting possibilities that were never realized, joins the vast majority of species in extinction. So what? Most possibilities are never realized, and who will ever know the difference?

Arguments of this form lead me to the conclusion that biology's most profound insight into human nature, status, and potential lies in the simple phrase, the embodiment of contingency: *Homo sapiens* is an entity, not a tendency.

How are we to deal with this view of life and history emotionally? No design; no slime to man; no ultimate purpose; just endless, meaningless chance. Well, we begin by recalling that Bertrand Russell thought happiness depended more on good digestion than a view of life. And Gould wisely wrote that contingent history gives us maximum freedom to thrive, each in our own individualistic way. Myself, I prefer the declarations of the modern Russian writer Yevgeny Yevtushenko expressed in his poem "Thanks":

> *Say thanks to your tears.*
> *Don't hurry to wipe them.*
> *Better to weep and to be.*
> *Not to be is to die.*
>
> *To be alive — bent and beaten.*
> *Not to vanish in the dark of the plasm.*
> *To catch the lizard-green minute*
> *from creation's cart.*
>
> *Bite into joy like you bite a radish.*
> *Laugh as you catch the knife's blade.*
> *Not to be born, that's what's frightening*
> *even if it's frightening that you live.*
>
> *He who is — is already lucky.*
> *Life is a risky card.*
> *To be drawn — that's a cocky occasion.*
> *It's to draw a straight flush.*
>
> *In the sway of wild cherry blossoms,*
> *drunk on all, drunk on nothing,*
> *don't shake off the large wonder*
> *of your entrance upon the scene.*

The application of contingent history should again be expanded to include the formation and development of our solar system. To date, astronomers have discovered 130 extra-solar planets ranging from the size of Saturn to more than 10 times Jupiter's mass. Some have highly elliptical paths and take so few days to orbit their star that they must be roasting. Astronomers dubbed them "hot Jupiters." The wide sweep and immense gravitational pull of these giants would be deadly to all planets in the so-called habitable zone of liquid water. No chance for carbon-based life there.

If viewed from a distance in space, our solar system appears gem-like in its perfection with near-circular orbits. Only Pluto's path is clearly elliptical to the naked eye — and many astronomers believe it to be an errant asteroid. This clockwork stability over geological ages has provided the necessary time for life on earth to develop. And our Jupiter, at a safe distance and in a virtually circular orbit, has long since gravitationally vacuumed up most of the solar-space junk to save earth from fatal impacts. Recall the comet Shoemaker-Levy bombardment of the Jovian planet in July 1994.

Without such fine-tuned points of planetary position, orbit, and size, we could not exist. But present evidence indicates our solar system is not the standard model. Astronomers believe that for life to develop on earth an orbiting moon of the mass of ours is essential. A moonless earth would rotate three times faster than it presently does, and our tides would be much lower. Storm force winds would prevail like those on Jupiter prohibiting "tall" mammals such as humans from evolving. . . . The patient reader may feel as I do that this list of contingencies is long enough — we will go no further.

ON THE FAR HORIZON we see a black speck, a small dark dot. The rising heat causes it to simmer in the sunlight and wobble from side to side. Occasionally it dips below the horizon and disappears from view. We wait. But it resurfaces and grows larger. The fluctuations continue with maddening frequency for what feels like an eternity, but every time the small black dot re-emerges ever larger and closer. We wait! Our anticipation increases with the dot's size. It simmers, wobbles, dips, and resurfaces like an immortal. Again we wait. Ultimately, we discern some shape. It appears to be a wild stallion with an ape-like being clinging to its back. As it approaches, the shape morphs more than the distance should allow. The stallion repeatedly attempts to dislodge the desperate hominid — but the bravery of the rider excites our admiration. We shout encouragement! He doesn't hear us. Again, the rider in the whirlwind appears to transform. Now he rides upright with free hands gripping the reins. Once again, he dips and disappears. And once more we wait! We cheer, as he miraculously breaks free of the horizon. Much closer now, so that this time he hears us yell. He's so close that we sense his determination and passion for survival. The whirlwind of events pummels him ever harder; we despair, but he remains mounted. Now we clearly see both horse and rider. He shouts words of recognition. And we know the rider in the whirlwind is us.

AB – SURD UNIVERSE

Pythagoras . . . was intellectually one of the most important men that ever lived, both when he was wise and when he was unwise.
Bertrand Russell (1872–1970)

PYTHAGORAS was a most paradoxical person. Two great rivers of western thought originated with him, the mystical, and the rational. Born on the Greek island of Samos around 560 BC, he could easily see Asia Minor only one mile away. And most of his mystical musings flowed from there into his work and on to his followers in later ages. But we shall paddle in the other stream.

He's the first person in history known to have proved anything. Of course, I'm referring to the great theorem that still carries his name. Oh, a thousand years earlier the Babylonians of Hammurabi's time knew hundreds of special examples of this theorem, but reasoning with numbers and diagrams in *complete generality* was new with Pythagoras. As Jacob Bronowski says in *The Ascent of Man*, ". . . Pythagoras raised this knowledge out of the world we should now call empirical fact into the world of what we should now call proof." We don't know his actual proof, but we suspect it was straightforward and simple — not the one in textbooks.

Recall the theorem. Given a right triangle, then the sum of the areas of the squares on the two shorter sides equals the area of the square on the longer side (hypotenuse). Consider Figure 3.17, a dissection, or look-see proof; possibly the one the legendary sage discovered. The surrounding outer squares in (a) and (b) are identical in size. The dark right triangle of (a) has shaded squares on its two shorter sides, while (b) has a large shaded square on the longest side of the same dark triangle.

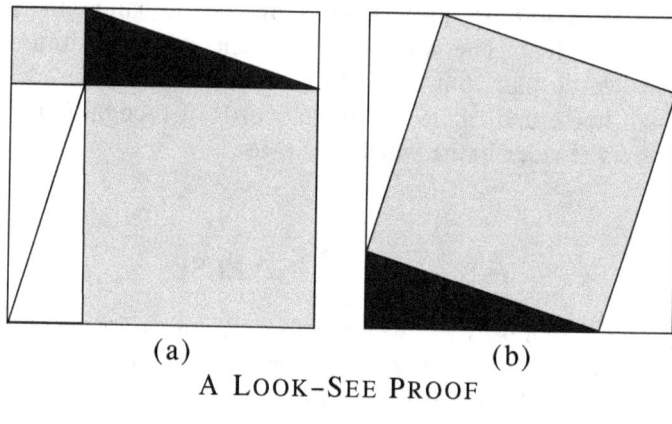

(a)   (b)

A LOOK-SEE PROOF

Figure 3.17

Both (a) and (b) have four identical right triangles, just differently arranged inside the same surrounding square. So, the leftover shaded area of (a) must equal the leftover shaded area of (b). Voilà! The theorem is true. Take note that since the dimensions of the dark right triangle were never stated, it can be any size whatsoever. So the theorem works for *every right triangle* not just the ones with sides of 3, 4, 5 or 5, 12, 13 units.

But the claim is stronger than that, much stronger. We now declare the theorem is true for *all time and everywhere in the universe*. It's part of the fabric of space itself, our space. In a world where "Man never bathes in the same stream twice," the feeling of power and permanence that deductive reasoning offers is intoxicating. With respect to this enchantment, Bertrand Russell says it better than anyone in the powerful "Prologue" to his three-volume autobiography: ". . . I have tried to apprehend the Pythagorean power by which number holds sway above the flux."

Greek deductive reasoning culminated with Euclid's *Elements*, one of the greatest books ever written. Mostly a compendium, it includes the Pythagorean theorem in Proposition 47 of Book I. Even after 2300 years, Euclid's scope and undertaking still excite our collective imagination. The entire structure, all 13 books, stands on the bedrock of five postulates and five axioms, or self-evident truths as he called them. For example, when you add equals to equals, you get equals.

This mathematics book has been so incredibly influential that when the Declaration of Independence says, "we hold these truths to be self-evident," it's modeling itself on Euclid. Franklin substituted the term "self-evident" for Jefferson's "sacred and undeniable." Abraham Lincoln could give any proposition in the first six books of the *Elements*. And President James Garfield, while still a Republican congressman from Ohio, discovered a novel proof of the Pythagorean theorem. Curiously, it involved exactly half of Figure 3.17 (b) — *see* the Chapter Notes for his proof.

"Number rules the universe," Pythagoras asserted. He had shown that that was so in the world of vision because the vertical and the horizontal formed a right angle — the premise of his theorem. History also records that he did a similar thing for the world of sound by discovering a basic relationship between musical harmony and whole numbers. By "number" Pythagoras meant what we would today call "rational quantities." That's either whole numbers themselves or ratios of them. And he believed he had demonstrated the dominion of these numbers in the worlds of both vision and sound.

Let's look more closely at his numbers. We will use decimals, something he didn't have, because they'll get to the core problem directly. Every whole number can be written as a terminating decimal (period of zero):

$$3 = 3.000000\ldots = 3.\overline{0}$$

So can terminating fractions:

$$\frac{1}{2} = 0.5 = 0.5000000\ldots = 0.5\overline{0}$$

We can say that every whole number and all terminating fractions have a period of "0." Now consider the following list of non-terminating rational numbers. Each has a period, at most, of one less than its denominator. Again, by analogy to our earlier logistic function, each has an attractor of one or more digits. For instance, our old friend 1/7 cycles through six attractors, i.e. 1,4,2,8,5,7. Hence, all Pythagoras' numbers — all rational numbers — are periodic.

$$\frac{2}{3} = 0.666666\ldots = 0.\overline{6}$$

$$\frac{1}{7} = 0.142857142857\ldots = 0.\overline{142857}$$

$$\frac{3}{11} = 0.27272727\ldots = 0.\overline{27}$$

$$\frac{4}{13} = 0.3076923076092\ldots = 5.\overline{307692}$$

$$\frac{1}{17} = 0.05882352941176470582\ldots = 0.\overline{0588235294117647}$$

As with the logistic function, the length of a period can be as long as you please. Before modern computers, dedicated arithmophiles found extended periods by incredible persistence and a few tricks. An Englishman named William Shanks, suffering from a terminal addiction to long division, found the 17388 digits in the period of 1/17389. Few remember him today — *sic transit gloria mundi!* His degree of dedication is reminiscent of Indian holy men who roll across the entire subcontinent to demonstrate their faith, or Catholic saints who sit atop poles for decades to be closer to their god. The most renowned of these, Simeon Stylites, reportedly spent 35 years at his craft.

The period need not begin directly after the decimal point. In fact, the division can jump about erratically through any number of digits before settling down on its final set of attractors, its period. The two examples below have been specially constructed. All this behavior recalls the logistic function.

$$\frac{11111111}{900000000} = 0.01234567\overline{8}$$

$$\frac{130898383259}{416662500000} = 0.31415926\overline{14142}$$

To continue the comparison, writing rational numbers as decimals is an iterative process just like the logistic rule. Recall how periods are generated by long division. For example, with 1/7, you recursively divide by 7 until a remainder repeats and then you go through the entire cycle again. To see this as a recursive equation refer to the Chapter Notes.

Without exaggeration, we conclude that rational quantities imitate all the behavior of the logistic rule up to the "point of accumulation." Beyond this limiting value, periodicity transforms into chaos for the logistic function. But with rational numbers, there is never any chaos, just the perfect order found in every period.

THE PROBLEM AND THE POWER of great discoveries is they open unexpected doors to new truths. Pythagoras' theorem unlocked one such door. Truth wears no clothes, and her stark nakedness alarmed the sage and his followers. What they saw was a new kind of number, one they couldn't express as the ratio of two whole numbers. Because of this, it never has a repeating group of digits — a period. Today we call them "irrationals" and that name still echoes the Pythagorean shock.

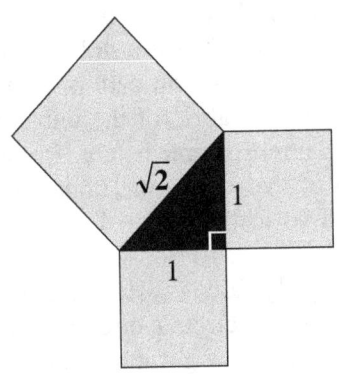

THE ROOT OF THE PROBLEM

How they found them is no mystery, but rather a clear-cut application of their theorem. Consider the dark isosceles triangle to the left with short sides each one unit. Since this triangle is right-angled, the theorem tells us the longest side must be $\sqrt{2}$ units. This was the first irrational number to be discovered — and the taproot of the problem.

All our knowledge about Pythagoras comes from the hinge of history and legend. He wrote nothing himself. And those who wrote about him lived long afterward, some many centuries. He is said to have had a golden thigh; he believed in the transmigration of souls; he refused to eat beans. As Bertrand Russell said, "He may be described as a combination of Einstein and Mrs. [Mary Baker] Eddy."

After the discovery of $\sqrt{2}$, his secret society grasped at the hope that a rational fraction would eventually be found that when multiplied by itself equaled 2 (that is $\frac{a}{b} \times \frac{a}{b} = 2$). But it was not to be. Someone, perhaps the sage himself, proved by deduction that no such rational number could exist. Euclid's *Elements*, Book X, Proposition 9, is a more general proof of this. After 2,300 years, his reasoning still speaks clearly to us. When we read a mathematician like Euclid, we look him straight in the eye; our minds meet, and we understand. On the other hand, when we read a Greek dramatist or philosopher, we are always gazing at his feet — the nuances of words change with time and the social context is forever lost. And we don't fully comprehend.

Little time passed before the deep significance of what the Pythagoreans had found was fully realized. Then the feeling among them must have been similar to that of the founders of modern chaos theory. In both situations, the irrational was uncovered in the temple of certainty among their most treasured formulas. In the modern case, it sprang from Newton's equations; for the ancients, it arose from the theorem of Pythagoras.

If the Pythagorean Society had been a scientific rather than a religious gathering of men and women, they would have at least modified their credo that "rationals rule." They did no such thing — much the opposite. Members swore an oath of utmost secrecy never to reveal this flaw in the fabric of the universe. Of course, the truth was revealed almost immediately. Proclus (about AD 450) records that one of their society, Hippasus by name, uttered the unutterable and so justly perished in a shipwreck — or so they say.

Some irrational numbers are rightly famous. Here are three of the most celebrated: pi, the ratio of a circle's circumference to its diameter; e, the base of the natural logarithms; and phi, the golden number of art, architecture, and nature.

$\pi = 3.14159265358979323846264338327950288419...$
$e = 2.71828182845904523536028747135266249775...$
$\varphi = 1.61803398874989484820458683436563811772...$

Great hope takes its loftiest seat when it has the least chance of realization: in asylums, among religious zealots, in the midst of circle-squarers, angle-trisectors, and cube-duplicators. *Expand pi far enough* some declare — the current record is 500 billion digits — and you will see its period. Proofs to the contrary are like art in the valley of the blind.

In an earlier age my teachers referred to irrationals as "surds," a word whose etymology is similar to its sound. Surds are everywhere. Randomly pick a point on the number line, and you get a surd. They're more abundant than rationals even though both are infinite. All measurements — no matter how accurate your measuring devices — are surds. And this inexactness is the root source of the Butterfly Effect. We live in a surd universe.

In describing these numbers, you must choose your words with care. All the *known digits* of pi occur with equal frequencies even when tallied to billions; yet this is not so for all irrationals. And although surds never have periods, they may have a pattern — as strange attractors do. Consider two examples:

$$0.101001000100001...$$
$$4.737337333733337...$$

Once again, this reminds us of the logistic function that for $3 \leq m \leq 4$ has the wondrous form of the Feigenbaum fig tree but nothing much outside that interval.

The graphs of real numbers, rational and irrational, are just dots on a line — not aesthetically pleasing. Yet how they're placed on that line is bizarre. Between any two rationals, however close they may be, is another rational or surd. And between any two irrationals is another rational or surd. Moreover, there are an infinite number of each. Nevertheless, compared to the surds, rationals are very rare!

A marvelous numerical method exists for finding the roots of any polynomial function. Although the method is named after Isaac Newton and his contemporary John Raphson, the Greeks used a version of it for finding *square roots*. The process is iterative. Begin with a guess, and iterate to a better guess. The whole method is similar to a dynamical system swiftly zooming in on a steady state. If the root is a surd, the Newton–Raphson method will allow you to calculate as many decimal places as your patience permits. See the Chapter Notes for the formula and a worked example.

The history of mathematics can be seen as one continuous extension of the number system starting from counting numbers, progressing to fractions, then to surds, and on to negatives, and ending with complex numbers. These are "complex" only in the sense of having two parts: the *real* part shows your distance east or west of zero, and the *imaginary* part (always indicated by *i*) shows your distance north or south. The two parts give each complex number a unique address on the plane. The real number line is only a subset of this plane when the imaginary part equals zero.

As I have said, mathematics is burdened with bad names for good numbers. "Real" numbers are not real; "imaginary" numbers are not imaginary. "Rationals" are not particularly well balanced and reasonable, but they do have a period; "irrationals" are not in need of psychoanalysis, they merely have no period. And although I have been observing "improper fractions" for years, they have yet to do anything unseemly, but I keep hoping.

Complex numbers are usually written as $a + bi$, where "a" is the *real* and "b" the *imaginary* part. All the basic operations can be performed on them, including powers and roots. The "i" equals $\sqrt{-1}$ and is best visualized as a counterclockwise rotation of 90°. For that reason, $3i$ means rotate 3 ccw a quarter turn.

The Newton–Raphson method, although designed for reals, works equally well on complex numbers. And exactly the same formula is used. With this extension, the graphs of each iteration — whether rational or irrational — will be released from the monotony of just points on the real number line and spill out into the stupendous richness of the complex plane.

Karl Gauss proved that the number of roots of any polynomial equation equals its highest degree. So, $x^4 = 1$ must have four; clearly 1 and –1 are two; the other two are the complex roots i and $-i$. All four roots are on a circle of radius 1, centered at the origin (*see diagram to the left*). And now for the order and surprise.

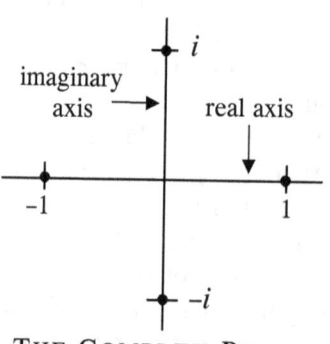

THE COMPLEX PLANE

Start the Newton–Raphson method on any point in the complex plane. As it is iterated, observe which root this initial point goes to. Now ask the right question and, open sesame, you conjure up a world of wonders. Here the magic words are "What is the pattern of these basins of attraction?" To find out we literally paint a picture. To each point on the complex plane assign a color depending on which root it terminates at: red for 1, blue for –1, yellow for i, and green for $-i$. Have the computer scan the visible plane at the pixel level of resolution coloring the points as it goes. Now these basins of attraction will be clear. See for yourself — *look down*.

BOUNDARIES OF THE NEWTON–RAPHSON METHOD FOR $x^4 = 1$

Figure 3.18

In the above figure, the complex plane is divided into four pie-shaped wedges around each root of the equation. The roots are at the bottom of the dark basins. This image represents the way the Newton–Raphson method solves equations by leading from different starting points to one of the four possible solutions. Of interest are the regions between any two roots — where the fractals grow. Along these diagonals are pools of red, blue, yellow, and green — each a distance from its main attractor but nonetheless destined to go there. On closer inspection, you will notice that where any two colors seem to meet, the other two colors separate them. In the nature of fractals, this action goes ad infinitum, and as such, no two colors actually touch each other!

The Newton–Raphson method behaves like a dynamical system with, in our case, four attractors. Some starting guesses go quickly to an attractor while others on the fractal boundary leave a trail of values indistinguishable from chaos. Figure 3.18 is one of an infinite variety of similar patterns that exist for all higher degree equations of the form $x^n - 1 = 0$. Each has n roots equally spaced on a unit circle. *And any two colors are always separated by all the other colors.*

No classical mathematician opened the door to this fractal world — they lacked words of enchantment. Had they done so their amazement would have been profound. Even Newton would have danced, perhaps in horror, perhaps in delight. Such surprises, hiding in well-ordered formulas, are the signature of chaos. Ancient myths preach about creating order out of chaos. Modern science speaks of simulating chaos out of order.

THE CHAOS THEORISTS have guided us out of a long night's journey into light. We have traveled very far from Genesis 1:2, "... without form and void" and Newton's clockwork universe to a distant and happier land.

They began by demonstrating that the classic pendulum, the very paradigm of order, had chaos at its core. And this chaos was not boundless, but rather cabined, cribbed, and confined within a strange attractor. Armed with the logistic function and its fantastic fractal form, they ventured out from the dull study lamps to successfully apply their hypotheses to the natural world. We explored cobweb diagrams and endless self-similarity. Edward Lorenz introduced us to the Butterfly Effect, the emblem of this new science. Steven Jay Gould applied this contingency to evolutionary theory through the lens of the Burgess Shale. The rider in the whirlwind brought everything down to the human level and the mystery of personal consciousness. With the "Ab-Surd Universe" we learned that all the elements of chaos theory already existed in our number system. We have been like Dante as his guide Virgil led him through the various circles of *The Inferno*. At the end of our journey, these scientists have shown us a way out of the repetitive cycles of Ecclesiastes 1:9 where, ". . . there is no new thing under the sun" into a new universe. We have emerged from the earth as Dante did and realized that every day is Easter Monday — when all is reborn fresh and new.

The American movie industry has never been kind to scientists. Hollywood portrays them as either Faust or fool: lunatics in lab coats, or dribbling misfits with coke-bottom glasses, inarticulate, and barely able to walk. So when Steven Spielberg's *Jurassic Park* came out, it was a new path for these moguls. Although chaos theorist Ian Malcolm, one of the protagonists, is largely two-dimensional, it was nonetheless a breakthrough depiction.

A much deeper and more sympathetic examination of the way modern science looks at the world is found in the English dramatist Tom Stoppard's dazzling play *Arcadia*. There is enough thoughtful reasoning here to gladden the heart of any beleaguered rationalist. The characters — some from the present, some from 1809 — discuss mathematics, poetry, history, landscape design, and chaos theory. In the passage quoted below, one of Stoppard's characters crisply summarizes chaos theory. His last sentence is motivated by the essence of the scientific method that values the journey over the destination in the search for truth.

Two characters from the present are Valentine Coverly, a biologist and computer scientist, and Hannah Jarvis, a writer and historian. Valentine uses hunting records to model chaotically fluctuating populations of game birds on the grounds of the estate. He speaks to Hannah (Act I, scene 4):

> The unpredictable and the predetermined unfold together to make everything the way it is. It's how nature creates itself, on every scale, the snowflake and the snowstorm. It makes me so happy. To be at the beginning again, knowing almost nothing. People were talking about the end of physics. Relativity and quantum looked as if they were going to clean out the whole problem between them. A theory of everything. But they only explained the very big and the very small. The universe, the

elementary particles. The ordinary-sized stuff which is our lives, the things people write poetry about — clouds — daffodils — waterfalls — and what happens in a cup of coffee when the cream goes in — these things are full of mystery, as mysterious to us as the heavens were to the Greeks. We're better at predicting events at the edge of the galaxy or inside the nucleus of an atom than whether it'll rain on auntie's garden party three Sundays from now. Because the problem turns out to be different. We can't even predict the next drip from a dripping tap when it gets irregular. Each drip sets up the conditions for the next, the smallest variation blows prediction apart, and the weather is unpredictable the same way, will always be unpredictable. When you push the numbers through the computer you can see it on the screen. The future is disorder. A door like this has cracked open five or six times since we got up on our hind legs. It's the best possible time to be alive, when almost everything you thought you knew is wrong.

CHAPTER — 4

FORM AND FRACTAL

Fractal Geometry will make you see everything differently. There is danger in reading further. You risk the loss of your childhood vision of clouds, forests, flowers, galaxies, leaves, feathers, rocks, mountains, torrents of water, carpets, bricks, and much else besides. Never again will your interpretation of these things be the same.
Michael F. Barnsley

WE UNLOCKED THE ENTRANCE to the fractal universe in the previous chapter and glanced at a few of its extraordinary forms. Now we will walk through the portal and tour this unique landscape called Frackinos Island, jewel of the Aegean.

As we stepped off the tour boat, we saw a Koch snowflake close by the harbor's entrance. Named after Helge von Koch, a Swedish mathematician, he was the first (1904) to describe fractals. So bizarre are its properties that mathematicians of his time called it *monstrously pathological*.

To construct such a snowflake do the following simple procedure: start with an equilateral triangle of 1 unit per side for a perimeter of 3 units. In the middle of each side place another equilateral triangle $\frac{1}{3}$ the perimeter of the original and pointing outward — see Figure 4.1. With the addition of these 3 attached triangles, we now have the Star of David with its six points. The star's boundary has 12 segments for a total perimeter of $3 \times \frac{4}{3}$ or 4 units.

In the next step add a smaller triangle — again $\frac{1}{3}$ the size — to the middle of each of the 12 sides of the star. Since all of these new segments have a length $\frac{4}{3}$ times the original, the resulting perimeter is $\left(3 \times \frac{4}{3}\right) \times \frac{4}{3}$ or $\frac{16}{3}$ units. So, if each piece is increased in length by $\frac{4}{3}$, then the *entire perimeter* is increased by the same factor. When this procedure is expanded to infinity, we create the Koch snowflake. And its final perimeter is $3 \times \frac{4}{3} \times \frac{4}{3} \times \frac{4}{3} \times \ldots$, which is infinite! Because of its construction, any portion of the curve under magnification by 3 is identical to the original — the quality of self-similarity.

By extending this last property even further, we see that the partial perimeter between any two points, however close they may be, is also infinite. After all, the yet unseen pieces will be $\frac{4}{3} \times$ the old pieces.

The boundary of the snowflake is continuous, but not smooth — far from it. At *any* magnification, it has an infinite number of zigzags. This implies that although the snowflake is continuous no tangent can be drawn at any point: continuous but entirely non-differentiable as mathematicians say. No, it is not a paradox; it is a fractal.

As strange as the previous property may be, the following one seems impossible. Imagine you drew a circle around the Koch snowflake. Nowhere will this curve extend beyond the boundaries of the circle. Thus, the snowflake has a *finite area* since it's less than that of the finite circle. However, at every stage in building this figure, the perimeter is multiplied by $\frac{4}{3}$, implying it's always increasing. So, the ideal snowflake (ideal, meaning you go through an infinite number of stages in its creation) has an infinite perimeter, yet a finite area — exactly $\frac{8}{5} \times$ the original. No, it is not a paradox; it is a fractal.

Series of numbers (and both the area and the perimeter can be expressed as such) are open or closed, divergent or convergent, wander off or wink out. With the Koch snowflake, we get an example of each in the same figure: divergent perimeter but convergent area.

THE KOCH SNOWFLAKE

Figure 4.1

You could put the entire Koch snowflake on your smallest fingernail or in the palm of your hand. And with your mind's eye, it would be a way of seeing infinity. William Blake certainly did not have science or mathematics in his mind when he penned the introductory lines to his famous "Auguries of Innocence." Yet the very subjects he loved to loathe illustrate his imagery.

> *To see a World in a grain of sand,*
> *And a Heaven in a wild flower,*
> *Hold Infinity in the palm of your hand,*
> *And Eternity in an hour.*

LET US STEP INTO the fractal garden — a preternatural world — armed with our knowledge of the Koch curve. The keeper of this place is a tall elderly gentleman called Benoit Mandelbrot. Without my asking, he gives me his entire life's history. Born in Warsaw in 1924 of Lithuanian Jews, he moved with his family to Paris when he was 12. Once the war began they escaped the Nazis by migrating to the south of France. His schooling, like his fractals, was discontinuous and irregular: he claims never to have memorized the multiplication tables past the fives, or to have learned the entire alphabet. He regards these omissions as badges of distinction. And no interruption on my part slows this biography until I ask him the size of the island. Paradoxically, rather than telling me, we silently climb a steep hill and survey the entire terrain. To my surprise, it appears to be perfectly circular!

BENOIT MANDELBROT

"How large is Frackinos?"

"Exactly 153 acres, no more, no less," he replies.

"Since this island is a circle with a fixed area, its shoreline could be easily calculated."

"Yes," he answers in an unusual tone, "and you would get 9,152 feet or 1.7 miles. But you would be wrong, profoundly wrong." He clears his throat and affirms, "Clouds are not spheres, mountains are not cones, coastlines are not circles, and bark is not smooth, nor does lightning travel in a straight line."

After this outburst, we descend from our lookout and walk toward the beach. As we do, several small inlets and prominences come into view. Closer yet, other irregularities appear, and I remember the Koch snowflake and how magnifying it always revealed new details. Could this entire island be a type of Koch snowflake? Before I could explore this line of reasoning, my guide interrupted my thoughts. "What perimeter would you get if you strapped a pedometer to your leg and walked around Frackinos traversing all these inlets and prominences?"

"Well, common sense says it would be *longer* than 1.7 miles," I replied.

"And what would the distance be if a sparrow hopped from pebble to stone all about the island's rim?"

"I get the point. Of course, it would be much greater."

"And a microbe on a multi-generation voyage over every speck of rock and down each microscopic crevice would travel further still," he stated.

"Is there a limit to this ever-expanding perimeter?" I asked.

"Probably when you reach the atomic level no further increase would be possible. But remember, this measurement — any measurement — is an irrational, non-periodic decimal and hence can never be exact. And there is always some discrepancy between our theories and the real world — reality is so much richer and more complex than our imaginings. And it is the task of science to minimize these discrepancies. For example, fractals are light years ahead of Euclidean geometry in modeling reality."

"Are there other differences between theory and practice?" I inquired.

"One more. At all levels of magnification the Koch snowflake — a fractal island — is just equilateral triangles that are exactly similar. Nonetheless, on Frackinos Island or any real island, the similarity is statistical: the average proportions of prominences and crevices remain the same under magnification although their arrangements will vary. One is exact; the other is statistical."

"I can accept that and still admire how marvelously fractals model reality," I asserted.

"Good," said my guide, "you're learning."

"Let me see if I really understand all this," I responded. "My walking around the island's coastline is like measuring it with a yardstick. But the sparrow's measuring rod is only an inch or two long, while the poor microbe's is a millionth of that. So you're saying the perimeter depends entirely on the length of the measuring device."

"Exactly!" replied Mandelbrot. "In Revelation 11:1 John says, 'I was given a reed like a measuring rod and was told, "Go and measure the temple of God and the altar. . . ."' Certainly, John's measuring with a long reed would pass over all the cracks, bumps, holes, and doorways of the temple and be considerably less accurate than with a shorter stick. The accuracy of your answer depends on the length of your ruler. Length cannot be measured independent of scale."

MAURITS ESCHER'S *LIMIT CIRCLE III*

Figure 4.2

The modern Dutch artist Maurits Escher in his fascinating work *Limit Circle III* (*see* Figure 4.2) drew patterns of fish diminishing in size until they morph into a fractal boundary at the circle's circumference — a fractal lake. So, at this boundary both the sea and the land are fractals.

"All this regression to smaller and smaller pieces of whatever reminds me of Jonathan Swift's famous doggerel about larger fleas having smaller fleas on their backs to bite 'em."

"Yes," he said smiling. "That's a common ditty all over Frackinos. And the English logician Augustus de Morgan wrote an expanded version:

> *Great fleas have little fleas*
> *upon their backs to bite 'em*
> *And little fleas have lesser fleas,*
> *and so ad infinitum,*
> *And the great fleas themselves,*
> *in turn, have greater fleas to go on,*
> *While these again have greater still,*
> *and greater still, and so on."*

"Very entertaining verse but very bad entomology," I stated. "Fleas have a natural scale of existence, but as you've shown me, coastlines don't. Not every natural object scales like a fractal — certainly living organisms don't. If you magnified a flea 10-fold, its weight would increase a 1000-fold but its surface area only a 100-fold. It would be unable to breathe or feed itself, least of all jump. Remember, this was the theme of my book's first chapter 'The Measure of All Things.'"

"As to a flea's size that may be so," agreed my guide. "But what book are you referring to — I don't know any book with a chapter titled 'The Measure of All Things.'"

"That's the book you and I are characters in right now," I told him.

"Surely you're trying to trick an old man with verbal nonsense?"

"Not at all. See, I have a copy in my hand and we're both in the fourth chapter called 'Form and Fractal.'"

Growing more agitated, he demanded to know the book's title.

I showed him its cover: *Immortal Ideas: Shared by Art, Science, and Nature*. "Here, read about yourself."

He took the book and did so. After some minutes, he spoke again but in a quieter tone.

"So, you're saying I'm the Benoit Mandelbrot in this book who's reading about myself as a character in this book who's reading about myself as a character in this book and so on ad infinitum."

"Yes! In the mind's eye, you're now a kind of fractal yourself: the guide within the book, within the book, and so on. Like a real fractal, you quickly speed away to inaccessible places buried deep within this narrowing tunnel of regression. You're as a small boy in a barber shop standing between two parallel mirrors watching himself watching himself as his images march off into Alice's world."

"You make it sound attractive and inevitable," he replied, almost smiling again.

"It was inevitable," I responded. "And after all, *you* discovered and organized this hidden world, this fractal universe. You made sense of it, using it to explain large parts of the texture of reality. You have lived so long among your creations that you have become one of them. I'm told you coined the word *fractal* from the Latin adjective *fractus* and the verb *frangere* meaning to break, and from this you named Frackinos Island."

I glanced at him sideways as we strolled along the garden's zigzag paths. Yes, he was definitely smiling now. Soon we reached his home, and he invited me into his study for fractal tea. I accepted. Above the doorway's arched entrance, three bold letters were stamped: IBM. I wondered if they stood for **I'm Benoit Mandelbrot**.

Seated in his comfortable study, surrounded by dozens of fractal graphs, I asked the most crucial question. "You have told me these self-similar objects describe large parts of the natural world. Why should that be? Why should reality have fractal skin, bark, coverings, and coastlines? What possible purpose can they serve?"

"Many," he replied. "Consider the majestic elephant. It has a huge volume for what appears to be a relatively small surface area. This is ideal for heat retention, as your book's initial chapter shows, but exactly the situation you don't want in equatorial Africa. *Look* at Figure 4.3. The enormous jumbo-size ears act as radiators to siphon off excess heat. Now look more closely. The skin of the beast has a fractal structure — at least for a few levels of magnification — perfect for increasing the surface area to dispel even more heat. If the skin were scaleable at all levels of magnification, then its surface area would be infinite. And in that ideal case, the infinite skin of the pachyderm would be analogous to the infinite perimeter of the Koch snowflake."

AFRICAN ELEPHANT DUSTING

Figure 4.3

"I appreciate the cogency of your reasoning for the elephant's skin, but what purpose does the fractal coastline of this island, or any island, serve?"

"Ah, now that's a more intriguing question, nonetheless with a simple answer," said Mandelbrot. "And the answer has broad applications in the natural world. Consider two opposite types of coastlines: one smooth, flat, linear — the other fractal. Picture a great wave, a rolling wall of kinetic energy striking both. In the former case, the force per unit area on the wall is formidable, and if continued would surely inflict damage and cause erosion. However, in the second case the force of the wave front is broken up into innumerable wavelets. The immense pressure is spread over a vast area and hence the force per unit area and accordingly the erosion of any part is diminished or extinguished. Divide and conquer wins the battle with the waves." He paused and smiled, but before I could interrupt he spoke again. "As a result you could say that only the 'fittest' coastlines survive. And these tend to be fractal in form.

This knowledge implies breakwaters around harbors and beaches should be jumbled piles of rocks of different sizes rather than ribbons of concrete." After this, he paused for an instant.

I jumped in. "Let me state this thought in my own words to see if I truly comprehend it. So, you're saying the reason why fractal coastlines exist is precisely because they're best at dampening waves. As the coast dampens down the waves, the erosion is reduced — their fractal structure stabilizes them."

"Exactly so! I couldn't have said it better myself. Similarly, when a river meanders through a valley, its sinuosity reduces erosion by allowing it to slowly give up its potential energy. But a linear path would cause cascading torrents of kinetic energy tearing away at the earth. A skier's sinusoidal path is the same trick. Whenever you smooth the energy out over a larger surface, you diminish its destructiveness. Fractal coastlines do this best"

We talked the entire afternoon until through his study window I noticed the sun setting. I thanked my wise guide for teaching me the geography of this terra incognita, and I left. That night in the island's only tourist hotel my mind ran about these 153 acres in delight. I revisited all of the places he had shown me and the ideas we had talked about, focusing at last on the following topic.

Fractal Dimensions

SCIENCE ADVANCES BY CRUMBS. A small increase in understanding here, a clarification there: never a full loaf, always a crust. For those who seek public adulation, fame, and fortune, science is a barren ground. Once in a long time, a very long time, a young scientist will have a larger idea than usual, a concept composed of many crumbs — a bread roll or biscuit if you like.

The imprint of the power and veracity of a novel idea is that it usually subsumes older concepts as *special cases*. If an idea denies entirely the work of previous scientists, then we suspect the recipes of this new cook and judge his ideas half-baked. Johannes Kepler's three rules can be derived from Isaac Newton's more general and powerful three laws plus his law of universal gravitation. In turn, Newton's formulations are just special cases of Einstein's equations — at slower velocities.

Consider the concept of dimension. We say lines have 1 dimension; planes have 2; cubes have 3; tesseracts have 4, and so on. But why is this? You can move in two directions on a line (forward and backward), and an infinite number of directions on a plane or in space. Nonetheless, our standard definition of dimension is legitimate, just not rigorously expressed. Let's explore each:

LINES: All line segments are self-similar regardless of their length. A line segment can be cut into 3 identical pieces and when each is multiplied by a factor of 3, this reproduces the

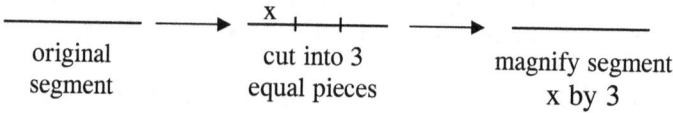

| original segment | cut into 3 equal pieces | magnify segment x by 3 |

original segment. More generally, we can cut a line segment into N self-similar pieces, each with magnification factor M. The exponent "1" in the following equation is the dimension.

$$(\text{Magnification factor})^1 = (\text{Number of pieces})$$

SQUARES: All squares are self-similar regardless of their area. We can decompose any square into 4 equal, self-similar sub-squares, but now the magnification factor will be 2.

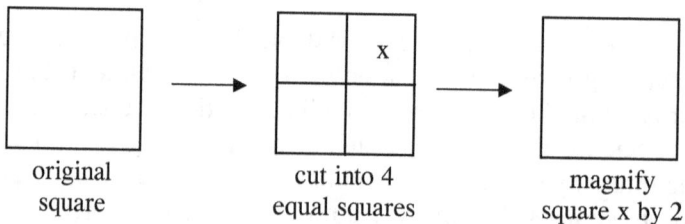

| original square | cut into 4 equal squares | magnify square x by 2 |

On the other hand, we could also divide a square into 9 equal sub-squares with magnification factor 3.

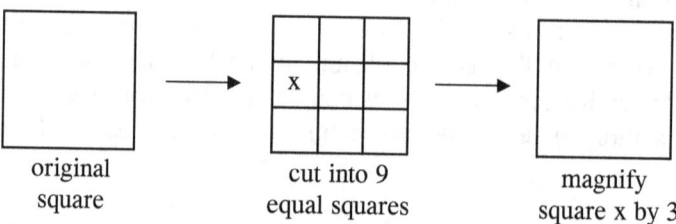

| original square | cut into 9 equal squares | magnify square x by 3 |

Generally, we can cut a square into N equal self-similar squares, each with magnification factor M, such that $M^2 = N$. The exponent "2" is the dimension.

CUBES: Any cube may be cut into 8 self-similar sub-cubes each with magnification factor 3, so that $2^3 = 8$. The exponent "3" is the dimension. As before, this can be stated with complete generality: a cube can be decomposed into N self-similar cubes, each with magnification factor M.

$$(\text{Magnification factor})^3 = (\text{Number of pieces})$$

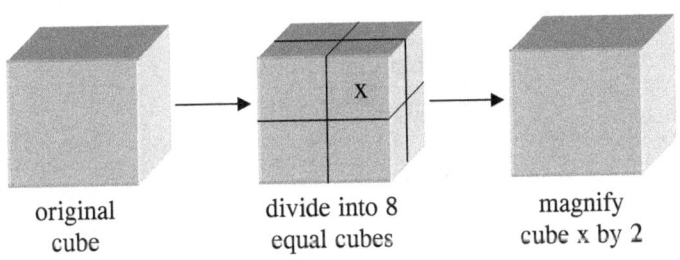

| original | divide into 8 | magnify |
| cube | equal cubes | cube x by 2 |

ANY DIMENSION: Now we see an alternative way to specify the dimension of all self-similar objects. The "D" below is the exponent of the magnification factor that equals the number of self-similar pieces into which the figure may be broken. The universal equation for all dimensions follows:

$$(\text{Magnification factor})^{\text{Dimension}} = (\text{Number of pieces})$$
$$(M)^D = (N)$$

By using the logarithm laws, this formula can be explicitly solved for "D," the dimension:

$$D = \frac{\log(\text{Number of pieces})}{\log(\text{Magnification factor})}$$

This simple yet powerful equation (*see* Chapter Notes) produces *fractional dimensions* as well as the ordinary 1, 2, 3 of our experience, thereby stamping it with credibility. But what, you might ask, can the purpose or use of fractional dimensions be? The concept is wholly anti-intuitive! One answer says they offer a way to calculate the degree of roughness of a fractal curve. My guide Mandelbrot likes to say they measure the complexity or degree of wiggliness. As an example, consider the Koch snowflake. The diagram below recalls one loop in the endless process for making this fractal. And with this knowledge, we find its dimension to be 1.26 — more than a line but less than a plane.

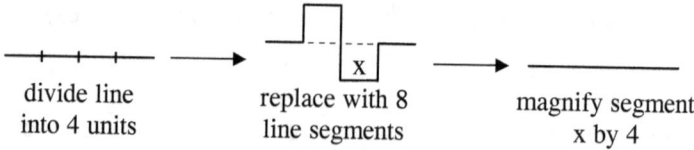

| original segment | build a triangle on the middle third creating 4 pieces | magnify piece x by 3 |

$$D = \frac{\log(\text{number of pieces})}{\log(\text{magnification factor})}$$

$$= \frac{\log(4)}{\log(3)}$$

$$= 1.26 \text{ dimensions of wiggliness}$$

In his 1977 book *Fractals: Form, Chance, and Dimension*[2] Mandelbrot wrote about a squarish asymmetrical snowflake created à la Koch. To construct this object, take one side of a square and mark it off into 4 equal units. Next, replace these 4 with line segments of 8 units as below. Do this forever.

| divide line into 4 units | replace with 8 line segments | magnify segment x by 4 |

The dimension is easily found using our formula, and it's larger than the Koch curve. By comparing Figures 4.1 and 4.4, it's apparent that the latter curve is rougher or wigglier in outline.

$$D = \frac{\log(8)}{\log(4)}$$

$$= 1.5 \text{ dimensions of roughness}$$

At each step, the same portion of area is added and subtracted, so it remains fixed at the original amount. Ah, but the perimeter is the wayward property here as was its Koch cousin before. If the original perimeter were 4 units, then the first transformation would double this to 8 units, the second to 16 units, and so on. At each nth transformational stage, the perimeter would be 4×2^n. And as n goes to infinity, so does the perimeter. This boundary will not spread much beyond what you see at the right below. At the limit, the maximum width of the snowflake is $1\frac{2}{3}$ units.

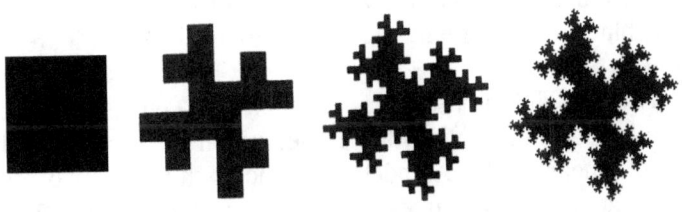

THE SQUARE SNOWFLAKE

Figure 4.4

THE SIERPINSKI TRIANGLE

THE NEXT MORNING I chose to explore more of Frackinos Island but without my guide. Before I had traveled far from the hotel I noticed a grand archway engraved with gold lettering saying: "Gather hope, all ye who enter here." Encouraged by this anti-Dante greeting, I passed through into the island's most astonishing place — heaven's half-acre.

Almost immediately an animated, nerdish gentleman greeted me, introducing himself as the *shade* of mathematician Michael Barnsley. He grasped my arm and led me to a table spread with a large sheet of plain white paper.

"Here, let's play the *chaos game*," he said. "First you mark 3 points on this paper at the vertices of a triangle. Any triangle will do — right, equilateral, isosceles, scalene or whatever. Color one vertex red, the second blue, and the third green."

His whole demeanor was so singularly earnest that I did as he asked by placing the colored points at the vertices of an equilateral triangle. After I had done this, he immediately continued.

"Next, take a die and paint two of its faces red, two blue, and the last two green. Now, randomly choose a point or seed anywhere on the plane. It could be miles away, or even on the other side of the galaxy. For now, keep it on this sheet of paper. Then the die is cast. Depending on what color comes up, mark a dot on the paper exactly halfway between the seed and the appropriate vertex. For example, if blue comes up mark a dot halfway from the seed to the blue vertex. Move to this new point. Roll the die again, note the color, and mark a second point midway from the new point to that vertex. Continue in this fashion for 6 rolls and you could get the following orbit: seed, blue, blue, green, red, red, green." *See* Figure 4.5 below.

"This is a dull amusement," I stated. "If you're sufficiently compulsive about playing this game, you'll just end up with a random smear of dots. And by continuing for 1,000s of dots, those among us who are sufficiently brain dead will fill the entire triangle."

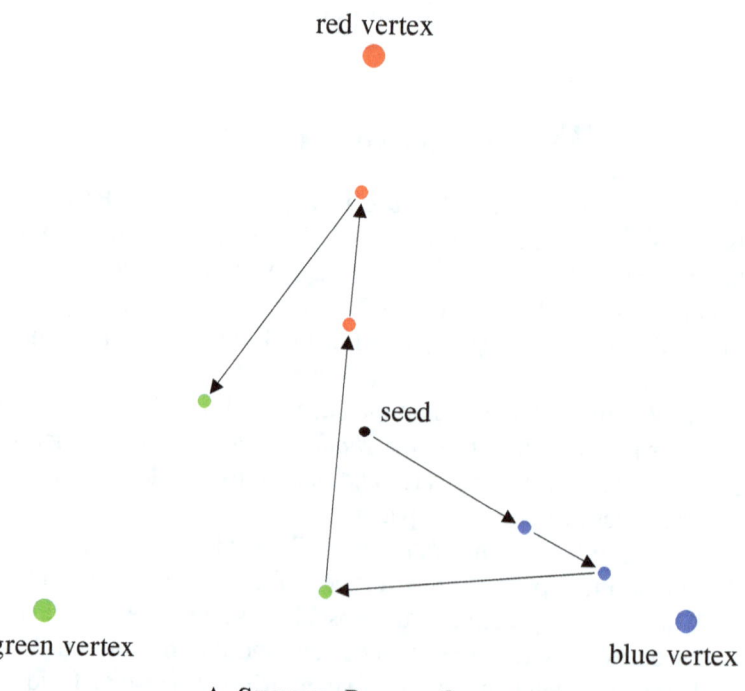

A Seven–Point Orbit

Figure 4.5

He smiled like a magician about to pull a rabbit out of a hat. "Your guesses are quite natural considering the random nature of this game. But they're wrong, completely wrong! I have programmed a computer to graph a million or so points, and the resulting image is the graceful, delicate tapestry of points you see on this page." (Interested readers should go to the Chapter Notes for a simple QBasic program for this fractal.)

THE SIERPINSKI TRIANGLE

Figure 4.6

At this moment Barnsley's shade noticed two elderly Greek gentlemen passing under the archway. Straight away he bolted from my side in their direction, and I was left to explore Sierpinski's garden in privacy.

If, in a second game of chaos, you choose a different initial condition (seed), and you have an infinite number of choices, the pattern will be the same. Always the same! In the language of chaos theory, this tapestry is a strange attractor for the iterative process. The first few points may be off the pattern but those following quickly scurry into position like soldiers at the bark of a drill sergeant. The colors are mainly decorative indicating the proximity of the vertex with the same hue.

Oddly enough, the pattern is dependent on the die being "fair." If the die were weighted in such a way that red always came up, then Sierpinski's triangle (called S for convenience) would never materialize. Most people are surprised by this outcome, but they shouldn't be. Randomness plays an essential role in many types of order. Our very existence is the result of an evolutionary process with key random elements — recall Stephen Jay Gould's concept of *contingent history* from the previous chapter. For anyone who still believes chaos and order are diametrically opposed, all this is a dilemma. For the rest of us, it's just another dance in the vast repertoire of this inseparable couple whose pas de deux unfolds here as a tapestry.

THE SELF-SIMILAR FEATURE of fractals comes in the two varieties seen in Figure 4.7. For the highly stylized tree at the left, it resides in the canopy, for Sierpinski's triangle it exists throughout the structure. (We must be careful here because a *real* tree's trunk is like the triangle, self-similar everywhere.) For instance, with the Koch curve the self-similarity was in its boundary, never in its interior. This distinction of types is useful.

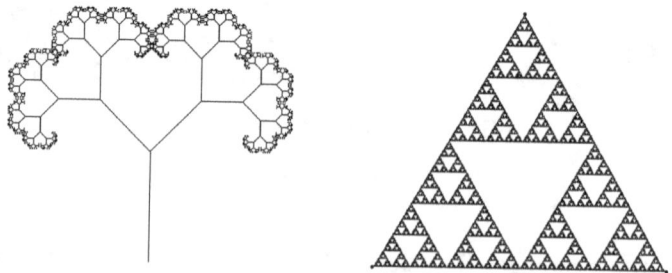

TWO TYPES OF FRACTALS

Figure 4.7

With this in mind, the dimension of S may be quickly found. Consider the sub-triangle at the top of S in the above figure. Multiplying it by 2 generates the original 3-piece figure. That is, S consists of 3 self-similar copies of itself, each with magnification factor 2.

$$D = \frac{\log(\text{number of pieces})}{\log(\text{magnification factor})}$$
$$= \frac{\log(3)}{\log(2)}$$
$$= 1.58$$

We can gaze deeper into S and see additional copies of it. For the Sierpinski triangle also consists of 9 self-similar copies of itself, every one with a magnification factor 4. Or we can cut S into 27 self-similar pieces, every one with magnification factor 8. In general, we can divide S into 3^n self-similar pieces, each of which is congruent, and each of which may be magnified by a factor of 2^n to generate the entire figure. If we now calculate the dimension using these general terms for the number of pieces and the magnification, we get the identical answer. In other words, the dimension is universal at all levels — micro- or macrocosmic — precisely identifying this fractal.

$$D = \frac{\log(3^n)}{\log(2^n)}$$
$$= \frac{n \log(3)}{n \log(2)}$$
$$= 1.58 \text{ as above}$$

A MYSTERY LIVES HERE: why does the chaos game generate the Sierpinski triangle? Learning mathematics and science — unlike the humanities — is often a series of "aha" experiences, moments of epiphany where understanding comes abruptly. Scientific facts accumulate by small steps, but its larger comprehension is a succession of giant strides. The recent discovery of an antibody that blocks live SARS viruses from entering cultured human cells was one of those eureka experiences, says its discoverer Wayne Marasco.

Although the midpoints scurry away as quickly as cockroaches from the light, the reason why they map out S is straightforward and an "aha" experience. Color the seven-point orbit of Figure 4.5 red; superimpose it on a black copy of

Sierpinski's triangle to get the result shown below. Carefully examine the orbit from the seed to the last red dot. Observe that the seed is approximately at the geometric center of the largest empty triangle. And the following dot is in the identical relative position in the next smaller empty triangle. Why? Because the next 3 smaller triangles represent *all the points that are half the distance to the 3 vertices from any point in the largest empty triangle.* As you continue around the orbit — using the same reasoning — all the dots are in the same relative position but in successively smaller empty triangles (halfway houses). By the 5th or 6th dot, their position on the computer screen is indistinguishable from the actual Sierpinski triangle. This outcome was inevitable.

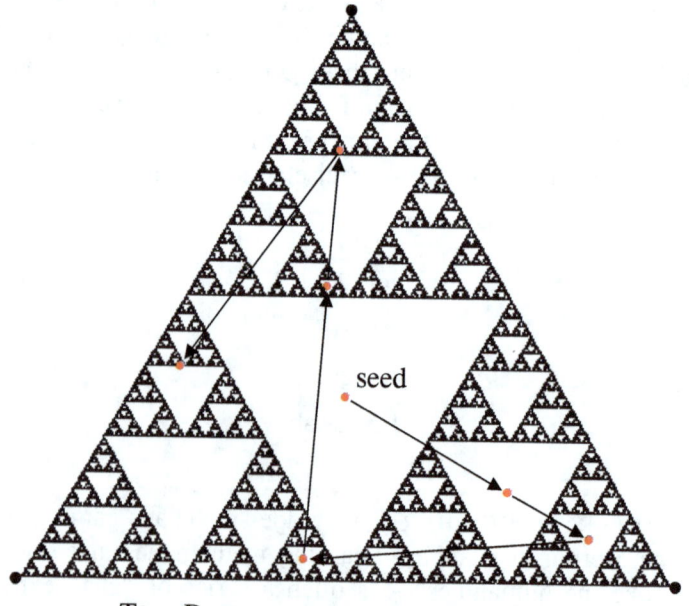

THE DANCE OF THE CHAOS GAME

Figure 4.8

Throughout the garden, symbols and representations of S abounded. One enormous tricolored flowerbed of impatiens conformed strictly to the pattern. It was so immense that, like the Nasca lines of Peru, the whole was visible only from the air. As well there were wall displays running computer programs. One large-screen video program descended forever into the infinite recesses of Sierpinski's triangle.

Off to one side I observed the two Greek men who had allowed my earlier escape from Barnsley's specter. Evidently free themselves, they were about to enter a smaller pavilion; so I followed. They seemed a strange pair: one huge and rugged looking, the other small and refined. Moreover, they usually conversed in an ancient Greek dialect.

IN CONTRAST TO the previous random algorithm, there exists a second path to constructing Sierpinski's triangle — a completely deterministic way. Begin with any triangle and follow the three steps below as they unwaveringly march to the drumbeat of Sierpinski:

1. Divide every dark triangle into four congruent sub-triangles by drawing line segments joining the midpoints of the three sides.

2. Remove the middle triangle of each.

3. Repeat steps 1, 2, and 3 for all new dark triangles.

Note that this process is recursive because step 3 always requires more repetitions of itself. The *residue* after infinitely many steps is the Sierpinski triangle.

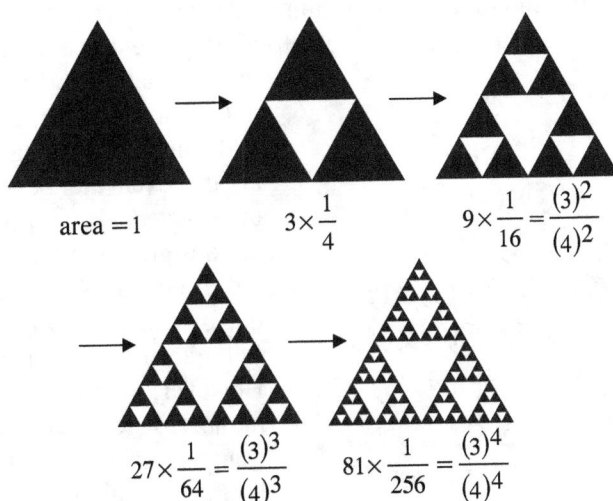

area = 1 $3 \times \dfrac{1}{4}$ $9 \times \dfrac{1}{16} = \dfrac{(3)^2}{(4)^2}$

$27 \times \dfrac{1}{64} = \dfrac{(3)^3}{(4)^3}$ $81 \times \dfrac{1}{256} = \dfrac{(3)^4}{(4)^4}$

DETERMINISTIC CONSTRUCTION OF S

Figure 4.9

Set the area of the first and largest triangle at 1 square unit. Let's see how much of this area remains after iterating. Following the first, there are 3 dark triangles left each with an area of $\frac{1}{4}$ the original. The second leaves 9 triangles each with an area $\frac{1}{16}$ of the original. Beneath every iteration in Figure 4.9, I have shown the surviving area. In general, the amount remaining after n iterations is $\left(\frac{3}{4}\right)^n$. Consequently, as n gets very large, $\left(\frac{3}{4}\right)^n$ gets very small (try it on your calculator). In the limit when n is infinitely large, the total residual area is zero. Since all the area has vanished, what's left? Just the exquisite tapestry of Sierpinski's pattern — a never-ending recess of triangles composed exclusively from dust, fractal dust. All form and no substance!

As soon as I left this pavilion, another blocked my path. Its signs about Pascal's triangle promised even greater surprises. So like a child running from booth to booth at the local country fair, I went in.

EVERYTHING COMES IN THREES, a saying probably as old as Aesop's fables. And I was about to discover a third way to construct Sierpinski's triangle.

In Chapter 2, we learned something of Pascal's triangle (*see* Figures 2.14 and 2.15), and its unusual history and uses. Incredibly, this same array of numbers, when correctly viewed, is Sierpinski's triangle. Where? How? Represent every number in the triangle by a small dot. Then darken all the odd numbers leaving clear all the even numbers, and unexpectedly you produce the design to the left — our old friend S. Well, maybe not precisely S because the full fractal form is lacking. However, when an infinite number of rows of Pascal's triangle are included, the limiting pattern is fully fractal, and as we say, self-similar or scale invariant.

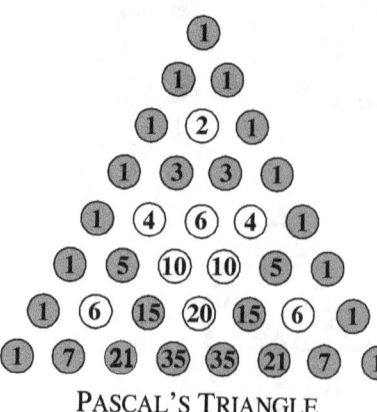

PASCAL'S TRIANGLE
— the first 8 rows —

Figure 4.10

Just so that this pattern is perfectly clear, I have reproduced its top 64 rows with all the dots and shading.

THE PASCAL-SIERPINSKI TRIANGLE
— the first 64 rows —

Figure 4.11

When we speak of even and odd numbers, we're really thinking of divisibility by 2: evens exactly so, odds with a remainder. Dividing by 3, 4, 5, etc., will also produce striking patterns when you darken all the remainders correctly and leave clear all those with none. The properties of Pascal's triangle are exceedingly numerous, and anyone studying it long enough will discover new ones, but these are unlikely to be original. In his famous treatise on the triangle, Pascal said, "It is a strange thing how fertile it is in properties."

This fertility extends beyond the walls of the study into the natural world and its realm beneath the waves. We prize the pearl, but the most remarkable product manufactured by a mollusk is its shell. Consider the tent olive shell below, an image of Sierpinski with all his self-similar triangles plus the roughness of nature we have come to expect when several forces interact. Imagine that the shell's logarithmic spiral were unrolled and pressed flat into a rectangle. Like a tree's tips, shell growth occurs only on its leading outside edge. So this rectangle is a kind of space-time diagram: its width for position and its length for time. With these two coordinates, any shell triangle or chevron can be located.

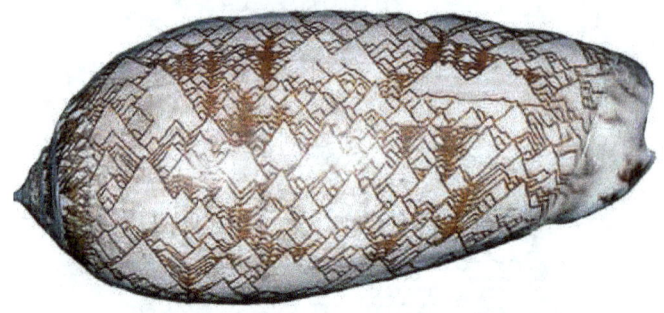

TENT OLIVE SHELL, *Oliva porphyria*

Material is deposited by a row of cells on the leading margin of the mollusk's mantle. Through hormonal activators and inhibitors, this pigment forms patterns that are rarely altered thereafter.

The tent olive shell is one of the most treasured finds of collectors — a true glory of the seas — and a favorite topic of mathematical shell modelers. The one above comes from the Gulf of California, but similar and related shell designs are found in most of the world's oceans. To uncover them on Frackinos Island, you need only prospect at the hotel's dock for a few minutes. The study of pattern formation in organisms has long been one of the more mathematical areas of biology. D'Arcy Wentworth Thompson set the standard in his thousand-page book *On Growth and Form*.

Let's return to the study and consider an extension of Pascal's triangle to a tetrahedron. You can see in Figure 4.12 (a) that each of the three upright faces is the well-known array, but beyond this note the developing internal structure. By analogy with Pascal, every number is the sum of the *three numbers* directly above. Granted this may sometimes be difficult to see, but picture yourself standing on the "6" in level four. Looking up you see the three 2s that form your sum. Face values of the pyramid are the sum of the two numbers above and an unseen zero in the surrounding space. Some readers may wish to find the next level; the answer is in the Chapter Notes.

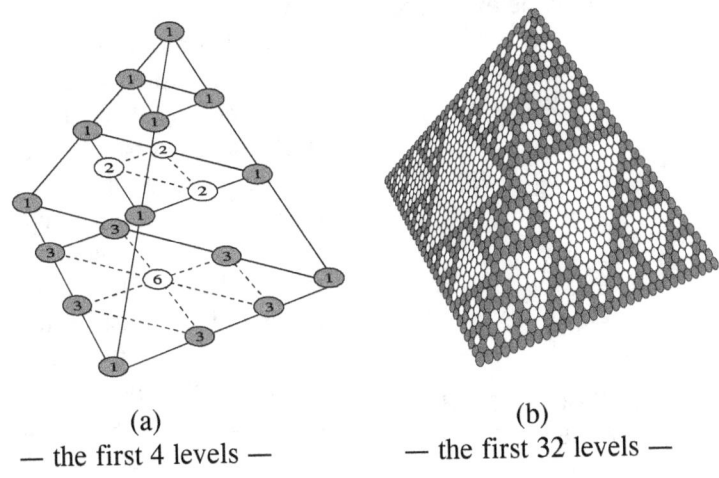

(a)
— the first 4 levels —

(b)
— the first 32 levels —

THE PASCAL-SIERPINSKI PYRAMID

Figure 4.12

The exterior pattern of the pyramid is self-evident, but what *form or design* do we have when all the interior points are included? You could easily be forgiven if you imagined it to be a series of space-filling tetrahedrons, but you would be mistaken! Even the great Aristotle in his work *On the Heavens* proclaimed that tetrahedrons could fill space. But of all the five perfect solids (*see* Chapter 2), only the cube can do this and leave no empty pockets. On the other hand, the void may be packed by numerous convex polyhedrons, yet only five (again) have regular faces — not necessarily all the same:

1. Cubes
2. Triangular prisms
3. Hexagonal prisms (the honeybee's comb)
4. Truncated octahedrons
5. Gyrobifastigiums (8 faces: 4 squares, 4 triangles).

What is behind the largest inverted triangle of even numbers in Figure 4.12 (b) or behind any such triangle? Presuming a pattern exists, plainly more levels must be found before it emerges. Expanding this design is messy and difficult on paper, but building a 3-dimensional model of toothpicks or straws, with at least the first 16 levels, will reveal the internal structure. Fortunately, this is not a homework assignment. You need only look below to see the answer.

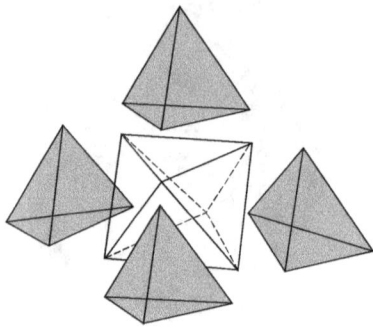

OCTAHEDRON–TETRAHEDRON
SPACE PACKING

Figure 4.13

Neither the octahedron nor the tetrahedron by itself fills space, but together they could pack the universe. In the exploded view at the left, the central octahedron, composed entirely of even numbers, is surrounded by four tetrahedrons, which must be further subdivided in the familiar self-similar fashion. Continue these subdivisions to infinity and Pascal morphs into Sierpinski. The ratio of the volumes — octa to tetras — is 1 to 1.

So, an octahedron of even numbers lies behind every clear (empty) triangle on the face of Pascal's pyramid. This is the answer to the question, and it's a good one because it generalizes the previous results of the renowned array.

Unexpectedly, since the pyramid appears 3-dimensional, you would suppose its fractal dimension to be more than 2 and less than 3. But that's not the case. Magnify the crowning solid in Figure 4.13 by 2, and then you have 4 tetrahedrons.

$$D = \frac{\log(4)}{\log(2)}$$

$$= 2 \text{ dimensional}$$

Perhaps at this point you don't wish to call 2 a *fractal dimension* since it's not fractional. But it must be because by definition it fits the formula!

In the true Sierpinski pyramid, all the evens are *removed*. So, as the number of transformations goes to infinity, the pyramid's volume zooms to zero, while its area grows beyond all bounds. Similar transformations were done on the cube decades ago generating the Menger "sponge" (*see* the figure below). Also, by recursively removing the appropriate volumes from every part of an octa-, icosa-, and dodecahedron, their Sierpinski analogue can be produced. All these objects have zero volume but an infinite area, much like the infinite perimeter of the Koch snowflake. These denizens of the island's fractal forest may seem strange at first glance, but you probably have one lurking in a dark corner under your kitchen sink and another creeping about in your bathroom.

THE FIVE PLATONIC NON-SOLIDS

Figure 4.14

THIS PAVILION OF PUZZLES AND PROBLEMS housed one last booth that I noticed was manned by Barnsley's shade. Clearly the two Greek tourists had escaped his frothy enthusiasm.

He who hesitates is lost, they say, or at least captured. "There you are. I've been looking for you for hours. Thought maybe you'd become lost in the infinite hallways of some fractal or other," Barnsley chuckled. "So you've been studying all the 3-dimensional Sierpinski generalizations. Great stuff, but only *part* of the story."

By way of evading capture I responded, "Just on my way back to the hotel for a rest before supper." Instinctively, however, I knew resistance was futile, and I was about to be assimilated. He clutched my arm and I surrendered without a struggle, whispering, "Show me."

"It's all quite simple," he enthused. "Instead of allowing only 3 vertex points in the chaos game, why not have 4, 5, 6, or as many as you please? Let me show you on the computer the spectacular designs that accumulate. This game is played as before. Plot 6 points, say, at the vertices of a regular hexagon, and number them 1 to 6 — like a normal die. Choose a seed anywhere on the plane and roll the die. Mark the point $\frac{2}{3}$ of the distance from the seed to the appropriate vertex. Then iterate this procedure. The ultimate effect is superbly intricate, almost Alhambresque, and as earlier the colors are decorative." As the pattern magically grew, he turned directly to me and asked, "Do you recognize anything in the self-similar *white spaces* that the fractal dust outlines?"

CHAOS GAME RESULTS FOR SIX POINTS

Figure 4.15

"They seem familiar somehow," I observed. . . . "Yes, yes, it's the Koch snowflake I first saw when I arrived on the island. Besides the one precisely in the center, your monitor shows four more size levels. Intriguing!" *See* Figure 4.1.

"The Sierpinski hexagon, as it's sometimes called," he said, "appears 'rougher' than the Sierpinski triangle; its fractal dimension reflects this quality. If you multiply any sub-ring of garlands by 3, the original-sized figure is reproduced with its 6 rings. As a result, the dimension is $\log(6)/\log(3) = 1.63$, which is somewhat higher than the triangle's 1.58."

The specter handed me a sheet of paper, "Here's the QBasic computer program that generated this design. Type it into your home PC to create the chaos game for any number of vertices you please." I took the program. (Interested readers will find it in the Chapter Notes.) For a while we talked about the beauty of this pattern and others he had shown me and how such delicate order arises from such robust chaos.

After a few minutes Barnsley's face developed an unusually thoughtful appearance, he said: "In previous centuries, mathematicians and scientists studiously avoided nonlinear equations. These were equations behaving badly, but they explained the world as it is. Later, and for similar reasons, mathematicians and scientists also avoided fractals, labeling them pathological monstrosities. These were geometries behaving badly, but they described the world as it is. What is the relationship between nonlinear equations, and fractals? Simple! Fractals are the geometrical structure of chaos, the paths of nonlinear equations over time. They make the mathematical world gloriously visible. Fractals are a big tent — you will find more than computer drawings of chaos here — they shelter the creations of nature (clouds and coastlines) and of man (Koch curves and Sierpinski triangles)." Following this we stood in silence for some time. Then I thanked him and walked toward the pavilion's exit. As I opened the door, he called after me, "Tomorrow be sure to visit the garden in the center of the island. Maybe you'll be lucky enough to meet the old groundskeeper."

THE GARDEN OF SIMPLE COMPLEXITY

The mathematician lives long and lives young;
the wings of his soul do not early drop off,
nor do its pores become clogged with the earthy
particles blown from the dusty highways of vulgar life.
James Joseph Sylvester (1814–97)

THIS WAS TO BE my last full day on Frackinos. With Barnsley's advice in mind, I asked the hotel's desk clerk for directions to the garden. She told me, adding that it was elliptical in layout, walled on the outside, and with only a single entrance at its easternmost vertex. Upon arriving, I was greeted by a cherubic young man who informed me there were two gardens: a real one in the ellipse's interior and a virtual one on its border, and I should visit both. Only two trees grew in this garden: an *almond* at one focus and a *fig* at the other.

On either side of the entrance were two immense pillars — as if to guard against intruders — with a formula and a diagram engraved on each. The one on my right was an old friend from the previous chapter: the logistic function $x_{n+1} = kx_n(1-x_n)$. The other formula I had never seen before, although it was simpler still: $z_{n+1} = z_n^2 + c$. As to the drawings, they seemed to be the two trees, the fig on the right and the almond on the left pillar.

Pointing to the second equation, I asked the guard, "What does this mean?"

"Wish I could tell you. I've been here as long as I can remember, and I tour the garden daily, but I still don't *fully* understand it. Even my many talks with the old groundskeeper haven't helped."

"You can't have been here long."

"Longer than you can imagine," he said smiling curiously.

With the early morning sun on my back, I stepped across the threshold and realized I was alone in the Garden of Simple Complexity. Glancing about the enclosure with its magnificent floral displays, I decided to survey the boundary first. The walls were hung with psychedelic posters of incredible forms and colors, and at intervals large computer monitors were creating similar designs.

GASTON JULIA

The posters were of floral patterns made by the former groundskeeper Gaston Julia (1893–1978). As a soldier in the First World War, Julia was wounded in an attack on the French front. He lost his nose and for the rest of his life wore a black leather strap across his face. Between painful operations, he heroically carried on with his mathematical researches.

The first poster's caption — *see* the figure on the next page — stated "Julia Set for c = - 0.13 - 0.85i." This constant was a value for the c in the recursive formula $z_{n+1} = z_n^2 + c$. And z, like c, is complex and of the form a + bi where a and b are real numbers (*see* page 120). Every new c creates a different pattern; some values for c are "better" than others.

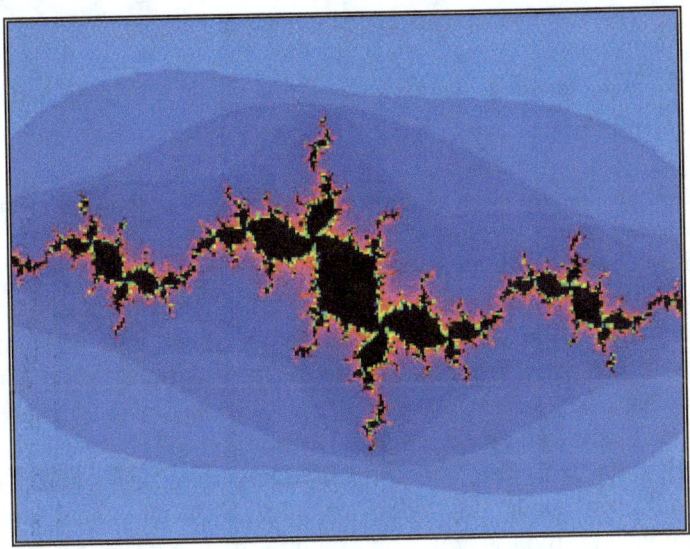

JULIA SET FOR C = − 0.13 − 0.85i

Figure 4.16

Rather than the particular quantities used, the creative force resides with the *iteration*. This is analogous to the logistic function, where although the choice of the constant k is important, without iteration there is no pattern. The full method to generate any Julia set follows:

1. Choose a complex value for c. (You soon learn that all choices are not created equal.) Keep it fixed for a given Julia set.

2. From the complex plane, pick any starting value z_0. Again, you soon discover which selections aren't part of the pattern. Roughly, any number whose distance from the origin is greater than 2.

3. Iterate:
$$z_1 = z_0^2 + c$$
$$z_2 = z_1^2 + c$$
$$z_3 = z_2^2 + c$$
$$\vdots$$

4. If after 50 or more iterations (this is why you need a computer), the initial value z_0 has escaped to infinity, discard it. Now go to step 2 and pick a new starting value. Otherwise, proceed to step 5.

5. If z_0 remains bounded (a prisoner), then color it and go to step 2.

When do you stop? Although the process is endless, after a while the final pattern — the strange attractor — emerges. If all the dots escape, then the attractor is the single point at infinity. Accordingly, all points on the complex plane are divided into two groups: the *escape set* and the *prisoner set*. Well, not exactly two sets. The Julia or *keeper set* is defined as the common boundary between the other two. These sets are different for each new value of c.

If Julia sets were tangible to touch, then *some* could be picked up whole by any tendril or bulge, although others would crumble like moldy bread or a lamp's mantle. These sets are either wholly connected or wholly disconnected; no hybrids exist. Graphs don't always display these qualities due to the tenuousness of the tendrils or the density of the dust. The black areas below are all prisoners; the Julia set forms the boundary between these and the escapers.

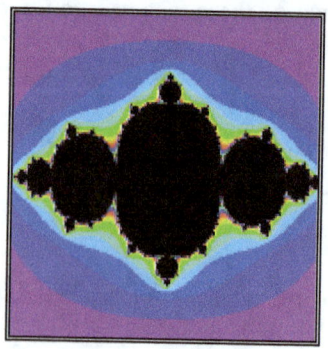

Connected Julia Set
c = − 0.75

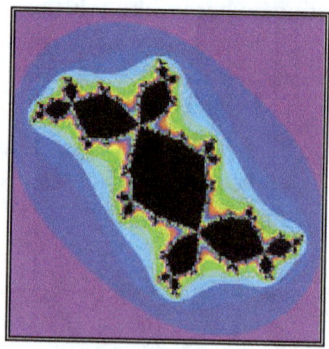

Connected Julia Set
c = − 0.125 + 0.75i

The next two sets are Julia dust; they take no prisoners.

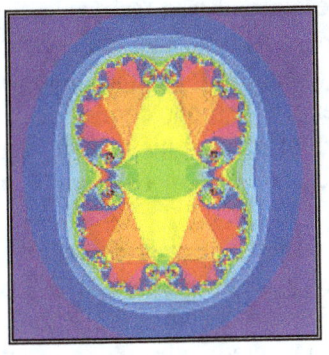 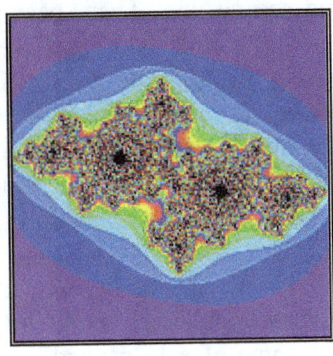

Julia dust for
c = 0.275

Julia dust for
c = −0.687 + 0.312i

The colors surrounding these strange attractors are mostly decorative, but they do serve to indicate how quickly the escapers flee to infinity — that is, the number of iterations needed. Prisoner sets, normally colored black, could also be given a full palette since they're attracted to an interior point — a sink hole. Conscious of this, we now understand that every *connected* Julia set dances tiptoe on two basins of attraction: the edge of infinity and the rim of the abyss.

Since both prisoners and escapers flee from Julia points as if from a medieval plague of pox, perhaps you think it's a mistake to label these figures "strange attractors." But by finding the inverse transformation to $z_{n+1} = z_n^2 + c$ (i.e. square z_n and then add c), the situation is easily remedied:

$$z_{n+1} = \pm\sqrt{z_n - c}$$ (Subtract c and then take either square root.)

Because the points are now traveling in opposite directions, the subscripts were interchanged. The above equation does the inverse operations of the first one, and it does them in the reverse order. This means, if putting on your shoes and tying your laces is the first transformation, then untying your laces and removing your shoes is the inverse transformation: *inverse* operations in *reverse* order. Iterates of the second equation for *any point* in the complex plane approach a Julia set (choose either root randomly). The entire process is dynamically discrete, resulting in the exquisite strange attractors we call Julia sets.

I BECAME AWARE of an extra presence in the garden, yet I could see no one. Then, near the almond tree, someone stood up — it was the old groundskeeper. All this time he had been on his hands and knees studiously tending his plants. I spoke a greeting. He responded in a friendly fashion, coming to welcome me. Stepping carefully over or around his colorful creations as he approached, I realized we had met before — the groundskeeper was Mandelbrot.

Before I could even ask a question, he was describing in detail his section of the garden around the almond tree. "My predecessor Julia created an infinite number of possible flowerbeds: some closely packed and connected, others just isolated islands of color. As for myself, I have created only a single unique form, the one named after me (*see* Figure 4.17). Nevertheless, it is a kind of index, an anthology of every *connected Julia set* (*see* the Chapter Notes). Once upon a time, when I first discovered this pattern, mathematicians christened it the world's most complicated object. Now we have a virtual bestiary of such creatures."

"Why do you think it became so popular?"

"In a word, computers," he answered. "As complicated as it is, this entire form can be graphed on a home PC with ten or so lines of programming. It was a profound shock to see such a simple program produce something so incredibly complex — wholly unforeseen. Everyone had his or her own copy shimmering on the monitor waiting to be explored. Suitably selected fragments of Julia's sets are precisely self-similar; my set is quasi-self-similar, no mindless duplication here. Nevertheless, the enigmatic 'little man' appears now and then, resplendent with every blob and tendril. More compelling is his occurrence in other fractals, elevating him to a universal symbol for the computer age — the über-form."

I broke in, "What can you tell me about the fig tree at the other focus?"

"You'll have to ask the other gardener yourself," Mandelbrot rejoined. "He's a part-timer, and this morning is one of his many times off, but you can gather a lot of information from the displays at his end of the ellipse."

I thanked the old groundskeeper, and later I did as he suggested. First, however, I explored his personal gallery of images.

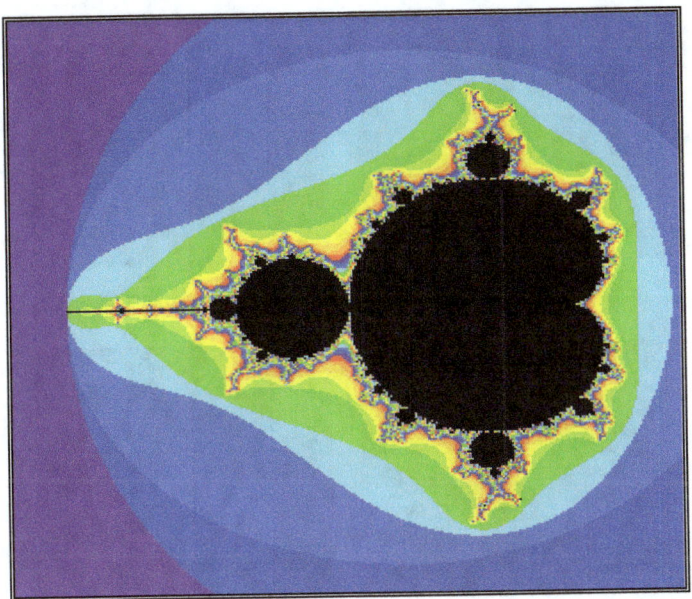

THE MANDELBROT SET

Figure 4.17

The cleft where the "head" joins the "body" on the reclining Mandelbrot man (above) is called Seahorse Canyon. Many Internet websites run Java applets to explore this chasm. By just pointing and clicking on anything in the valley, you descend to the next level of magnification. A few such programs will magnify the original image to beyond the diameter of Pluto's orbit. And that's just a beginning, the atrium to a fantastic mansion. This descent is a voyage to infinitesimal infinity. Along the way you pass seahorses, waves, cacti, pinwheels, arabesques, starbursts, whirlpools, sprouts, swirls, zigzag lightning, Julia sets of all descriptions, and every so often, hidden away, another Mandelbrot man (*see* following page).

This virtual voyage echoes the first few lines of Samuel Taylor Coleridge's phantasmagorical poem *Kubla Khan*. The poet, by his own account, composed it while in an opium-induced reverie:

> *In Xanadu did Kubla Khan*
> *A stately pleasure-dome decree:*
> *Where Alph, the sacred river, ran*
> *Through caverns measureless to man*
> *Down to a sunless sea.*

| PINWHEELS | SEAHORSES |
| MANDELBROT MAN | NIGHT VISIONS |

This landscape appears strange yet at the same time familiar. Some basic outlines are recognizable, but the detail and the everlasting abundance are alien — strange objects in a stranger land. At least, that was the opinion of most when these figures were first discovered. But that was then. Today, everywhere you look in our culture you see them. They're on calendars, book jackets, seminar covers, store windows, advertising logos, and so forth. Advertisers — those heralds of consumerism — have pocketed them for their private monetary purposes. On the other hand, Ian Stewart in his book *Does God Play Dice?* sums up their true importance and usefulness:

> But there's much more to fractals than just a few buzzwords. "No one will be considered scientifically literate tomorrow who is not familiar with fractals," says the physicist John Wheeler. Fractals reveal a new regime of nature susceptible to mathematical modelling. They open our eyes to patterns that might otherwise be considered formless. They raise new questions and provide new kinds of answers. "Fractals," says the science writer Jeanne McDermott, "capture the texture of reality." [3]

AFTER GATHERING SOME LITERATURE on the fig tree, I left the resplendent Garden of Simple Complexity. Lingering regrets crossed my mind that I would never have an opportunity to speak to the keeper of the second tree. Walking eastward, passing the two pillars, I said goodbye to the young guard and went to the hotel. It was time to prepare for the final dinner planned by the tour guide and the captain. Everyone was to be there, some of whom I hadn't met.

By the time I went down to the dinning room, several people were already present. In the center of the room was a massive wooden table carved in the shape of a Koch snowflake. It was spread with liberal quantities of Greek appetizers: saganaki shrimp, pita bread, tzatziki, fish soup, keftedes, pickled calamari, and a huge bowl of Greek salad sprinkled with feta cheese and dotted with black olives. Bottles of red and white retsina were everywhere. Tonight we'd dine like Olympians.

I mingled. I met. I munched. I met everyone except the two older Greek men who hadn't yet arrived. But just as we were sitting down to the main course of roast lamb, moussaka, and a variety of souviaki, the last two tourists came in. The larger, rough looking man said his name was Epos, a retired boxer and engineer. The other was a more refined person called Femos, a poet and composer.

We ate and drank too much, the affluent affliction, and the conversation turned to everyone's favorite topic — him- or herself. The ship's captain was a clever man for he encouraged this egotism by suggesting we each had two minutes to relate some unique personal accomplishment. I spoke of the book I was writing and that whatever anyone said in their two minutes could be recorded for all time. Eventually it was the turn of the two late arrivals. Epos told how as a "young" man working with Heinrich Schliemann (1822–90) at Hissarlik Hill (Troy), he had found the bronze latch to the hidden door on the Trojan horse. Femos related a parallel story about excavating on Odysseus' home island of Ithaca and uncovering an ancient lyre, which he just happened to have with him. These tall tales produced smiles all around. The captain spoke last. He claimed to have sailed the Aegean his entire life but always kept to the well-traveled routes directly between the islands. This sea is like a carpet; some areas are worn threadbare while others are never touched. A few years ago during a terrible storm, he was blown far off course and found Frackinos Island.

Following baklava and coffee, we mingled again and talked. I stood near the two Greeks to listen to their conversation. Femos was extolling the mind-bending beauty of the Mandelbrot man. Epos concentrated on the mathematical economy and grandeur of the equation that generated it: $z_{n+1} = z_n^2 + c$. By way of joining their dialogue I said, "It's a trivial seed, but it holds an entire world of wonders — a homunculus that grows to become the Almond-bread man."

"Whatever its strange genesis, it's still great art," affirmed the poet.

His friend jumped in, "And this art was created in the bowels of a computer."

I remarked that "If you can get great art from a machine, then the ancient argument between science and art is hollow — an empty barrel of hot air not worth the price of its cork bung." My two listeners mused on this for some time.

The captain called on the poet-minstrel for an example of his art. And with the consummate ease of the professional, Femos asked the audience what they would like to hear. Upon request he recited some Greek lyrics from Callimachus and Sappho and sections of Aeschylus and Sophocles accompanied by his unearthed lyre. The clarity of his Pythagorean cords, the artfulness of his skill, and the power and the glory of the words hushed the dinner guests to a reverie. Out from this silence, came the voice of Epos asking him to conclude by reciting "The Funeral Games for Patroclus": the penultimate book of the *Iliad*. Without hesitation, the poet began his narration, which continued for almost an hour, all this without a single written note. As he finished, I glanced about the room to see several people wiping their eyes, while Epos sobbed quietly in a corner. The poet went to console his friend. There was nothing more to be said; so, we dispersed to our individual rooms leaving the artist and the scientist alone.

Late the next morning I collected my belongings for the long boat trip to Athens. When I went to the dock, it was empty, so I found a chair, sat in the sunshine, and thought about everything I had seen and learned and talked about the previous evening. It felt good. The memories, the sun, and the world seemed more wondrous; the mysteries deeper, and subtle things mattered again.

In a daydream of musings on the beauty of Seahorse Canyon and its myriad complex designs, my mind drifted to an unexpected place, a sonnet by Edna St. Vincent Millay:

> Euclid alone has looked on beauty bare.
> Let all who prate of Beauty hold their peace.

And a memorable passage from Bertrand Russell that echoes the identical sentiment:

> Mathematics possesses not only truth, but supreme beauty — a beauty cold and austere, like that of sculpture without appeal to any part of our weaker nature, sublimely pure, and capable of stern perfection such as only the greatest art can show.

Within the discoveries of Julia and Mandelbrot we see no austere beauty; we find no Elgin Marbles, but we do uncover incredible abundance, psychedelic images, and opium-like visions. Neither beauty bare, nor truth naked! Here the Lady is resplendent in infinite ballroom gowns. What are we to make of these contrary versions of beauty, one classical the other romantic? Perhaps, like music lovers have done in their art, we should accept both, value both. Beauty cannot be fully quantified and defined in a legal manner; those who try this reveal more about themselves than the subject they attempt to legislate. In either version, whether Russell's or Mandelbrot's, *truth is beautiful*. All this neither says nor implies anything about *truth* being relative or a cultural artifact.

The boat trip to the mainland turned out to be uneventful and long, but my island mood persisted as I relaxed in a deck chair near the stern and talked to a few fellow passengers. I gazed up at the passing parade of cumulus clouds: large puffballs of water vapor in a variety of sizes — self-similarity in the sky. They seemed harbingers of good weather and good things. Higher still, and moving more slowly, cirrus clouds strolled across providing a backdrop of finer fractal forms.

After some time, I glanced over the rear handrail at the boat's turbulent wake disturbing the perfect symmetry of the calm sea. This turbulence had several forms: tiny waves on larger waves on bigger waves still. Another more noticeable pattern near the wake's edge caught my eye: spirals alternating clockwise and counterclockwise diminishing in size and spinning to extinction.

WHIRLPOOLS CAUSED BY A PADDLE OR A PROPELLER

By now it was early evening, and since we were only a few degrees north of the equator, the sunset was over almost before it began. No tiring the sun with talking and sending him down the sky here. Darkness descended and hidden lights appeared on the great dome of the sky. The stars rose out of the sea at our stern while others plunged beneath our bow. Here too, in the deep star fields of space, modern astronomy has revealed a fractal universe of galactic clusters. And since the Hubble Space Telescope has been upgraded to higher resolutions, we see this fractal form persists universally. Hubble is a modern cyclopean eye looking forward in space but backward in time.

The world is not as we imagined in the days of our scientific youth among the Greeks. Euclid was a towering intellect, but we have surpassed him — his geometry is not of the natural world. Rather it belongs to the *ideals* from Plato's allegory of the cave in the *Republic*. Nowhere in nature do we find perfect lines, planes, cones, spheres, or anything else. These are visions from another place. Everywhere you look the world around you is self-similar: the bark of a tree, its leaves, rock falls, glaciers, streams, coastlines, the clouds above your heads, the craters within craters on the moon, and the galaxies beyond. Fractals are the very stuff of the universe. Sometimes in life you turn a corner and a new vision of the natural world emerges, unclear at first, but gradually coming into focus as you study it. I now knew Barnsley was right; I never would see the world the same again.

CHAPTER — 5

NUMBERS: SUPERNATURAL

I believe that mathematical reality lies outside us, that our function is to discover or observe it, and that the theorems which we prove, and which we describe grandiloquently as our "creations," are simply the notes of our observations.
G. H. Hardy (1877-1947)

A GROUP OF PEOPLE at a bar are discussing the usual topics when the young bartender, a mathematics student, asks, "What are numbers?"
Politician: "I have no idea."
Voter 1: "Three's a number."
Voter 2: "I bet I can think of a larger number."
Bill Gates: "The only numbers you need are 0 and 1."
Baptist Minister 1: "My favorite number is 7."
Baptist Minister 2: "Of all the numbers 8 is the holiest because 6 and 9 and sometimes 2 and 4 have one hole each while 8 always has two holes."
Satanist: "My lucky number's 666."
Voter 2: "I bet that's a larger number."
Philosopher 1: "Seven's an odd prime number."
Philosopher 2: "That's an interesting statement. I'll have one of my research assistants look into it."
Quantum Physicist: "All numbers are equally prime and non-prime until observed."
George W. Bush: "What's an odd prime? Sounds like a terrorist! Call Condi! And put that 666 on the enemies list. I'm gonna get that numba dead or alive. If I can't get even, I'll get odder."
Bartender: "Ladies and gentlemen, if you can't tell me what numbers are, perhaps you can tell me where they hang out?"

If nature has numbers, where do they reside? Do they have validity independent of the human species, or are they residents of our nervous system? Are mathematical quantities and truths discovered or invented? All these questions pose the same core problem: where do numbers reside?

It might seem obvious that they reside either in the external world or in the minds of *Homo sapiens*. But a third possibility exists that they live exclusively in our cultures. And since this globe has many many cultures, this would imply many many fundamentally different types of mathematics. So, at least three answers are possible. Someone might even suggest another preference, proclaiming each individual creates his or her own version of reality. This fourth option is an extreme version of the cultural reality option when society has shrunk to a single individual. So now we have four:

1. the external world
2. the human species
3. every distinct human culture
4. each individual.

Alternatives 2, 3, and 4 can be viewed as size diminutions on a common theme of centrism: species to tribe to individual. But the last of these can be quickly dismissed: things that are too stupid to be spoken should be sung, and this option encompasses an entire opera. Choices 2 and 3 — size variations on a "collective solipsism" — have had more supporters than you might suspect. Dr. Leslie Alvin White, past professor of anthropology and ardent cultural relativist, ably expounded and defended this view in his essay "The Locus of Mathematical Reality," reprinted and widely read in James R. Newman's popular four-volume anthology *The World of Mathematics*[1].

At the conclusion of his essay, White says mathematics is a kind of primate behavior like languages, musical systems, and penal codes. He believes mathematical concepts are fabricated like our traffic laws. He says they're not part of the structure of the universe but nonetheless lie outside us individually, existing entirely within our culture.

In defense of option 2, Newman himself with Edward Kasner expressed similar thoughts near the close of their co-authored book *Mathematics and the Imagination:*

> ... we have overcome the notion that mathematical truths have an existence independent and apart from our minds. It is even strange to us that such a notion could have ever existed.[2]

To counter this and deliver the coup de grâce to all those afflicted with either 2 or 3, read Martin Gardner's insightful essay "Mathematics and the Folkways." In the postscript, Gardner notes:

> I will here say only that almost all mathematicians today agree with Hardy [*see* the opening quotation] that a mathematician discovers truths that are independent of his culture and that those truths are qualitatively different from the conventions of traffic regulations or codes of etiquette.[3]

People like White, Newman, and Kasner correctly point out that different cultures do have different number systems, and these are deeply and emotionally entrenched. The similarity of Indo-European *number words* is strong evidence of this entrenchment. The following chart illustrates this curious stability for *seven* down through the centuries:

	Hebrew	Sanskrit	Greek	Latin	German	English	French
7	*shevah*	*sapta*	*hepta*	*septem*	*sieben*	*seven*	*sept*

Yet whether you say *shevah* or *sept* is trivial; whether you champion the metric over the imperial system is of no matter; whether you sing the praises of meters over yards is pointless; whether you count by 5s, 10s, 20s, 60s, or 2s or 12s or whatever is of no consequence, and whether you use dots, lines, bars, squiggles, cursive forms, or any combination of these is meaningless. Computers utilize a binary system; even so, they perfectly represent our decimals. As Gardner points out, variations in these schemes are just variations in symbolization. All number systems use the same laws. Every earthly counting method — and we believe every unearthly variety as well — has different *numerals* for the same *numbers* all based on the same *logic*.

We can see these different numerals within our own culture. Look at the chalkboard (next page) marked with a single *number*. All these numerals for the identical number can be erased with a few swipes of a chalk brush. Yet the number lives on! We first encounter numbers as an abstraction from a diversity of particular, concrete examples. The very existence of so many terms for a single concept in our culture suggests that historically we were centuries making the transition from concrete to abstract. As Bertrand Russell once remarked, it took a long time for humans to realize that a pair of apples and a brace of partridges were both instances of the number two.

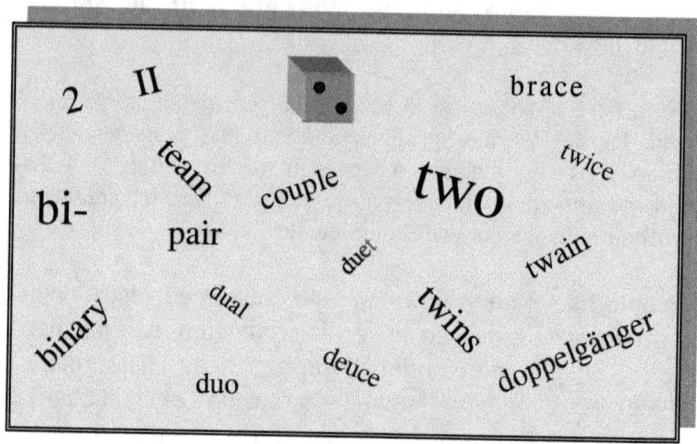

PERISHABLE NUMERALS — IMMORTAL NUMBERS

Figure 5.1

The universe is so subtle that the butterfly net of our senses gathers only a few truths in a lifetime. Yet no need exists to slide into some form of solipsism for imagined alternative mysteries. Believing mathematics is grounded in culture is the collective solipsism George Orwell satirized so effectively in his novel *Nineteen Eighty-Four* with the scene of Winston and the four fingers. Fleeing from ethno-, species-, and ego-centrism, science has made a long forced march up country to the landscape of objective truth. The spiral in the great Andromeda galaxy is truly *in* Andromeda and *not in* our neurons. Reflect on the close of Chapter 2 and the scene at Deep Space Nine. When the extraterrestrials and the humaaaans gather at Quark's Bar to play tongo, the nature of three-space — not their neurons — allows the tossing of only five perfect dice. Stand down, Protagoras!

Mathematics exists in the external world, not in our species or tribe, and certainly not within each individual. But where in the world might this be? The answer is more startling and comprehensive than you could expect. Once again, we listen to Martin Gardner:

> Mathematics is not only real, but it is the only reality. That is that the entire universe is made of matter, obviously.

And matter is made of particles. It's made of electrons and neutrons and protons. So the entire universe is made out of particles. Now what are the particles made out of? They're not made out of anything. The only thing you can say about the reality of an electron is to cite its mathematical properties. So there's a sense in which matter has completely dissolved and what is left is just a mathematical structure.[4]

THE SIGN OF THE FISH

One of the most striking differences between a cat and a lie is that a cat has only nine lives.
Mark Twain (1835-1910)

THE CLICHÉ ADVISES that there is more than one way to skin a cat, some ways more bizarre than others. And at least two methods exist to describe one. You can list its binomial name (genus and species) and write a detailed description, but it's often clearer just to show a photograph. In the previous section, I purposely neglected to give a comprehensive mathematical "picture" of what White calls "primate behavior." Now I intend to display a large canvas of one product of this ethnocentrism, to show its beguiling qualities, and its ultimately deceptive nature.

By "The Sign of the Fish" I do not mean Pisces the zodiac sun symbol found in your daily newspaper's "horrorscope" column. Rather I'm referring to the early Christian religious symbol (*see* below). However, even this fish has links to astrology.

Everything that follows in the next ten pages — those between the two fish symbols — is an example of White's primate behavior in mathematics. (Recent best-selling books on this and related topics have sold hundreds of thousands of copies and developed a near-cult following.) Although the arithmetical properties to be shown are mostly real, the associations and extravagant claims are imaginary or invented.

MUCH OF THE LIFE of *Jesus* of Nazareth involved fish, fishing, and fishermen. His ministry began and ended at the Sea of Galilee with a miraculous catch of fish. But *Christ* is normally symbolized by the lamb, which derives from the Passover sacrifice of Exodus 12:1-11. Furthermore, John's gospel speaks of the Jesus as the lamb that takes away the sin of this world. The significance of the lamb is clear, but what are the origins and meaning of the fish symbolism? Why this association?

The writers of the Old Testament had an inordinate fondness for acrostics. Psalm 119 is an extraordinary acrostic poem consisting of 22 stanzas of 8 verses each. In the original Hebrew, the first 8 verses begin with the letter *aleph*, the second 8 with *beth*, the third 8 with *gimel*, and so on. In this fashion the psalmist continues through the entire Hebrew aleph-beth.

Perhaps unknowingly the KJV translators of Matthew 7:7 achieved the same emphasis:

> **A**sk, and it shall be given you;
> **S**eek, and ye shall find;
> **K**nock, and it shall be opened unto you:

The world's most famous acrostic cannot be found in either testament, yet it flourished in the early church; it lives on in the hearts of Christians everywhere.

Ιησους Χριστος Θεου Υιος Σωτηρ

This translates as "Jesus Christ, Son of God, Savior."

The acrostic ΙΧΘΥΣ (ichthus) is Greek for fish.

And *ichthus* was further shortened to just ⊂×. For 1st century Christians, it served a special purpose: secret symbols for your faith meant personal safety. Since Roman emperors demanded worship "as gods," they ruthlessly slaughtered all competition. To be discovered a Christian in ancient Rome guaranteed a bad day.

Although this example inescapably associates the image of the Fish with Jesus, it's not the only connection. Let's turn our eyes from the earth and water to the sky and the stars.

The legendary Star of Bethlehem announced Christ's nativity. But is there a corresponding astronomical event surrounding Jesus' death and Resurrection? What would be appropriate? Possibly a great occasion in outer space, somehow celebrating rebirth, renewal, and resurrection. For Christians, what historic event could be more significant than the Resurrection?

Early studies of the heavens revealed a rare astronomical event that takes place once every 2,160 years. Like a child's gigantic top, the earth wobbles on its axis with one complete wobble (precession) every 25,920 years. This precession means the sun's position moves backward through the zodiac. So spring (the vernal equinox) comes 20 minutes earlier each year causing the sun to move through one constellation every 2,160 years (25,920 ÷ 12). Today, spring/rebirth takes place in the constellation of Pisces. It seems remarkable that at the time of the Resurrection the sun was leaving Aries the Ram and entering Pisces the Fish[5].

God referred to these great astronomical signs in Job 38:32 when he declared, "Can you bring forth the constellations in their seasons?" This implies he designs events to bring forth these star groupings when he pleases, even Pisces the Fish at the time of the Resurrection.

As noted in the previous chapter, "everything comes in threes" is an ancient proverb, an old saying even in the days of King Solomon. A third association between Jesus and the fish merits our investigation.

At the end of John's gospel, Jesus encounters seven of his disciples fishing on the Sea of Galilee. Although they had been casting nets all night, and it was now near dawn, they had caught nothing. Jesus told them to spread their nets on the right side of the boat. And when they did:

Simon Peter climbed aboard and dragged the net ashore. It was full of large fish, 153, but even with so many the net was not torn.
John 21:11

The Miraculous Draught of 153 Fishes
by Gustave Doré

The above verse, John 21:11, pointedly states the exact number 153 and emphasizes that the net was not broken — implying this quantity was significant. Every detail of this miraculous catch has great importance and should be intensely studied. I find the exact counting of the fish a curious and unusual event. Fishermen don't count their fish — at least not catches of this magnitude. If the gospel writer had rounded the quantity of fish to 150, say, we would have given it little thought. The only possible reason for this precise enumeration is to introduce the number 153. Consequently the outstanding question is why 153 fish and not some other number?

Augustine of Hippo, in his *Tractates on the Gospel of Saint John,* gave one of the first explanations. He began with 10, the number of the Commandments and a symbol of the old Mosaic dispensation. To this he added 7, the number of the Gifts of the Holy Spirit and the new dispensation. Therefore, by adding the old and the new he arrived at 17, and the sum of the first 17 numbers is 153:

$$1+2+3+4+5+6+7+8+9+10+11+12+13+14+15+16+17 = \mathbf{153}.$$

Curiously, the famous bishop missed the crucial point of his own reasoning:

> **153** is triangular,
> one of the Bible's foundation numbers

What are these numbers?
Why are they called triangular?

Numbers that can be arranged as shown in the figure below are called triangular. The reader can see from the diagrams that these are just the orderly sums of the counting numbers. With a handful of pennies you will quickly find these patterns.

THE FIRST FOUR TRIANGLE NUMBERS

For future reference, here's a table of these numbers:

1	66	231	496	861
3	78	253	528	903
6	91	276	561	946
10	105	300	595	990
15	120	325	630	1035
21	136	351	**666**	**1081**
28	**153**	378	703	.
36	171	406	741	.
45	190	435	780	.
55	210	465	820	

THE FIRST 46 TRIANGLE NUMBERS

Figure 5.2

We have actually met these numbers previously. The third column of Pascal's triangle is entirely composed of them — *see* Figure 2.13. Although infinite in quantity, they're not common. Up to 1000, only 4 percent are triangular. When the numbers get larger, these special ones become extremely rare. For example, in the first two million numbers, fewer than 2000 are triangular — on the average that's only 1 in 1000.

Why are **45**, **666**, and **1081** emphasized among this already extraordinary set of numbers? It's understood that the words of the Bible are special — but they're special in a way you may not know, a way foreign to the English language. The key to this uniqueness is revealed in Revelation 13:17-18 when John writes:

> *Which is the name of the beast or the number of his name. This calls for wisdom. If anyone has insight, let him calculate the number of the beast, for it is a man's number. His number is 666.*

Simply stated, each character in the ancient Hebrew and Greek alphabets did double duty as both a letter and a number. So, *every word was also a number*. This practice of summing up the numerical values of the letters in a name — called gematria (mentioned in Chapter 1, page 14) — is familiar to everyone from Roman numerals. For example, "CIVIC" would sum to 100+1+5+1+100 or 207. The Romans, however, unlike the Hebrews and Greeks used only the letters I, V, X, L, C, D, and M as numbers. In Hebrew, 45 is the gematria value of Adam's name and 1081 of God's Holy Spirit. The Scriptures provide the authority for gematria's use. From Genesis to Revelation, it links many outstanding biblical verses.

The intense triangularity of 153 reveals itself in many directions. Consider the following: instead of writing $1 \times 2 \times 3$, mathematicians write 3! (called factorial 3). With this type of notation it's possible to express our number as the sum of factorials:

$$
\begin{aligned}
1! &= 1 &&= 1 \\
2! &= 1 \times 2 &&= 2 \\
3! &= 1 \times 2 \times 3 &&= 6 \\
4! &= 1 \times 2 \times 3 \times 4 &&= 24 \\
5! &= 1 \times 2 \times 3 \times 4 &&= +120 \\
& && \ 153
\end{aligned}
$$

Therefore, 1! + 2! + 3! + 4! +5! = **153**.

Mathematicians refer to numbers expressed in the above manner as super-triangular. Happily, the Antichrist's quantity, 666, is not among them.

Not only is 153 triangular but its reverse, 351, is also (*see* Figure 5.2). Numbers with different digits that remain triangular when written backward are exceedingly rare. (Of course, if the triangle number has identical digits, like 666, then it's automatically reversible.) In the first six million numbers, only four exist. To see a list of these refer to the Chapter Notes.

Truncating 153 (cutting it into smaller pieces) further emphasizes its intense triangularity. To truncate means to slice off without rounding up or down. For example, 7.821 rounds to 8 but truncates to 7. Let's show why 153 can be truncated. The symbol "/" means "a cut":

$$\left.\begin{array}{l} 1/53 \text{ gives } 1 \\ 15/3 \text{ gives } 15 \\ 153/ \text{ gives } 153 \end{array}\right\} \text{ all triangle numbers.}$$

From a set of infinite possibilities, only six numbers prove to be truncated triangle numbers (again *see* the Chapter Notes). Amazingly, 153 is one of these. And 666 is also.

UP TO THIS POINT we have seen a few of the unusual properties of 153 and some of its parody by 666. Now we enter an expanse forever barred to the Antichrist's number and every other quantity except 153 — Jesus' number. Nevertheless, the next three characteristics don't exhaust what might be written about 153.

We'll use an extraordinary function to reveal what we could otherwise only dimly perceive. All mathematical functions have the same simple form. First, choose a number (the pre-image), and then do some operations on it (for example, +, —, x, ÷, raise to a power, take a root, or flip) to get a second number (the image or answer). That's it! The figure below introduces this rule in full formal attire, while the examples following show how it works.

> **THE TRINITY FUNCTION**
>
> $$abc \longrightarrow a^3 + b^3 + c^3$$
> pre-image   image

If you're unfamiliar with mathematics, this must appear a most unusual procedure. Its outstanding feature is the general *threeness*. Also, no matter how you scramble the pre-image digits, the image will always be the same. This function allows us to see clearly through a dark glass into a world of marvelous symbolism and beauty. As we continue, you will begin to appreciate the unbelievable clarifying value of this rule and the profound insights it allows into God's Holy Word.

The wonderful attributes of triangularity, reversibility, and truncation already discussed are just a preamble to our main idea. Consider the outcome from applying the Trinity Function to 666:

$$666 \longrightarrow 6^3+6^3+6^3 = 6\times6\times6+6\times6\times6+6\times6\times6 = 648$$

Nothing surprising. However, when we use this function on 153, the result is truly remarkable:

> **153** $\longrightarrow 1^3 + 5^3 + 3^3$
> $= 1\times1\times1 + 5\times5\times5 + 3\times3\times3$
> $= 1 + 125 + 27$
> $= \mathbf{153}$
>
> **THE RESURRECTION PROPERTY**
>
> Figure 5.3

The Trinity Function reveals the constancy of Jesus' Number. It's neither transformed nor changed. Reapplying the rule simply shows how 153 resurrects itself. In contrast the Antichrist's number, as we saw above, *is transformed* though not by much: i.e. $666 \longrightarrow 648$. If a 2 had been taken from the 8 and given to the middle 4, then this continuing parody would have been complete: i.e. 666 ⟶ 666. Satan may imitate Christ, but cannot equal Him.

> *"In the whole land," declares the Lord,*
> *"two-thirds will be struck down and perish;*
> *yet one-third will be left in it.*
>
> *This third I will bring into the fire; I will refine*
> *them like silver and test them like gold.*
> *They will call on my name and I will answer them."*
> Zechariah 13:8-9

In this quotation, the prophet paints an apocalyptic picture of the *end times* and the establishment of the Messiah's Kingdom. This "one-third" will include that last remnant of the Jewish People who recognize and follow Jesus as their Messiah at His Second Coming. As we shall learn, this one-third is also an outstanding attribute of 153. Mathematics mirrors reality.

To explain the preceding, choose any number you wish. There is only one condition; it must be a multiple of 3 (e.g. 6, 24, 336, . . .). Suppose you pick 1776 (= 3 x 592). Now, apply the Trinity Function to it and to each image:

$$1776 \longrightarrow 1^3 + 7^3 + 7^3 + 6^3 = 903 \quad \text{(1st image)}$$
$$903 \longrightarrow 9^3 + 0^3 + 3^3 = 756 \quad \text{(2nd image)}$$
$$756 \longrightarrow 7^3 + 5^3 + 6^3 = 684$$
$$684 \longrightarrow 6^3 + 8^3 + 4^3 = 792$$
$$792 \longrightarrow 7^3 + 9^3 + 2^3 = 1080$$
$$1080 \longrightarrow 1^3 + 0^3 + 8^3 + 0^3 = 513$$
$$513 \longrightarrow 5^3 + 1^3 + 3^3 = \mathbf{153} \quad \text{(final image)}$$

All the above can be summarized as follows:

$$1776 \to 903 \to 756 \to 684 \to 792 \to 1080 \to 513 \to \mathbf{153}$$

Amazingly, it doesn't matter how large a number you start with — it could be a googolplex (*see* the Chapter Notes).

The number need only be divisible by 3, and that's a third of all quantities. By continuously reapplying the Trinity Rule to each image, the process quickly ends at 153, the center of the net. So, a third of all numbers do go to 153.

I encourage the reader to use a hand calculator to confirm these unusual statements. Here are three examples:

$$99 \to 1458 \to 27 \to 351 \to \mathbf{153}$$
$$1125 \to 135 \to \mathbf{153}$$
$$375 \to 495 \to 189 \to 1242 \to 1080 \to 513 \to \mathbf{153}$$

No other number, be it 666 or whatever, can copy or counterfeit this outstanding property. So totally singular is this characteristic that out of the whole of God's created quantities none can imitate it. The sacred uniqueness of 153 makes it the most fabulous number in the Bible or anywhere else.

It's easy to prove that every third number eventually ends at exactly 153. Obviously we can't check all possible multiples of 3, even with the world's fastest computer. So we should look for a method to drastically shorten our work. When the Trinity Function transforms any number larger than 2000, something curious occurs. The image (answer) is always smaller than the pre-image (original number). For example:

$2,001 \to 2 \times 2 \times 2 + 0 \times 0 \times 0 + 0 \times 0 \times 0 + 1 \times 1 \times 1 = 9$
And 9 is much smaller than 2,001.

$2,899 \to 2 \times 2 \times 2 + 8 \times 8 \times 8 + 9 \times 9 \times 9 + 9 \times 9 \times 9 = 1,978$
And 1,978 is smaller than 2,899.

$9,999 \to 9 \times 9 \times 9 + 9 \times 9 \times 9 + 9 \times 9 \times 9 + 9 \times 9 \times 9 = 2,916$
And again, 2,916 is smaller than 9,999.

This shows that quantities greater than 2000 will steadily descend — by repeated application of the Trinity Rule — to less than 2000. So we need test only those that are both smaller than this and multiples of 3. How many numbers fit this description? Exactly 666! Here's another link between opposites: between the number formed from a single repetition of the first 3 odd digits — 153 — and the number forged from 3 repetitions of the 3rd even digit — 666. Significantly, 2000 is the watershed quantity! Coincidence? Not likely! God creates order, not chaos. Beware the Millennium.

Figure 5.4 mirrors many biblical passages. Note the 1080 in the third circle. Previously I wrote about using gematria to find the numerical value of a Hebrew or Greek word. It's surely no accident that the gematria value of "the Holy Spirit" is exactly 1080. On many occasions the Scriptures tell us that by way of the Holy Spirit we're led to Jesus, 153 — as the diagram shows.

To summarize, the Trinity Function uncovered the following three extraordinary properties:

- Property 1: 153 is fixed, constant, and unchangeable.
- Property 2: A third of all numbers transform to 153.
- Property 3: All quantities larger than 2000 behave differently from those smaller than 2000.

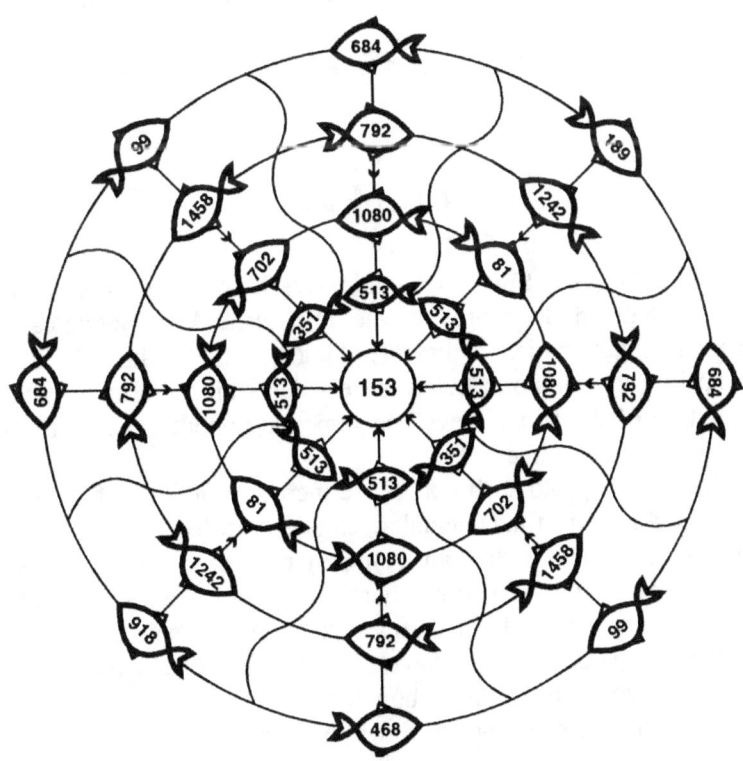

THE 153 FISHES IN THE NET

Figure 5.4

The 153 Fishes in the Net illustrates these three ideas in a unique diagram. The constant 153 is at the center of the net. It catches a third of all numbers. Only multiples of 3 are used, and only quantities below 2000 are shown or needed. These properties mark 153 as the most phenomenal quantity

Centuries ago, St. Augustine summed up man's condition and the nature of the Fish symbolism:

> Now if you connect the initial letters of those five Greek words, Ιησους Χριστος Θεου Υιος Σωτηρ *(Jesus Christ, the Son of God, the Saviour)* you have the Greek word ichthus, which means 'fish', and the allegorical meaning of the noun is Christ, because he was able to remain alive — that is, without sin — in the abyss of our mortal condition, in the depths, as it were, of the sea.[6]

FISH FRY

THE PREVIOUS SECTION, the pages between the fish symbols, is a detailed example of what Dr. Leslie Alvin White calls primate mathematical behavior. Although the author* of this material (*see* footnote) believes he's describing the underlying structure of the Bible as well as something about external reality, he's doing neither.

Those who hold such beliefs are members of Christian, Jewish, or Moslem fundamentalist cults — ecclesia, meaning a *called-out group.* Called out from general society and protected from it by having their own churches, schools, and even compounds, lest a scintilla of reality disturb their absolute "truth." When groups believe they have perfect knowledge, with no test in the external world, this is how they behave.

Let's look closely at the flaws in their reasoning. Let's fry some fish! Despite all the earlier ravings about the *unique*

---

* I wish to thank Isaiah Publications for allowing me to quote extensively from James Harrison's book *The Pattern & The Prophecy: God's Great Code* published in Toronto, Canada, 1995.

qualities of 153 — in particular the Resurrection Property shown in Figure 5.4 — these are false, totally false. Three other such numbers exist: 370, 371, and 407. *See* for yourself:

$$153 = 1^3 + 5^3 + 3^3 = 153, \quad 371 = 3^3 + 7^3 + 1^3 = 371$$
$$370 = 3^3 + 7^3 + 0^3 = 370, \quad 407 = 4^3 + 0^3 + 7^3 = 407$$

ODD FACTS

And even this list doesn't exhaust the many pathways down which the iteration of the Trinity Function could lead us. Sometimes it will escort us through a cycle of numbers (e.g. 55 to 250 to 133 and then back to 55, endlessly) and sometimes to unity. The outcomes are many; the uniqueness is nil.

It gets worse! When the Trinity Function is repeatedly applied, the second "extraordinary property" (page 177) says that a *third* of all numbers terminate at 153. Is that significant? Not when you realize that *half* of all numbers terminate at either 370, 371, or 407. The percentage of all numbers ending up as poles (individuals) or cycles (groups) are shown in the following table:

POLES		CYCLES	
153	33%	55 - 250 - 133 - 55	3%
370	17%	919 - 1459 - 919	6.5%
371	27%	136 - 244 - 136	1.5%
407	6%	160 - 217 - 352 - 160	2.5%
1	3.5%		

TERMINAL PERCENTAGES

The third "extraordinary property" (about the numbers above and below 2000) results from the Trinity Function, and so has no more relation to 153 than any other number. And the passage of time has shown that the apocalyptic deep breathing about the significance of the year or the number 2000 was nonsense. Recall the Y2K hysteria. The Millennium was of no more significance than your car's odometer rolling over from 99,999 to 100,000.

So there you have it. Of these three "extraordinary properties," the first isn't unique, the second isn't important, and the third didn't refer to 153.

In the modern era, when England's greatest number theorist G. H. Hardy wrote about 153 in his celebrated autobiography, *A Mathematician's Apology,* he commented on trivial number properties that neither generalize nor relate to theorems. But I'll let Hardy speak for himself:

> These are odd facts [*see* the table on the previous page], very suitable for puzzle columns and likely to amuse amateurs, but there is nothing in them which appeals much to a mathematician. The proofs are neither difficult nor interesting — merely a little tiresome. The theorems are not serious; and it is plain that one reason (though perhaps not the most important) is the extreme speciality of both enunciations and the proofs, which are not capable of any significant generalization.[7]

It gets worse! At the risk of dancing on the rim of overkill, I can cite another fatal problem with such "mathematics." All the supposed unique properties of 153, all but one that is, vanish when it's written in a different number base. For example, 153 base 10 (i.e. 153_{10}) equals 1103 base 5 (i.e. 1103_5). Proceeding left to right, in base 5 the columns have the following values: $5^3, 5^2, 5^1, 5^0$ or 125, 25, 5, and 1. So 1103_5 means 1x125 + 1x25 + 0x5 + 3x1 equaling 153_{10}, as expected. If you iterate 1103_5 with the Trinity Function, you quickly enter a cycle of three numbers where you spin for eternity:

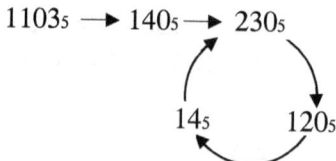

No resurrection, no net, no one third, no truncates, no super-triangularity. These vaunted properties are now seen as nothing more than artifacts of the number base. Change the base and you change the properties.

Is base 10 the best of all possible bases? Of course not! Historically societies have mostly used bases 5, 10, or 20 depending on whether they counted on one hand, both hands, or all fingers and toes. And why do we have 5 fingers? Only because our distant saurian ancestors suffered from pentadactylism.

As I mentioned above, a single property out of many survives the changing of the bases. E pluribus unum, so to speak. Can you guess what it is? If 153, in whatever base, is represented by a collection of black dots, these can be arranged in a triangle, hence 153 is a triangle number. So, of course, is 1103_5. But so is 666. Is this surviving property significant? Not really. Every number, *yes, every number,* has some unique properties. You doubt this? Read on.

Earlier chapters noted that famous stories hover about the lives of the great mathematicians: Archimedes in his bath, Newton and the apple, Gauss and the sum of the first 100 numbers. Undoubtedly the greatest self-taught mathematician of the 20th century was the Hindu genius Srinivasa Ramanujan. G. H. Hardy had the wisdom and foresight to arrange his passage to England and to obtain a special scholarship for him at Cambridge where collaboratively they did brilliant work.

Unfortunately his health was precarious and in the damp English climate he developed tuberculosis. As Hardy recounts the story, he arrived by taxi at the hospital at Putney where Ramanujan lay dying. Hardy — always awkward at starting a conversation and without the least greeting — said, "I thought the number of my taxi-cab was 1729. It seemed to me a rather dull number."

To which Ramanujan replied: "No, Hardy! No, Hardy! It's a very interesting number. It's the smallest number expressible as the sum of two cubes in two different ways." This impoverished Hindu from the streets of Madras, this unschooled professor of mathematics, this member of the Royal Society, this genius, had every integer in the first few thousand as a personal acquaintance. What follows is one of his last recorded friends, 1729, as the sum of two cubes in two ways:

$$1729 = 1^3 + 12^3 = 9^3 + 10^3$$

Still unconvinced that diligence and/or creativity can discover interesting properties about any number? What follows is an old mathematical yarn. Give me a number you think has no redeeming qualities whatsoever, the dullest of all numbers. Put it in a set. I'll now ask many others for their choice for the all-time dullard. These I'll put in the same set. Now, by popular ballot we will rank them in order of their dullness. Among all such, there must be the chief dullard — the all-time, universal, most boring number. By virtue of that fact alone, it has some appeal and so doesn't belong in the set. Then a new number steps up as the winner in the dullness sweepstakes, and by the same argument, it also doesn't belong in this set. And so it goes until the entire set is empty. Hence, there are no dull numbers!

INVENTION OR DISCOVERY

ON THE WALL BEFORE YOU is a huge map of the world, but you can't see it since you're blindfolded. In one hand you grasp a single dart that you must throw at the map. Where it lands will determine your birthplace, religion, longevity, culture, and native language. This is a crucial moment! You hurl the dart. It drills through the air and lands on the Arabian Peninsula. So, you grow up speaking an Arabic dialect and practicing some form of Islam. . . . Suddenly your whole body shudders and you awake and know you were having a nightmare. Or were you?

(Reality has more possibilities, more choices, than any mere nightmare can conjure up. Most of your beliefs and your strongly held opinions — please not all — are the result of your place of birth which is itself an accident, a dart in the dark.)

You're a dutiful son, the best that parents could hope for. Studying the glorious Koran is your special passion; you have memorized entire suras (chapters). With your parents' blessing and their abundant wealth, you pursue your youthful passions into adulthood. You develop a scholarly reputation that spreads to other villages. And then it happens — you begin to become aware of hidden patterns in the Koran. Since the Koran was dictated directly to the illiterate Mohammed by the archangel Gabriel, these must be the very words of Allah, the beneficent, the merciful. You research the nature of the patterns as well as their meaning. You outgrow the meager intellectual resources of your village; you move to a larger town and then on to Medina, and finally to Mecca itself. After a few years even the heart of the sacred land and the sight of the Kaaba prove inadequate, and so you journey to the great centers of Koranic learning in Cairo, Istanbul, and Tehran. We'll call you Ishmael.

One day you come across the writings of Dr. Rashad Khalifa, the self-proclaimed modern messenger of Allah. The good doctor is also a graduate of the Cairo schools, but presently he's working as a research assistant at the University of Arizona. Without hesitation you pack your belongings, mostly books, and travel to meet him in Tucson, Arizona, and sit at his feet to learn more about the phenomenal numerical patterns in the Koran — and particularly the number 19. Yes, 19!

In 1972, Dr. Khalifa privately printed a 60-page monograph titled *Number 19: A Numerical Miracle in the Koran*. He proclaimed this book offered the first physical proof in history of god's existence. Why? Because the number 19 is found everywhere in the Koran:

- Verse 1:1 has 19 letters.
- Verse 1:1 occurs 19x6
- 19x6 suras
- 19x334 verses
- 19x17,324 letters
- 19x142 appearances of Allah
- First revelation has 19 words
- First revelation has 19x4 letters
- Last revelation has 19 words
- ...

THE NUMBER 19 IN THE KORAN

And this numerology doesn't even touch the gematria of the Arabic letters which adds a further dimension.

To Muslims the number 19 is as mystifying as 153 and 666 are to Christians and about as important. In *The Magic Numbers of Dr. Matrix*, Martin Gardner discussed Khalifa's patterns with his numerological friend Dr. Irving Joshua Matrix (note the 6 letters in each of his names). The old scoundrel commented:

> "It's an ingenious study of the Koran," said Dr. Matrix, "but it would have been more impressive if Khalifa had consulted me before he wrote it. Nineteen is an unusual prime. For example it's the sum of the first powers of 9 and 10 and the difference between the second powers of 9 and 10."

Dr Matrix was not giving the full story on this: he always held something back, possibly for an encore or a rebuttal. This sum and difference of 9 and 10 can be marvelously expanded (*see* below and the Chapter Notes):

$$10^1 + 9^1 = 19$$
$$10^2 - 9^2 = 19$$
$$10^3 + 9^3 = 19 \times 91$$
$$10^4 - 9^4 = 19 \times 101$$
$$10^5 + 9^5 = 19 \times 8371$$
$$\cdot \quad \cdot \quad \cdot \quad \cdot$$
$$\cdot \quad \cdot \quad \cdot \quad \cdot$$

POWERS OF NINETEEN

His conversation was embellished with a string of pearls about the peculiarities of 19. "Did you know that Ramanujan's 1729 is the product of 19 and its palindrome (i.e. 1729 = 19 x 91)! Yet of all occurrences of this number, the strangest of these spirals around Dr. Rashad Khalifa's assassination in 1990. Do you see where? Note that the sum of the digits in the year of his death is 19. You might say his number finally came up," concluded the old numerologist.

Ishmael heard or read none of this. He was fixated on the Koranic designs as *invented* by Khalifa and other cabalists. Despite his patronym, he finally ceased his wanderings in search of the truth. But you can find him still wherever minds are closed and knowledge is absolute.

These two examples of culturally biased "mathematics," one Christian and the other Islamic, have profound similarities between them. A measure of your parochialism is the degree to which your "logic" disagrees with your neighbor's. Both Christian and Islamic fundamentalists believe they are absolutely right, and that's just another way of saying everyone else is absolutely wrong. Bertrand Russell wrote that all great crimes are committed by groups that know they are utterly correct.

> *The first of April is the day we remember what we are the other 364 days of the year.*
> Mark Twain (1835-1910)

"PHEMIUS, DIDN'T YOU TELL me about a book published a few years ago that concerns hidden texts?"

"You're probably thinking of Michael Drosnin's *The Bible Codes*. Simon & Schuster launched this book with a full-page advertisement in the New York *Times* and an initial printing of a quarter of a million copies."

"How does it differ from the gematria in the Koran and Bible?" Epius asked.

"Drosnin arranges the 304,805 Hebrew letters of the Torah into a single, large array. Spaces and punctuation marks are omitted, so that the "wordsruntogether." A computer then searches this array for names and words by skipping to every 2nd, 5th, 351st, or nth letter. You can do this by starting at the beginning or the end — this maneuver doubles your chances. Because of these skips, the series of letters found are called equidistant-letter sequences (ELS). If any group of these letters from the various skips is a word or a name, Drosnin cries 'hit,' otherwise he ignores it and moves on. His match for *Yitzhak Rabin* had a skip value of 4,772."

"Do you know a simple ELS?"

"Sure, consider one of your favorite words, ge*n*er*a*li*z*at*i*on. It contains a skip three sequence spelling out *Nazi* — all this within a single word."

"Phemius, I've heard enough. This is nonsense. With billions of step sequences for every array the opportunities for 'hits' are endless — you can find whatever you want. It's like buying all the tickets on a lottery — you have to win. This makes Drosnin's entire technique claptrap."

"Aren't you too skeptical?" Phemius asked.

"I doubt it," replied his friend. "Let's put this ELS business to the test by using only our two names, side by side, and see how many words we can find."

E P̱ I U̱S P̱ H E M I U S

Skip 1 letter: *pi, us, he, hem, mi, us, pius*
Skip 2 letters: *up, is, pup*

"And, Phemius, isn't *pup* [underlined above] your favorite word? You talk about wanting one every day."

"Yes, yes, ... but you've missed the prefixes *epi* and *hemi*."

"If we can get nine different words and two prefixes from 12 letters, think what we could get from the 300,000 or more letters of the Torah. I'm going downstairs to the common room to check out ELS and Drosnin online."

Since few of the elderly residents of Happy Acres were computer conversant, Epius got immediate access and did a Google search, coming up with 888 sites. The pages of David Thomas (physicist) at www.nmsr.org/biblecod.htm and Brendan McKay (mathematician) at http://cs.anu.edu.au/~bdm/dilgim/moby.html were excellent. McKay, while searching *Moby Dick*, found ELS assassination "predictions" for Indira Gandhi, Leon Trotsky, Rev. M. L. King, Abraham Lincoln, and John F. Kennedy — this beats Drosnin's meager list of one (Rabin). Melville would marvel at these findings in his masterpiece, since all the time he thought he was writing an epic about the struggle between man and nature, Ahab and the whale. Thomas discovered unexplained coincidences involving the number 19, not in the Koran, but in Ted Kaczynski's *Unabomber Manifesto*. And the list goes on. As Thomas says on his site, "*Any* message can be derived from *any* text."

"Why is it, Phemius, that Moslems never seriously consider the gematria or ELS patterns in the Bible, and on the other hand, Christians completely ignore similar 'designs' in the Koran?"

"Because, my old friend, the mind knows only that which lies near the heart. I understand that Drosnin has recently published a sequel, *Bible Codes II*, but I suspect this second tome won't make quite the slash of the first. The minds of the millions may grind slowly, but I believe they move inexorably forward."

"The ad on the cover of his new book should read 'Come believe the unbelievable; be a fool with me.' But enough of this! Let's go and flatter the kitchen help into giving us jam and peanut butter sandwiches before we go to bed," grinned Epius.

"Now that's a plan."

"Perhaps when we're back in our rooms you'll play your lyre and recite something from Pindar or Homer."

BEYOND ALL RELIGIOUS DISAGREEMENTS among Christian, Moslem, and secular groups — at least those of Western and Middle-Eastern origin — one number prejudice is common. Everyone in these cultures perceives something unusual about the number seven, even if they can't explain what it is. The Hebrew word for *seven* is *shevah* from the root *savah* meaning to be full or satisfied. This root dominates the usage of this word. On the seventh day, god rested because his works were complete and full. The derived word *shavath*, meaning to desist, cease, or rest became *Sabbath*, the day of rest.

From the Homeric age, the Greeks have revered the number seven — for them it always implied completeness:

- The 7 Wise Men
- The 7 Wonders of the World
- Odysseus spent 7 years as a prisoner of the witch Circe
- The 7 Against Thebes
- The 7 Sisters (daughters of Atlas)
- Even their horses' iron shoes were fastened with 7 nails.

From time to time, the keepers of lists changed particular Wise Men and certain Wonders by deletion and addition. But, the remarkable point is that the number of each was always kept constant. This implies the primacy of the number 7 over who or what might actually be in the list.

The Romans had their own 7 Sages, not to be confused with the Greeks' 7 Wise Men. Legend says 7 followers of Romulus raped 7 Sabine women and afterward took them for brides. (This is the basis for the Broadway musical *Seven Brides for Seven Brothers*.) Since the Romans built their city on 7 hills, they could hardly avoid this number. You may recall the passage about the Harlot of Babylon, a code phrase for Rome, from Revelation 17:9: "This calls for a mind with wisdom. The 7 heads are 7 hills on which the woman sits."

Our culture is adorned with a great variety of 7s. We see 7 colors in the rainbow. One beautiful constellation is called the 7 Sisters or Pleiades though the naked eye can discern only 6 stars. On more earthly matters, biologists divide the animal realm into 7 parts: kingdom, phylum, class, order, family, genus, and species. Musicians divide the world of sound into 7 tones: do, re, me, fa, so, la, ti and start the scale over with another do, an octave higher. Writers pen plays about the *Seven Ages of Man*, sailors speak of the 7 Seas, and movie producers say, and legends affirm, it's always *Snow White and the Seven Dwarfs*. An old Russian proverb says "Measure seven times before you cut." Seven is considered a lucky number. To be born the 7th son of a 7th son of a 7th son is said to be a triple blessing. Almost before memory we have had 7 days in our week; any other length seems unnatural. The French at the time of their revolution metricated the week into 10 days; however, the tormented citizens only briefly tolerated this outrage.

The Islamic people, the descendants of Abraham through his son Ishmael, have also inherited this number tradition. The Koran commands them — at least once in a lifetime — to go to Mecca and circle the sacred Kaaba exactly 7 times. Those unable to walk are carried. In the second sura, Allah is said to have created 7 heavens. Believers say that when Mohammed died he ascended directly to the 7th of these on the back of his white horse.

Nevertheless there is nothing divinely ordained about any of these 7s. We could choose to "see" 6 colors in our rainbow as the Cree and Natchez Indians did, or even 3 as some other cultures have done. Many societies divide their musical scale into more than 7 parts. Yet if Western or Middle Eastern civilization is your heritage you unconsciously insist on 7 different notes in each scale. In China 6, 8, and 9 are considered lucky. However, the Chinese word for death sounds like 7 — their most unlucky number.

The rainbow's divisions are in our eyes, the bell's tones are in our ears, and the lucky numbers are in our hearts. Different eyes, ears, and hearts have different numbers. This is an exploration of our cultures, not the natural world. We can correctly infer that *these "supernatural" numbers are arbitrary,* and therefore culturally determined. The scientific road to knowledge is observation of things as they are.

Curiously, Shakespeare knew all this centuries ago. In his celebrated play *King Lear* there is a scene where the Fool and the King discuss the number of stars. These are the seven — as it turns out — main bodies of the constellation Orion/Nimrod noted in Revelation 1:16: "In his right hand he held seven stars."

> The Fool speaks first: *The reason why the seven stars are no more than seven is a pretty reason.*
> And the King replies: *Because they are not eight.*
> The Fool says: *Yes, indeed: thou wouldst make a good fool.*

By accepting the universe as it is, the wise Fool and the King got it right. We should not force our cultural predispositions on the natural world, which is better served by other numbers — as we shall see.

CHAPTER — 6

NUMBERS: NATURAL

God made the integers; all else is the work of man.
Leopold Kronecker (1823-91)

FAMOUS ARTISTS FROM THE RENAISSANCE are known by their first name: Dante Alighieri, Michelangelo Buonarroti, Leonardo da Vinci; similarly, famous numbers are recognized by a single Greek letter.

Let's look at the world around us through three of nature's numbers. One is from the distant past, another from the Middle Ages, and the last from just yesterday. We know them by their Greek names π (pi), φ (phi), and δ (delta).

These are universal "words" representing profound ideas about the structure of nature, so universal that if we wished to communicate with the slimy-green mathematoids from the Trifid Nebula, any one of them should be enough to gain us an audience. Real mathematics, the mathematics of scientists, is the language of the cosmos — no universal translator required. Even music can't make such a claim — our Trifid friends could be insensitive to sound.

Not only numbers but also the theorems and forms of mathematics and geometry are believed to be part of a universal vocabulary founded on logic. In the 19th century when humanity speculated there might be humanoids on Mars, Karl Friedrich Gauss suggested we communicate with them by outlining a gargantuan drawing of the Pythagorean theorem on the Sahara desert.

We will entertain each number in chronological order. The first of our triad has been known, at least by approximations, since pre-biblical times.

PI IS THE MOST FAMILIAR of numbers; it's defined as the ratio of a circle's diameter to its circumference. We have all heard this boring definition since childhood. What, if anything, is novel or new about π? As it turns out, quite a lot since we live in the paramount age of mathematical discovery.

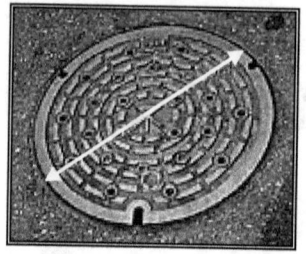

MANHOLE COVER

Consider manhole covers! Why are they circular? One excellent reason is so they can't be passed through themselves. Therefore, you can't accidentally drop one down the hole on a co-worker's head, nor can vandals wreak havoc in the streets by doing the same. We say circles are *curves of constant width* — their diameter. (But you say some manhole covers are square? Those are all hinged to prevent just such an accident.)

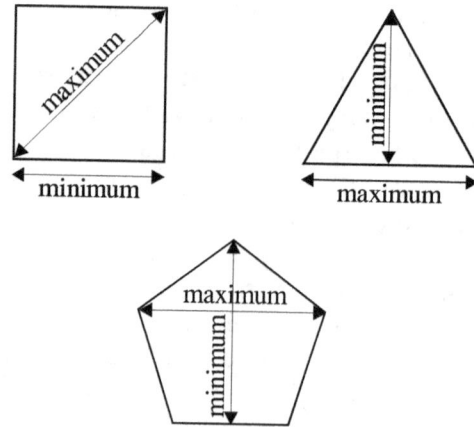

ALL POLYGONS PASS THROUGH THEMSELVES

Figure 6.1

In contrast, here are three polygons. These figures have necessarily different maximum and minimum widths, which allows them to pass through themselves. In fact, *every polygon* will do this whether equilateral or not. For the circle, the maximum equals the minimum in the diameter and this makes all the difference.

No object with straight sides has this non-passing property. The question is whether the circle is the *only closed curve* that does. You may think the question is bizarre; yet this is an example of where one's mathematical intuition spectacularly fails. Remarkably, an infinite number of curves of constant width exist — all related to polygons with an odd number of sides.

The simplest such figure is called the Reuleaux triangle after Franz Reuleaux (1829–1905), a German mathematician and teacher who discovered its constant width property. Its construction is quite simple (*see* the figure below). First, draw an equilateral triangle ABC. Put your compass at vertex A and draw arc BC. In a similar manner, draw the other two arcs. A moment's reflection will convince you the "curved triangle" at the right in Figure 6.2 has a constant width equal to the side of the "straight triangle" opposite it. By constructing arcs you extend the three minimum distances into maximums and hence make them equal as in the circle. This process allows the creation of an infinite number of curves of constant width each built on an odd-sided polygon. Why odd? Only these have a vertex opposite every side, which permits the drawing of the arcs.

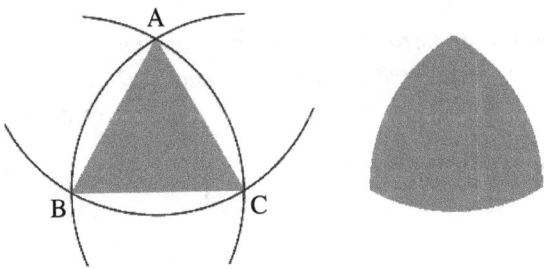

THE REULEAUX TRIANGLE

Figure 6.2

These curves of constant width — measured between two parallel lines — may be employed in the following way. Ancient engineers probably used circular rollers, perhaps under wooden platforms, to move the huge building blocks for Herod's temple and the monoliths of the pyramids — although no hard evidence of this survives. Wheels with axles could easily snap or buckle, but rollers double as both. As the cargo is pulled forward, the rollers left behind are picked up and moved to the front; the load progresses parallel to the ground without up or down movement. A log with a Reuleaux cross-section (*see* the top of the following page) works equally well, as would any curve of constant width. Though presumably the Egyptians contented themselves with the simple cylindrical rollers.

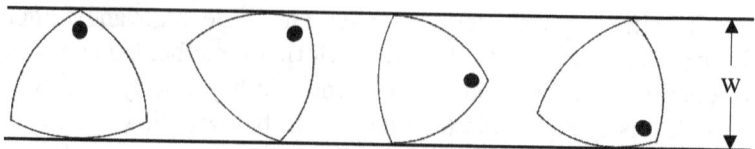

Take a Reuleaux triangle bounded by a pair of parallel lines. Next, take a second pair of parallel lines and place them perpendicular to the first pair and surrounding the triangle. The bounding lines necessarily form a square (*see* Figure 6.3). Like a circular drill bit in its round hole, the Reuleaux triangle rotates snugly within the square, at all times touching every side. The property of turning snugly and smoothly within a square is the basis for the Harry Watts drill bit. This unique device is a modified Reuleaux triangle with three pieces removed to create a cutting edge and to allow the shavings to escape.

The bit cuts *straight edges* but leaves the corners very slightly rounded. Nevertheless, it removes 98.7 percent of the material — as good as a circular bit boring a round hole.

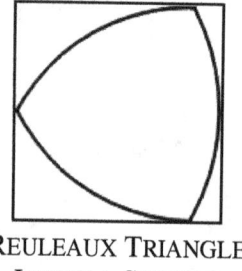
REULEAUX TRIANGLE
INSIDE A SQUARE

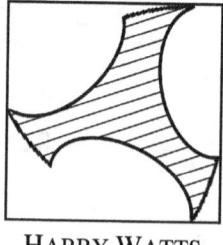
HARRY WATTS
DRILL BIT

Figure 6.3

As the drill rotates, the centroid (balance point or center of mass) doesn't remain fixed but travels on a path composed of four arcs of an ellipse. (This path can be closely approximated by Piet Hein's superellipse.) To allow for this movement, Harry Watts also invented a "free floating chuck" to hold the revolving and rotating bit.

It's even possible to draw unsymmetrical curves of constant width as seen below in the "curved heptagon" placed between the circle and the "curved pentagon" (Figure 6.4). For most individuals these figures are strikingly anti-intuitive.

All this is fascinating, but we have seen little or nothing of the subject of this section, the number π. That's about to change! Consider a set of curves of constant width w, as in Figure 6.4, arranged in order of ascending area. A startling characteristic, not easy to prove, is that all curves of the same constant width have the same perimeter. Since the circle is one of this group, then the perimeter of these curves must be πw — the same as the circumference of the circle with diameter w. This remarkable result is called Barbier's Theorem.

More to the point, this allows for a new depiction of π. Forget the ratio and diameter stuff! Now we see that π yields the perimeter of an entire class of objects — those of constant width w. The circle is just one among an infinite host, although it's the one with the largest area. The paramount property that teases π from one of its many secret lairs is this concept of constant width.

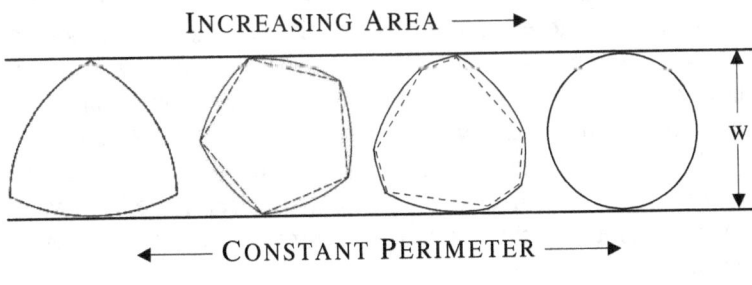

Figure 6.4

Even to the untrained eye it is self-evident that π occurs in nature. Pi is ubiquitous but never to the accuracy demanded by mathematicians. Fifty decimal places is more than enough to circumscribe the known universe to the accuracy of a proton's width. But that barely begins to satisfy the digit hunters who now pursue π to billions of decimal places. Can a trillion be far away?

You can feel this number in the discs of the sun and moon and the rising arc of the rainbow or the spreading rings of a raindrop falling on a tranquil pond. Sine waves have a fundamental period of 2π, and so π is vital in spectral analysis (finding wave frequencies). Therefore π is used in television, radar, telephone, radio signals, and of course color and music. No escape from π is possible. It occurs in *death tables* in what we call the Gaussian distribution. From conception in the DNA double helix to the birth canal to the final exit chamber, π is forever with us.

Although humanity originally discovered π in the geometry of circles — and we have expanded this to all curves of constant width — it occurs repeatedly in a variety of settings. One of the more curious is in the meandering of streams. Previously (page 133), I mentioned an *energy model,* such as a skier's track down a slope, to explain how twists and switchbacks result in the slow change of potential to kinetic energy under gravity's influence and friction's drain. This is not half the truth; a larger portion was first given by Albert Einstein. Strangely, the master of the energy concept actually suggested a *mechanical model:* a battle between the terrain and inertia under the overseeing authority of gravity.

A bend in a river can result from minor irregularities in the surrounding terrain: a gathering of rocks here, a hillock there, or a harder more conglomerate soil elsewhere, and so on. These constrain the river to bend, but its inertia (Newton's first law) compels it straight ahead. As the current twists around a turn, inertia flings the water forward toward the concave bank. (This is why many racetracks have banked turns.) And this produces faster currents on the outside leading edge, which cause erosion creating ever-sharper turns — so a cascade effect ensues.

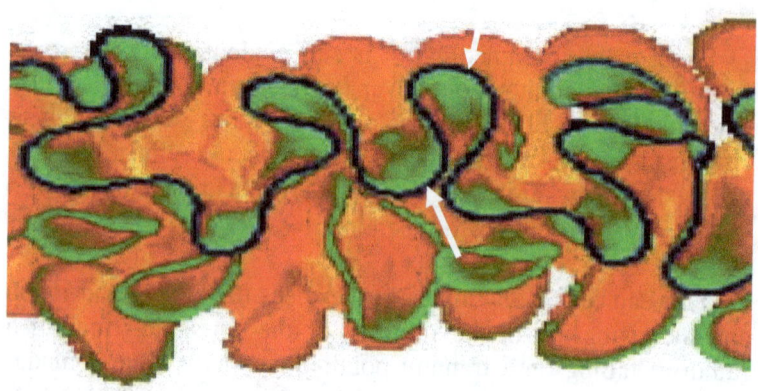

MEANDERING STREAM

Figure 6.5

You might expect this process to produce an exceedingly long river, yet that never happens. At all times and in all places, gravity opposes this by trying to straighten the waterway's sinuous switchbacks. In the previous figure, where the river flows left to right, we see an example of this. Two of many points of maximum erosion are indicated by arrows — these are just a little forward or downhill of the bend's apex. The black is the present riverbed; the adjoining green regions are the most recently eroded. This slightly displaced scouring — note *where* the green areas predominate — leads the bend downstream until eventually it intersects itself and thereby shortens its path by creating oxbow lakes and depressions.

To summarize: the terrain induces the original bending; inertial scouring magnifies this, and gravity takes the bend downstream until it intersects itself — all these processes keep the river's curved length in a kind of homeostasis.

Professor Hans-Henrik Strølum, an earth scientist at Cambridge University, has made a thorough study of this homeostasis. He calculated the actual curved length of rivers from source to mouth, as the canoeist goes, and the straight distance, as the crow flies. Their ratio, *curved to straight distance,* may vary somewhat from river to river but averages at 3.14. This is as close as nature usually gets to π.

The result is independent of the river's width or length, be it the mighty Mississippi or the humble Concord and Merrimack. Nonetheless, the ratio π is best found in rivers flowing across gently sloping plains like those in Brazil, Northern Canada, and Siberia. So, here we have uncovered another of π's subtle hiding places.

In the famous Rhind Papyrus written around 1650 BC, the Egyptian scribe Ahmes recorded the earliest known evidence of the pi ratio. One problem on the scroll — that of constructing a square equal in area to a given circle — implies pi is 3.16049. This misses the true value by less than 1 percent.

The chronology of pi is a staircase from the earth to infinity. With few exceptions, mankind has *traveled up* these stairs of decimals from Ahmes to the present record held by Kanada of Japan. But a thousand years after the Rhind Papyrus the Bible took a step downward. Chronicles and Kings both speak of a huge basin or cleansing font outside Solomon's Temple. The priests and supplicants used it to wash the blood from their hands and faces after slicing the pumping neck arteries of the sacrificial lambs. (Perhaps this is where the phrase "the innocence of the lambs" originates.) With respect to the basin's dimensions, I Kings 7:23 declares:

> *He [Huram] made the sea of cast metal, circular in shape, measuring 10 cubits from rim to rim and 5 cubits high. It took a line of 30 cubits to measure around it.*

With a circumference (= c) of 30 cubits and a diameter (= d) of 10, this clearly implies pi is 3. And this misses the true value by almost 5 percent.

$$c = \pi \times d$$
$$30 = \pi \times 10$$
$$\pi = \frac{30}{10} = 3$$

Now the "cubit" was an ancient unit of measurement equal to the distance from your elbow to the tip of your middle finger — on average about 20 inches. Let's get an idea of the magnitude of the biblical error here. Using pi as 3.14 and the given diameter of 10, the circumference is 31.4 cubits. This is 1.4 cubits — about 2½ feet — larger than the Bible's figure.

This particularly poor estimate of pi has bothered literalist Christians for centuries — presently one American out of three still believes the Bible is the inerrant word of god. But why should we expect a good approximation from an ancient people so uneducated that they had to import an architect (Huram of Tyre) to construct their most holy shrine?

Post hoc explanations for this biblical blunder are like mushrooms in the morning — everywhere; none of which I care to taste. Perhaps the reader might wish to concoct his/her own sautéed version. In the previous chapter, I gave a detailed example of primitive mathematical behavior — the material between the fishy symbols. Those *invented* associations swirled around 153, the number of Fishes in the Net from John 21:11. Let's do it again, but this time to find pi to a more accurate value than just 3.

Arguably, 153 is the most important quantity in the entire Bible. It has the ersatz property of resurrecting itself; besides this and its triangularity, one third of all numbers transform to it. Could it possibly produce pi also?

Draw a circle and write the number **153** curved around its circumference either clockwise or counterclockwise. Pick clockwise, say. Now, for greater accuracy write it again with each digit (shaded gray) beside the first as in the diagram to the left. At this point, reverse the procedure by writing the numbers in a straight line counterclockwise from the smallest to the largest pair, and group them as follows: (113)(355). After dividing the second group (**355**) by the first group (**113**), we get a stunningly accurate value for pi correct to six places of decimals. This degree of precision is sufficient to construct the hardware for an Apollo moon rocket.

THE PI CIRCLE

$$\frac{355}{113} = 3.141592\ldots$$

Nevertheless, this isn't a miracle god has wrought, but merely the consequence of luck and number juggling.

> *'Tis a favorite project of mine*
> *A new value of pi to assign.*
> *I would fix it at 3*
> *For it's simpler, you see,*
> *Than 3 point 14159.*

What's the personality of π? What kind of number is it? Some feel it's annoying and not well behaved: all those nasty, confusing decimals and its inability to be exactly expressed as a fraction are particular vices. Like the author of the above limerick, the Indiana state legislature was so vexed that in 1897 it attempted to pass a law making π's value exact. However, the local newspapers got wind of this, and their laughter blew House Bill 246 to never-never land.

It's difficult to know how to respond to these lawgivers; you might more easily communicate with the slimy-green mathematoids. Their misunderstanding is profound and identical to those who believe mathematics is an artifact like traffic laws or tennis rules. But the determination of pi's value is no cultural pastime; it's out there, everywhere, beyond our purview.

> *It's no pastime of mine*
> *To make pi asinine.*
> *You shouldn't fix it at 3*
> *For it's simple, you'd be*
> *If you don't make it 3 point 14159. . . .*

Closely related to these misguided lawgivers are those troubled souls who suffer from what Augustus De Morgan called morbus cyclometricus, the circle-squaring sickness. With only Euclidean tools — straightedge and compass — they have worked furiously but fruitlessly to construct a square or rectangle equal in area to a given circle (*see* figure below). Their dream was a nightmare!

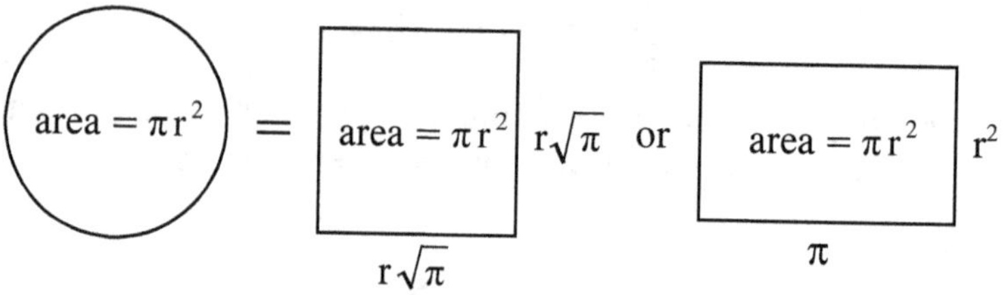

$$\text{area} = \pi r^2 \quad = \quad \text{area} = \pi r^2 \;\; r\sqrt{\pi} \quad \text{or} \quad \text{area} = \pi r^2 \;\; r^2$$
$$ r\sqrt{\pi} \pi$$

THE CIRCLE-SQUARER'S DREAM

Figure 6.6

Why was their quest doomed to failure? Well, that entirely depended on the nature of pi. As noted in the third chapter, "Without Form and Void," all real numbers can be classified as either rational or irrational; there are no hermaphrodites. The rationals behave well, can be written as fractions, and have periodic decimals — after all, that's how they're defined. The bar above the following decimals indicates the period, the part that repeats. This is the world of Pythagoras. Pi never lived here.

$$\frac{22}{7} = 3.1428571428\,5714... = 3.\overline{142857}$$

$$\frac{13}{12} = 1.08333... = 1.08\overline{3}$$

I can hear someone protesting they were taught that pi was 22/7 — it must have been a very hot, drowsy afternoon in class that day. As you can see in the above table, the fraction 22/7 equals 3.$\overline{142857}$; this is correct to only two decimals or three digits. An early Chinese mathematician, Tsu Ch'ung-chih (about AD 470), gave pi as 355/113, which — as we know from the previous page — is correct to an astonishing six places of decimals. (This is an easy fraction to recall: write the first three odd numbers twice each and then divide the second group of three by the first group.)

But 6 or 6 zillion decimals are not enough to make pi exact. In 1761 Johann Heinrich Lambert proved that this strange number was irrational in the full mathematical as well as the everyday sense. So no number of decimals will ever reward you with exactness. This knowledge — though not fatal — should have been discouraging to the circle squarers.

Now the nature of pi takes an odd twist. The real numbers, which were defined as the rationals plus the irrationals, may be classified in another way. It's this second way that opens new doors on old problems: any number that's the root of a polynomial equation with integer multipliers is termed *algebraic*. For example:

$$4x^2 - 12 = 0$$
which simplifies to
$$x^2 - 3 = 0$$
or
$$x = \pm\sqrt{3}.$$

Consequently $+\sqrt{3}$ *and* $-\sqrt{3}$ are algebraic numbers. This new group includes all the rationals plus *some* irrationals. Any quantity not captured by polynomial equations is termed *transcendental* — meaning it goes beyond. Accordingly, the real numbers may be separated into two branches, the algebraic and the transcendental; once again, there are no hermaphrodites.

And the algebraic numbers fork into two streams: those constructible with Euclidean tools and those not. Straightedge and compass allow you to add, subtract, multiply, and divide, and do a finite number of *square roots* — that's it. All the branches in this "real" river of numbers are displayed in the following figure.

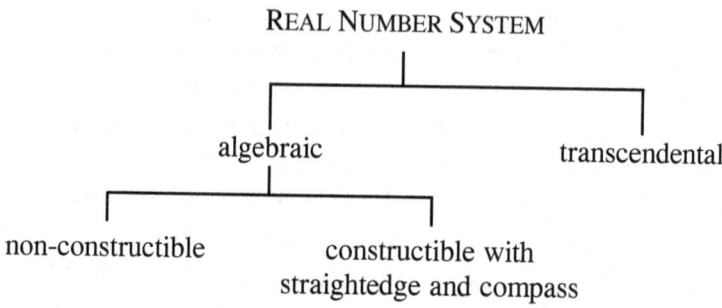

We have reached the crux of our investigation on circle squaring: "In which branch of the river does pi reside?" Well, it turns out that we know the answer to this historic question. The explorer who first discovered in which majestic stream pi floats through time and space was Ferdinand von Lindemann.

In 1882 Leopold Kronecker said to Lindemann, "What good is your beautiful investigation regarding π? Why study such problems, since irrational numbers don't exist?" That same year Lindemann proved π is a transcendental number; therefore, no finite polynomial equation has it as a root. This implies pi can't now or any time in the future ever be constructed with Euclidean tools. This epic discovery, built on centuries of mathematical knowledge, buried the chances of circle squarers forever. R.I.P!

Children and mathematicians are much alike: they both love to play games that appear meaningless. By the time we're young adults, most of us have left these Happy Isles, but number theorists never do. One of their games is trying to find a pattern in pi's decimal expansion. And, yes, it's possible to have a pattern without a period (*see* page 119). Presently we have uncovered 206,158,430,000 decimals of pi's transcendental meanderings. With *letter type* the size you're reading, those decimals would stretch more than eight times around the earth's equator. So far, all the digits occur about equally. When this happens, mathematicians say the sequence is *normal;* this surely implies some manner of randomness.

This randomness — whether valid or not — has fascinating consequences. In a bizarre twist of circumstances, *when no pattern exists then every pattern exists.* Whatever sequence of digits you want in pi's expansion you can find, providing you go far enough. For example: say, you want seven 7s in a row. As it happens, that sequence occurs twice in the first 10 million digits, once at digit 3,346,229 and again at digit 3,775,288. Since we presently know over 200,000,000,000 decimal digits (200 billion), and these digits occur with a normal frequency, we should expect to find the first 11 digits of pi itself twice (one digit for each zero and twice for the 2).

Write pi using only 0s and 1s (base 2). By arranging groups of these into square arrays you will eventually make a circle — à la Michael Drosnin (pages 184–5). The conclusion of Carl Sagan's novel *Contact* notes this weak geometric "self-similarity":

> In whatever galaxy you happen to find yourself, you take the circumference of a circle, divide it by its diameter, measure closely enough, and uncover a miracle — another circle, drawn kilometers downstream of the decimal point. . . . As long as you live in this universe, and have a modest talent for mathematics, sooner or later you'll find it.

The first few billion decimals are no more than a handshake in the life of pi. Search its expansion long enough, and you will discover all the digits of pi itself — infinity within infinity! The part can equal the whole — a strong arithmetic self-similarity.

All this begins to have echoes of chaos theory that we discussed in the third chapter "Without Form and Void." Recall that chaos was defined as probabilistic behavior within a lawful system. One system we explored in detail was the *logistic* defined by an equation, which could be clocklike in its precision or drunken in its progress depending on the initial values used. This marriage of pattern and patternlessness is exactly the point of the following dialogue from *Are Quanta Real?*[1] by J.M. Jauch. The conversation is modeled on Galileo's three questing friends in the famous *Dialogue on the Great World Systems*.

Salviati: Suppose I gave you two sequences of numbers, such as 78539816339744830961566084... and 1, −1/3, +1/5, −1/7, +1/9, −1/11, +1/13, −1/15, ...
If I asked you, Simplicio, what the next number of the first sequence is, what would you say?
Simplicio: I could not tell you. I think it is a random number sequence and that there is no law in it.
Salviati: And for the second sequence?
Simplicio: That would be easy. It must be +1/17.

Salviati: Right. But what would you say if I told you that the first sequence is also constructed by a law and this law is in fact identical with the one you have just discovered for the second sequence?

Simplicio: This does not seem probable to me.

Salviati: But it is indeed so, since the first sequence is simply the beginning of the decimal fraction [expansion] of the sum of the second. Its value is $\pi/4$.

Simplicio: You are full of such mathematical tricks, but I do not see what this has to do with abstraction and reality.

Salviati: The relationship with abstraction is easy to see. The first sequence looks random unless one has developed through a process of abstraction a kind of filter which sees a simple structure behind the apparent randomness.

To gauge its full significance, let's look more closely at this sequence by first expressing it as a series (*see* below). Discovered independently by James Gregory and Gottfried Leibniz, the regularity on the right of this equation is amazing — remember what we thought we knew about π. On the left of the equal sign we have a random decimal; on the right we have a perfectly patterned series. This is the very essence of the definition of chaos. But at an emotional level, we all agree with Simplicio's "This does not seem probable to me." We could go even further and say it's absurd. Nonetheless, there it is, and it's true! Intuition is often a false guide. Adjust your vision to the world as it is. This is the way of *Homo sapiens*.

$$0.785398163397448\ldots = 1 - \frac{1}{3} + \frac{1}{5} - \frac{1}{7} + \frac{1}{9} - \frac{1}{11} + \frac{1}{13} - \frac{1}{15} + \ldots$$

or

$$\frac{\pi}{4} = 1 - \frac{1}{3} + \frac{1}{5} - \frac{1}{7} + \frac{1}{9} - \frac{1}{11} + \frac{1}{13} - \frac{1}{15} + \ldots$$

The three dots at the right, implying the series is to be continued forever, is the salient feature. However, for actually calculating π this series is *useless*. Why? To get just four decimal places of accuracy, you need about 10,000 terms. The theory is perfect; the practice is impossible. Many series exist that converge on pi extremely rapidly, and these are used by the modern pi-hunters, but none have quite such a lovely design as the Gregory-Leibniz equation.

There are many ways to write numerals for numbers (*see* Figure 5.1). There are also many modes to represent numerals: fractions, different bases, logarithms, powers, and so on, or just words. Each is convenient for one purpose or another and each will be familiar to anyone who has done some mathematics at school. But, surprisingly, one of the most striking and powerful representations is completely ignored in school mathematics and it rarely makes an appearance in university courses, unless you take a special option for number theory. Yet *continued fractions* are a most revealing representation of numbers. They provide us with a way of constructing rational approximations to irrational numbers — called convergents.

For our purposes, we need a continued fraction expansion of π; fortunately there is one. The same man who first proved pi to be irrational, Johann Heinrich Lambert, also discovered the following extraordinary expression:

$$\pi = 3 + \cfrac{1}{7 + \cfrac{1}{15 + \cfrac{1}{1 + \cfrac{1}{292 + \ldots}}}}$$

If we truncate, that is chop off, this infinite continued fraction after a finite number of terms, we create a rational approximation (convergent) to the original transcendental number pi. For example, the first cut produces 3 — the biblical value; the second gives 22/7 — our schooldays' friend. The next truncate is less familiar but accurate to four decimals:

$$3 + \cfrac{1}{7 + \cfrac{1}{15}} = \frac{333}{106} = 3.1415$$

The fourth approximation is the brilliant Chinese discovery and the "fishy religious delusion" from a few pages back:

$$3 + \cfrac{1}{7 + \cfrac{1}{15 + \cfrac{1}{1}}} = \frac{355}{113} = 3.141592$$

Lambert's continued fraction expression for pi is something of an historic map of man's struggles to compute it. And the more terms we retain, the better the rational number approximation. In fact, his continued fraction for π is the best possible.

In 1910 Ramanujan discovered a Euclidean construction for 355/113 seen beside his portrait in the picture below which is something of a montage of pi's past and computer-driven future.

Pi à la Mode
by Ram Samudrala and Shriram Krishnamurthi

The patterns we find in π are messages from nature, not a conscious being. We can investigate pi long and hard, but it will always have unknowable elements waiting to be discovered — this is the joy of pi. It has been suggested that computers connected to the Internet be linked at night to calculate it to a trillion or more decimal places. Mathematician and pi researcher Peter Borwein has predicted that we will never know its 10^{51st} digit; after all, there are probably only about 10^{78} atoms in the entire universe. However, negative predictions in mathematics and science have a way of appearing foolish with the passage of time. As I've said, π's infinite expansion is not just outstandingly long, it's a different creature that lives in the undiscovered country.

When you were a tadpole and I was a fish . . .

WHEN THE EARTH WAS YOUNG and the summers green and carefree, my cousin and I would roam the fields in search of whatever we might find. We had no prescribed duties, except perhaps bringing the cows to the barn at dusk; we were rather the keepers of berry patches and bluebird eggs. Since we weren't in search of rare animals like moose, wolves, or bears, we were never disappointed. Hollow fence posts had to be inspected for birds' nests, ponds for tadpoles, swamps for turtles, pools for gilled salamanders, and special hidden places we alone knew for snakes and the occasional blue skink.

We were unfettered in our natural interests. As far as we knew, the adults in our lives had little interest or knowledge of the world outside. Rarely did they speak about it — never a bird's name or a flower's location. Infrequently an aunt would express some fear or other, especially about poisonous snakes (we never found one), or skinks that might run up inside your pant leg to do great damage, or the ever-vicious wolves. On one occasion, a not too likeable aunt with a goiter asked my cousin and me to capture a snake large enough to wrap around her neck twice, and then to release it before sundown. Local wisdom affirmed this would cause the goiter to shrivel up and disappear. Since she feared snakes, and we didn't particularly like her, we granted her wish. Trusting to adult wisdom, we fully expected the wretched disfigurement to vanish, but that never happened. So we learned grown-ups were not always wise.

The open fields were our special domain — the home of bumblebees, wasps, and innumerable wildflowers. We were young Pythagoreans for we loved to count. Oh, not as lovesick adolescents might with "she loves me, she loves me not" but just a straightforward enumeration of the flower's petals. At that time what we found was *consistency;* years later this became a *pattern,* and presently we have an *explanation.*

Let's begin at the beginning. The consistency lay in each healthy variety of flower always having a certain number of petals, even if the number was exceedingly large. As we shall see, this regularity also applies to leaves, branches, and seed spirals. Consider the examples on the following page picked from the field of boyhood dreams.

FIBONACCI IN THE FIELD

Figure 6.7

Every summer evening when we had tired the sun with exploring and sent him down the sky, we would wander out in search of the cows. These warm, placid creatures had to be sheltered in the safety of our barn. Since the farm was heavily forested, they were usually hard to find, even with the help of our faithful dog Pal. So we would call to them in a smooth comforting voice, "Cooo-bos, cooo-bos, cooo-bos." Typically they would give themselves away with deep lowing sounds. Well it just so happens that *bos* or $\beta o \varsigma$ is the ancient Greek word for *cow*. This word had come down through the centuries — to two skinny boys chasing cattle in the darkening hinterland of Ontario — *totally unchanged from the original Greek*. That's the wonderful consistency of a word through time and over space. But in the shadowy fields all around us was a design by nature that had endured millions upon millions of years and on every continent. And this is the pattern I wish to investigate.

Years passed and the dreams faded to the back of my mind for more pressing and immediate matters, but they *never* left me. I suspect we all have such special places, sacred in our memory. One afternoon all that changed! In my senior high school mathematics class the teacher wrote the following sequence of numbers on the blackboard and asked if anyone could discover the next term:

$$1, 1, 2, 3, 5, 8, 13, 21, 34, 55, \ldots$$

Without raising my hand or a moment's hesitation I blurted out "It's 89." These were the numbers my cousin and I had found a decade earlier — *see* Figure 6.7.

The teacher was more amazed than annoyed by my outburst; he asked how I knew the answer so quickly. I told him a much-shortened version because one is protective of such precious memories. It was now his turn to amaze me, for the numbers we had stumbled upon as boys were none other than part of the famous Fibonacci sequence named for the medieval mathematician Leonardo of Pisa, also called Fibonacci (about 1170–1240). And the secret for finding the next term in his sequence — which as boys we had overlooked — is marvelously simple: *add the preceding two terms to get the next*. For example, $3 + 5 = 8$, $21 + 34 = 55$, and so on. As my teacher gave instance after instance where this sequence occurs in nature, my surprise grew — a feeling that lives with me to the present moment. Come and share this wonder with me.

With composite flowers, meaning those with multiple florets on their heads, you can distinguish a *double* Fibonacci pattern by concentrating on the spirals. The head of the ox-eyed daisy below has 21 spirals going clockwise and 34 going the other way. Hence, at the edge you have 21+34 or 55 special florets we call petals (plucked off here). The 34 spirals are more easily seen near the edge of the head, the 21 nearer the center. Count them! Depending on the size of the particular seed head, there are different numbers of spirals; nevertheless, they're always neighboring pairs from the Fibonacci sequence. The pairs commonly found on sunflowers are 21/34, 34/55, 55/89, and occasionally a giant 89/144. These spiral patterns can also be discerned on the heads of other composite flowers.

OX-EYED DAISY: 21/34

The pineapple, that standard of the modern summer kitchen, has 8 rows of scales sloping to the right, and 13 sloping to the left. With it, and many other fruits and plants, the florets arrange themselves in spirals *up the stem*. Smaller pineapples can have 5 spirals one way and 8 the other — always-adjacent Fibonacci pairs. As with sunflowers, this difference is not the result of genetic variation but rather a matter of soil and sunshine.

PINEAPPLE: 8/13

The banana has 5 sides, 2 small and 3 large. All types of apples, apricots, pears, peaches, cherries, plums, and so on, have 5-petaled blossoms. The apple of myth and legend has a curious consequence from its association with the number 5. Consider the forbidden fruit from "the tree of the knowledge of good and evil" generally thought to be an apple. In countless paintings and cartoons, in earnest and in jest, artists have used the apple as a symbol of dark temptation. Every variety of this fruit has a striking property, known only to wannabe witches and warlocks. Take a knife and slice across the apple's middle instead of through its stem. The resulting cross section reveals the

THE APPLE'S CORE

core's perfect 5-pointed pentagram, an occult symbol since before the time of Pythagoras (*see* the above figure). Moreover, the pentagram has a profound connection, as we shall see, with *all* Fibonacci numbers.

Besides sunflowers and pineapples, nature's best-known example of the Fibonacci pattern is the pine cone. In Figure 6.8 below, we have a cone on the left with its spirals outlined on the right: *8 solid* black or white counterclockwise; *13 banded* black and white clockwise. Larger or smaller cones can have different pairs of numbers, but they're consistently adjacent terms from the Fibonacci sequence.

PINE CONE: 8/13   PINE CONE SPIRALS

Figure 6.8

Every spring, queen bumblebees fly erratically over the fields in search of abandoned mouse nests. All early spring bees are queens, large and vigorous. These are the true wild bees of America, not the puny domesticated honeybees from Europe.

The rodent's nest with its abundant insulation will provide an ideal site for her to reestablish her lost colony of the previous summer. She's the winter's sole survivor; her eggs — fertilized last fall by short-lived drones — will produce new workers (fertilized but non-reproductive females) and a few late-summer drones (from unfertilized eggs). The poor drone, one brief summer fling and then he dies; any of his brothers still in the nest will be driven out to perish. Drones have only a single parent, the queen, but she has the regular two. Thus, the number of his *predecessors* in each generation is a term from the Fibonacci sequence. Consider the family tree of the hapless drone.

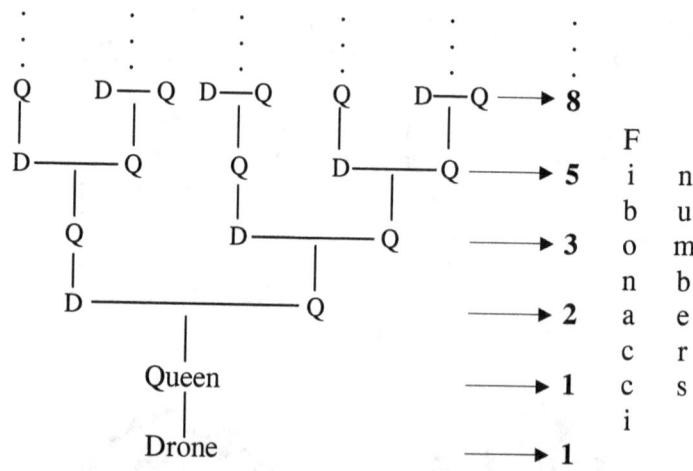

LINEAGE OF THE DRONE BUMBLEBEE

Figure 6.9

As we have just seen with the drone bumblebee, Fibonacci numbers have a habit of popping up in unexpected places. One of the most curious is in Pascal's triangle — the array we encountered in Chapter 2. Figure 6.10 below reveals that the sums of the numbers on the rising diagonals of this triangle produce the entire Fibonacci sequence.

Bee reproduction. Sunflowers. Flower petals. Daisies. Pine cones. Celery. Broccoli. Tree branching. Leaf arrangement. Thistle heads. Cactus spines. Pascal's triangle. These are a few of my favorite things — Fibonacci in the field and in our culture.

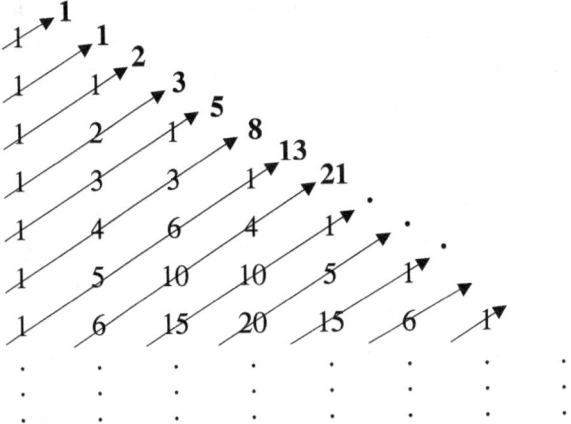

FIBONACCI NUMBERS IN PASCAL'S TRIANGLE

Figure 6.10

Extensive botanical observations confirm that more than 90 percent of all plants worldwide display these numbers. This is a confirmation of the *consistency* my cousin and I had found. But this is nature, and multiple forces are always at work, so outcomes tend to be a little messy. What about the remaining 10 percent? That's still an enormous quantity of plant species. H.S.M. Coxeter, in his *Introduction to Geometry,* comments on this divergence from Fibonacci:

> However, it should be frankly admitted that in some plants the numbers do not belong to the sequence of f's [Fibonacci numbers] but to the sequence of g's [Lucas numbers], or even to the still more anomalous sequences . . .

Named after the French mathematician Édouard Lucas (1842–91), the sequence below has the same method of formation as the Fibonacci just different starting values (2 and 1 rather than 1 and 1). Remarkably, it'll turn out that the *process of formation* — adding the previous two terms to get the next — is more important than the actual *numbers*. I once grew a sunflower with 29 spirals one-way and 47 the other, an adjacent Lucas pair as you can see:

$$2, 1, 3, 4, 7, 11, 18, 29, 47, 76, \ldots$$

The Fibonacci sequence (and occasionally the Lucas) is the grain in the stone, the path in the forest, the long sought after *pattern* for the *consistency* we would-be Pythagoreans had blundered upon in the dreamtime of youth.

OUR PURPOSE STILL HOLDS to investigate three of nature's numbers: π (pi), φ (phi), and δ (delta). To paraphrase Tennyson's "Ulysses," we have accomplished much, but much remains — some work of noble note is yet to be done. We have touched the happy isles of pi and seen the green fields of Fibonacci, but where is the land of phi?

According to the sometimes mystical Johannes Kepler:

> Geometry has two great treasures: one is the Theorem of Pythagoras; the other, the division of a line into extreme and mean ratio. The first we may compare to a measure of gold; the second we may name a precious jewel.

Let's see exactly what Kepler meant by "the division of a line into extreme and mean ratio" a phrase taken directly from Euclid. Consider a line segment divided so that the smaller length is 1 unit long and the larger is x units, and assume these lengths are in the following ratio:

$$1:x = x:1+x$$

or

$$\frac{1}{x} = \frac{x}{1+x}$$

In words: the smaller part, i.e. 1, is to the larger part x as the larger part is to the whole (1+x). The 1 and 1+x are called the *extremes* and x the *mean*. Cross multiply in the above equation and solve for x using the quadratic formula from secondary school and you get two values for x. We want the positive value (root):

$$1+x = x^2$$

so

$$x = \frac{1+\sqrt{5}}{2} > 0$$

$$x = 1.6180339\ldots = \text{phi} = \varphi$$

The number x is known by many names: the golden ratio, mean, section, or number; Luca Pacioli called it the divine proportion; modern mathematicians call it either phi or tau. By whatever name, this quantity has curious properties and an irrational decimal expansion:

However, it's not transcendental like π, but algebraic since it's the positive root of $x^2 - x - 1 = 0$. Not the least among phi's distinctive properties is its reciprocal and square:

$$\frac{1}{\varphi} = 0.618033\ldots = \varphi - 1$$

$$\varphi^2 = 2.618033\ldots = \varphi + 1$$

These surprising results can be easily checked with a calculator. The preservation of phi's irrational part after flipping and squaring is unique among numbers.

But the aspect of phi's personality that will prove most interesting and useful is its expression as a continued fraction. The interest resides in the stunning regularity of this representation for a number whose decimal is clearly chaotic. Whence this order? The multi-decked fraction can be developed from the initial ratio:

$$\varphi = 1 + \cfrac{1}{1 + \cfrac{1}{1 + \cfrac{1}{1 + \cfrac{1}{1 + \ldots}}}}$$

By inspection it's apparent that phi's expansion is the simplest possible for any continued fraction. Why? It's a noteworthy consequence of all the 1s. Consider its convergents below. Remarkably, they're all Fibonacci numbers; more precisely they're the quotients of consecutive Fibonacci numbers. This is no accident because this sequence provides the best possible rational approximations for phi ($= 1.61803\ldots$).

$$1,\; 1+\frac{1}{1},\; 1+\cfrac{1}{1+\cfrac{1}{1}},\; 1+\cfrac{1}{1+\cfrac{1}{1+\cfrac{1}{1}}},\; 1+\cfrac{1}{1+\cfrac{1}{1+\cfrac{1}{1+\cfrac{1}{1}}}},\; \ldots$$

which simplifies to $\frac{1}{1}, \frac{2}{1}, \frac{3}{2}, \frac{5}{3}, \frac{8}{5}, \frac{13}{8}, \frac{21}{13}, \ldots$

or as decimals 1, 2, 1.5, 1.$\overline{6}$, 1.6, 1.625, 1.615, ...

Dividing successive Fibonacci numbers gives closer and closer approximations to the golden number, and provably so. This is the link between phi and Fibonacci. Yet the multi-decked fraction for phi, because of all the 1s, has the distinct honor of being the *slowest of all continued fractions to converge* on its limit value of φ. And since the convergents provide a measure of how well phi is being approximated — and this is as bad as it gets — then the golden mean is *the most irrational of all numbers*. As we shall see, nature uses this property in intriguing ways.

There is a mystery surrounding phi's ubiquity. To experience a sense of wonderment over this, take your age (e.g. 33) and the number of freckles (e.g. 10) on the back of your left hand, say, or any two quantities you wish. Let these be the first two terms. Add them in the Fibonacci manner — each new term is the sum of the previous two — to produce an entirely new sequence (in this case 33, 10, 43, 53, 96,...). As before, divide adjacent terms (e.g. 53 ÷ 43). What do you notice? Voilà! The golden number rises from this new sequence like the phoenix from the ashes. Dividing consecutive terms from the Lucas sequence similarly produces phi. These aren't accidents. Any two numbers, when added in the Fibonacci manner, create a sequence such that the division of successive terms always approaches phi. *The gold is in the process not the numbers.*

Young Pythagoreans counted flower petals and noticed a consistency in their quantity. Youthful students were shown the underlying Fibonacci patterns to these numbers. And from the process of formation, we unearthed the golden number. We have come this far, yet one crucial question remains: "Why Fibonacci numbers?" Why not numbers from any of a thousand different sequences?

Where does this regularity of nature come from? Is it in the genes or the dynamics of growth interactions? Over time genes display enormous flexibility. Using "unnatural selection" man has in a few thousand years bred everything from Chihuahuas to Saint Bernards, all from the gene pool of wolves. On the other hand, chemistry and physics are mathematical in their uniformity. When hydrogen and oxygen are mixed, infinite combinatorial possibilities exist, yet with exceedingly rare exceptions we get no more than water (H_2O) molecules.

Let's attempt to explain just a single pattern: *the number of spirals* on a pine cone or a sunflower. Consider the computer-generated sunflower below. In both directions, the spirals are easily seen, and I encourage the reader to count them. (The QBasic program for this diagram is in the Chapter Notes. It's entertaining to enter it into your PC and play with the variables.)

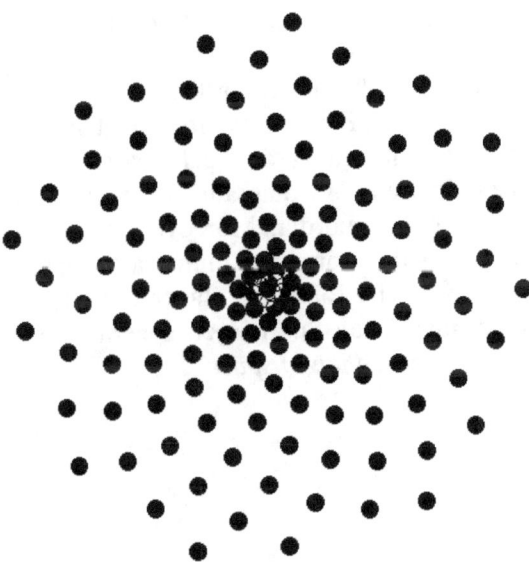

SUNFLOWER: 13 SPIRALS COUNTER-CLOCKWISE AND 21 CLOCKWISE

Figure 6.11

Plants grow from the tips of shoots, twigs, and branches, but grasses grow up from the ground. This allows people to mow lawns without killing the grass and animals to eat grass without slowing or destroying its growth. Had grasses lacked this simple characteristic, the earth's magnificent grazing herds of bison, gazelles, antelope, wildebeest, and reindeer could not exist. It's astonishing that something this minor — whether growth occurs at the tips or the roots — could have such devastating consequences.

216 / Numbers: Natural

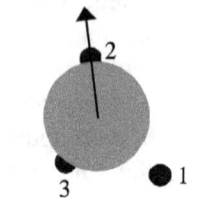

APEX WITH THREE PRIMORDIA

Basic knowledge of the spiral design is old and descriptive, but a deep understanding is current and remarkable. Like Janus, we gaze backward before looking forward. Consider the sunflower's apex: this is where new growth occurs. It's an undifferentiated area (colored gray in the figure at the left) where small bumps, called primordia (colored black), arise. As the plant grows, each primordium marches directly away from the apex and develops into a seed, leaf, or petal. Analogous to human stem cells, primordia are born undistinguished but mature into specialists.

But how is it possible for primordial buds to organize themselves into spirals — in Fibonacci pairs at that? Consider: since each advances radially outward, the firstborn will be furthest from the apex. In the above diagram, the numbers indicate the birth order of three primordia. Moreover, hidden on the sunflower's disk is another spiral not immediately apparent except to the mind's eye. The following illustration shows the birth order of 16 primordia on this unseen generative spiral: *it's a sequence in time rather than in space.*

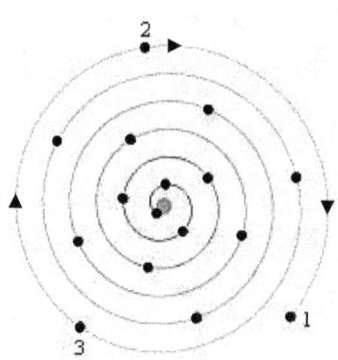

THE GENERATIVE SPIRAL

Figure 6.12

This is the fundamental spiral, the one that generates all the florets and seeds. The Fibonacci spirals are merely showy cast-offs lying on the surface of creation, picked out by the human eye because of their proximity in space. Normally primordia touch each other, but for clarity I've drawn them much smaller. The crucial quantitative element is the angle between consecutive buds. In Figure 6.12, join the apex and the first primordium with a line; do the same with the second. Measure the smaller angle between them; you'll see it's about 137.5°. Botanists call this the "divergence angle."

In the diagram below, the divergence angle is dark gray and its reflexive counterpart is light gray. Alongside the figure, I've reduced the ratio of these angles to their lowest terms — surprisingly, Fibonacci terms that in the limit equal the reciprocal of phi.

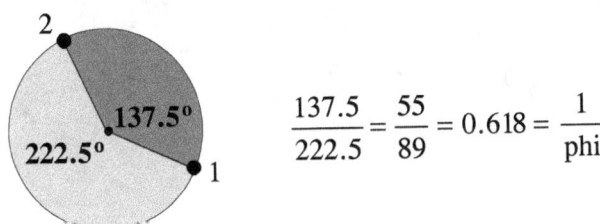

$$\frac{137.5}{222.5} = \frac{55}{89} = 0.618 = \frac{1}{\text{phi}}$$

Depending on which way you divide them, the Fibonacci numbers get closer and closer to phi (1.618...) or 1/phi (0.618...). Instead of starting with the divergence angle, we can reverse the process as follows:

$$360°\left(\frac{\varphi-1}{\varphi}\right) = 360°\left(\frac{0.618...}{1.618...}\right) = 137.507764... \text{ degrees}$$

Mathematicians call this the "golden angle."

In 1907 an early researcher G. Van Iterson plotted consecutive points separated by angles of approximately 137.5° along a generative spiral and created a diagram like 6.12. But since the real generative spiral (in time) is so tight and the primordia so much larger than I've shown them, the human eye doesn't see it, but rather we perceive two families (in space) of interlocking spirals. And because of the relationship between Fibonacci numbers and the golden angle — outlined above — the double spirals are consecutive Fibonacci pairs. With these figures, the number of such spirals is directly proportional to the number of primordia, nothing else — more primordia imply more spirals. *See* the following examples.

34/55 SUNFLOWER WITH TIGHTLY-PACKED BUDS

Figure 6.13

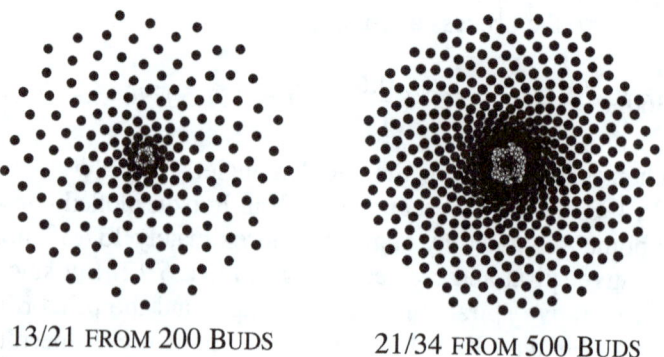

13/21 FROM 200 BUDS   21/34 FROM 500 BUDS

As you can see, the Fibonacci pairs you get are a function of the number of primordia/florets/seeds. And petals form on the outer edges of each spiral on only one of the two families. Nonetheless, we still don't have the complete story as to why these particular numbers occur. But as Ian Stewart says in his marvelous book *Nature's Numbers*[2], ". . . it all boils down to explaining why successive primordia are separated by the golden angle: then everything else follows."

The reason for the golden angle is a pretty one, and it issues from dynamics rather than genetics. Nature is forever maximizing her gains and minimizing her losses — she's the grand economist. This is well demonstrated in an indirect approach using divergence angles *other than* 137.5°.

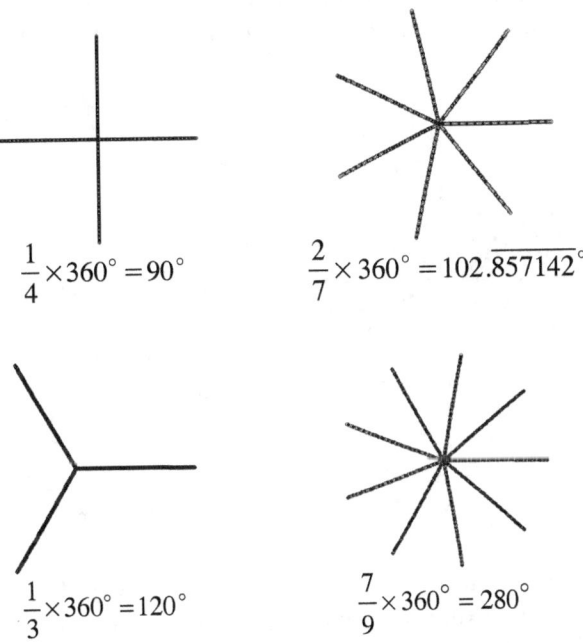

RATIONAL DIVERGENCE ANGLES

Figure 6.14

As you can see, these angles — *these rational angles* — leave gaps. And gaps mean inefficient packing, which is not nature's way. Furthermore, *all rational angles* leave gaps! Every such number, by definition, can be expressed as a fraction — with whole numbers in the numerator and denominator. For example, in this context 2/7 means that in 2 complete turns, 7 rays are laid down, or 1/3 implies that in 1 turn, 3 rays are laid down, and so on as in the figures above.

For these reasons, we need a divergence angle that's an *irrational* multiple of 360°. The question is which one? Like Cantor's paradoxical infinities, where some are larger than others, all irrationals are not equal; some are more irrational than others! We know the most irrational number; we were introduced to it a few pages ago. It is phi = 1.618033. . . . And the golden number yields the golden angle of 137.507764. . . degrees. The number or angle *most poorly* approximated by rationals — Fibonacci rationals — turns out to be the number *best* at packing primordia. And any deviation, by so little as a single degree higher or lower, will destroy the close packing — you can see this in the following two figures.

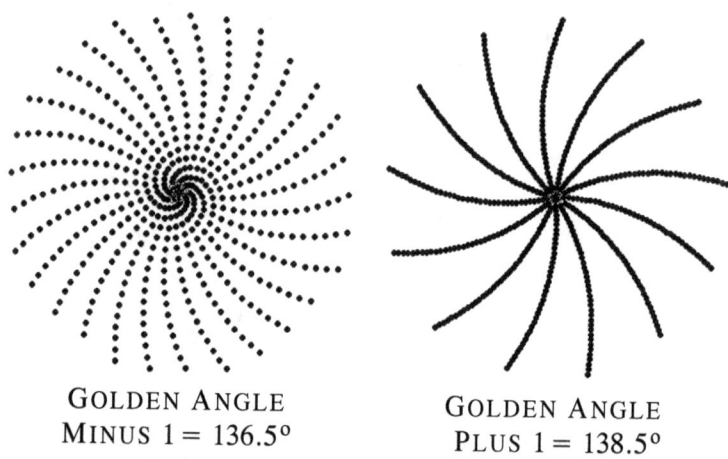

GOLDEN ANGLE MINUS 1 = 136.5°   GOLDEN ANGLE PLUS 1 = 138.5°

There you have it: the explanation as to why this Fibonacci frenzy, this exquisite pattern, fills nature's fields with the beauty of art and the designs of mathematics. It's all a matter of maximizing the seed head through close packing. And you do this by choosing the golden angle as the angle of divergence. It's phi — nothing more or less will do!

The plant, of course, knows nothing about the golden angle, and less of Fibonacci. Evolution has put it on a close packing quest: produce more seeds, have more progeny, be fruitful and multiply, or perish. When only a tiny portion of all your offspring (DNA) survives, it pays to produce hundreds, even thousands of copies.

Several times, I have referred to a debate between those who base all this on dynamics and those who believe every design issues from genetics. But it seems more likely to be a partnership among the dynamics of physics, chemistry, and the environment, plus genetics.

Two modern French researchers Stéphane Douady and Yves Couder have obtained the golden angle as a direct result of simple dynamics. Nevertheless, there could be more than one method of doing the same job. Perhaps it would have been constructive to the resolution of this dynamics versus genetics conflict had I planted the previously mentioned 29/47 Lucas sunflower to determine whether its progeny were Lucas or Fibonacci. Well I did plant the seeds but some squirrels ate them! However, recall that the result of dividing successive Lucas terms also approaches phi; hence, its primordia too are laid down along a generative spiral each separated by the golden angle.

The universe works in remarkable ways: it rarely goes directly from simple laws to simple patterns as in the case of gravity to the elliptical paths of planets. The very simplicity of the heavenly orbits was the reason for their early discovery. Oh, if only it were on earth as it is in heaven — but that was not to be. Instead, nature starts from simple laws, then runs down a rabbit hole and through a warren of complexity before resurfacing into simple designs again. Fibonacci numbers and spirals are unexpected patterns resulting from deeper processes that are either dynamic or genetic, but more likely both. In *Nature's Numbers,* Ian Stewart waxes almost poetic over this conclusion:

> Nature's patterns are "emergent phenomena." They emerge from an ocean of complexity like Botticelli's Venus from her half shell — unheralded, transcending their origins. They are not *direct* consequences of the deep simplicities of natural laws; those laws operate on the wrong level for that. They are without doubt *indirect* consequences of the deep simplicities of nature, but the route from cause to effect becomes so complicated that no one can follow every step of it.[3]

IT IS IMPOSSIBLE to think of sunflowers without mentioning the glorious creations of Vincent van Gogh. In a letter to his brother Theo he said, "You may know that the peony is Jeannin's, the hollyhock belongs to Quost, but the sunflower is mine." He painted eleven: four in Paris and seven in Arles.

Not only do his paintings reach out and touch us, but also we feel we can reach out and touch them. They have a tactile quality, a sense of being immediate, close and alive. Curiously, all the Paris paintings are cut sunflowers on a table showing spirals, yet all the Arles masterpieces are in vases showing few spirals.

Vincent van Gogh's *Two Cut Sunflowers*,
1887, Metropolitan Museum of Art, New York

Figure 6.15

Vincent van Gogh's *Two Cut Sunflowers*,
1887, Kunstmuseum, Bern, Switzerland

Figure 6.16

After these pages of detailed analysis on sunflower spirals, I can almost hear some aesthete proclaiming I have destroyed any appreciation of their beauty, that essence van Gogh so magnificently celebrated. That's false, foolishly false! And I'm not alone in my declaration. No less a person than the late Richard Feynman (as quoted from the "Best Mind Since Einstein," a NOVA Video) expressed identical sentiments; I'll let the last word be his:

> I have a friend who's an artist and he's some times taken a view which I don't agree with very well. He'll hold up a flower and say, "look how beautiful it is," and I'll agree, I think. And he says, "you see, I as an artist can see how beautiful this is, but you as a scientist, oh, take this all apart and it becomes a dull thing." And I think he's kind of nutty.
>
> First of all, the beauty that he sees is available to other people and to me, too, I believe, although I might not be quite as refined aesthetically as he is. But I can appreciate the beauty of a flower.
>
> At the same time, I see much more about the flower that he sees. I could imagine the cells in there, the complicated actions inside which also have a beauty. I mean, it's not just beauty at this dimension of one centimeter: there is also beauty at a smaller dimension, the inner structure . . . also the processes.
>
> The fact that the colors in the flower are evolved in order to attract insects to pollinate it is interesting — it means that insects can see the color.
>
> It adds a question — does this aesthetic sense also exist in the lower forms that are . . . why is it aesthetic, all kinds of interesting questions which a science knowledge only adds to the excitement and mystery and the awe of a flower.
>
> It only adds. I don't understand how it subtracts.

NUMBERS: CULTURAL

A lie gets halfway around the world before the truth has a chance to get its pants on.
Winston Churchill (1874–1965)

MANKIND HAS A LONG TRADITION of playing with phi, some of it admirable, most of it not. The dawn of history is rumored to have given us the first instance of this in the

Great Pyramid at Giza. Time hasn't been kind to either Khufu (about 2650 BC) for conceiving the pyramid or his slaves for building it. The Egyptians so hated him that for thousands of years they used his name as a curse. And the naturalist-writer Henry David Thoreau, in his American classic *Walden,* was equally harsh on the slaves:

> As for the pyramids, there is nothing to wonder at in them so much as the fact that so many men could be found degraded enough to spend their lives constructing a tomb for some ambitious booby, whom it would have been wiser and manlier to have drowned in the Nile, and then given his body to the dogs.[4]

More occult nonsense has been scribbled about the Great Pyramid than any other single subject. A quick run through the New Age section in any bookstore will convince you of this.

The founder of the religious sect called Jehovah's Witnesses, Charles Taze Russell, was deeply involved with pyramid mysticism. The third volume of his series *Studies in the Scripture* is pure pyramidology. Much of this silliness concerned various "measurements" of the tomb's secret chambers and hallways. It's on such measurements, especially in the Grand Gallery, that he based his prophecies. Repeatedly his dire dates passed with nothing happening, but undaunted Russell continued to churn out new predictions. On their door-to-door rounds, I have found that Witnesses are — without exception — unaware of their founder's occult obsessions.

The truth is we have no idea how long the basic Egyptian measuring unit was, nor are we ever likely to. Unless scholars can decide this, all these so-called measurements on the pyramid are spurious. Ratios, however, are a different matter: they don't depend on any unit whatsoever. If one block is twice as high as a second block, this says nothing about the actual height of either.

One legend — beloved by occultists — concerning the Pyramid of Khufu persists even to the present time. It's so pervasive that reputable mathematics books still repeat it. Allegedly, the ancient Egyptian priests revealed to Herodotus that the pyramid was constructed so the area of each triangular face was equal to the area of the square of its height.[5] In other words these areas are in the ratio of 1 to 1.

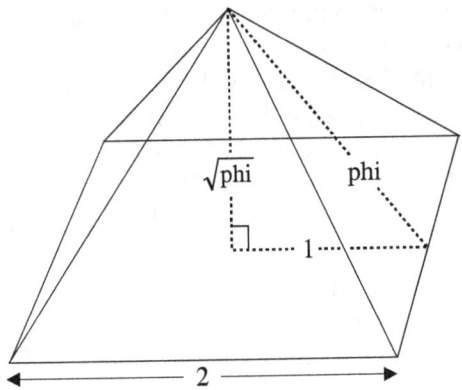

THE PHI PYRAMID OF KHUFU

Figure 6.17

The above diagram shows the precise implications of this. Without loss of generality we may let the pyramid's base length be 2 so the distance from its center to the edge is 1 — remember these are ratios. As a direct consequence of Herodotus' report (*see* the Chapter Notes for details), the ratio of the slant height to the vertical height is exactly phi : √phi. Only this number phi allows the area of the pyramid's triangular face to equal its height squared. From the Middle Ages to the present day, these pyramid ratios have been secrets of the Masonic Order.

This is a remarkable story resulting in some nice mathematics, but it's clearly false. Herodotus didn't report anything like that. How do we know? In section 127, Book II, of his classic *The Persian Wars*, we can read what he really did say and it's completely different. Here are the last three sentences of that passage:

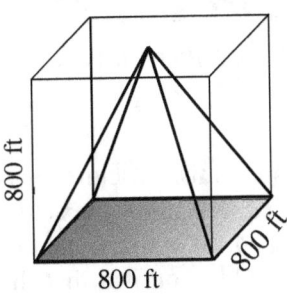

The pyramid itself was twenty years in building. It is a square, *800 feet each way, and height the same* [italics are mine], built entirely of polished stone, fitted together with the utmost care. The stones of which it is composed are none of them less than thirty feet in length.[6]

As shown, this pyramid fits inside a cube making its design quite simple — pyramids are very basic structures. Playing in wet sand on the beach, they're the first thing every child builds, but the child's motivation is very different from the pharaoh's. The latter's compulsion was to reach to the heavens like the Tower of Babel, the church spire, or Simeon Stylites on his pole. It's all about being "Nearer My God to Thee" as the old hymn says.

To build a high structure — without modern building materials — requires every new level or layer to be smaller than the previous one. The ancient Step Pyramid at Saqqara demonstrates this need, but all pyramids worldwide have steps whether large or small. Closer inspection of the Great Pyramid's steps belies the stories of legend and pen regarding the size of these building blocks and their perfect fit. Look for yourself at the following photograph. Note the jumble of rocks at the lower right and the places where you could easily thrust your fist between them. This picture also refutes Herodotus' statement: "The stones of which it is composed are none of them less than thirty feet in length."

A PILE OF ROCKS — THE GREAT PYRAMID

Figure 6.18

What were the original proportions of this structure? Although its smooth plaster finish is gone, experts agree the actual dimensions were 768 by 768 ft (234 by 234 m) with a height of 482 ft (147 m). Your calculator will quickly confirm that the phi pyramid accords well with these facts, but the cube pyramid existed only in Herodotus' imagination. If you work further with your calculator, you'll see that the phi-pyramid ratios are very close to those of the real pyramid. In fact, they differ by about 1 percent — an amount normally undetectable to the human eye.

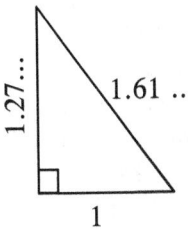 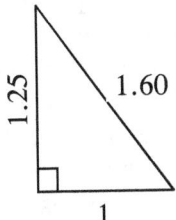

PHI-PYRAMID RATIOS   REAL-PYRAMID RATIOS

Consider these cross-sectional views of both pyramids. The one on the left is from Figure 6.17 where $\sqrt{phi} = 1.27...$ and phi $= 1.61...$; the one at the right is obtained from the real pyramid by dividing all sides by half the length of its base (*namely* 384). As is evident, these profiles are nearly identical. Perhaps now the reader may wish to reevaluate my dismissal of the veracity of the face area equaling the height squared, which produces the phi pyramid.

If you're unacquainted with number juggling, the ease with which you can get the answer you want rather than what the figures imply is startling. Lying with numbers is an ancient art. The following example of this deception requires no more than high school mathematics:

The profile of the cube pyramid below — the one Herodotus recorded — is quickly found by dividing its height and width by half its base (*namely* 400). And then use the famous theorem of Pythagoras to establish the pyramid's slant height of $\sqrt{5}$. Now for a little number juggling.

By taking certain ratios of angle A (the cotangent and cosecant) and adding them, we immediately produce phi:

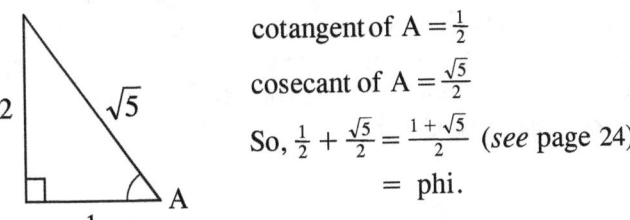

THE CUBE-PYRAMID OF HERODOTUS

Occultists will mistakenly jig with joy at this result, but it's mere playing with numbers since no such pyramid with these ratios ever existed. To recap: in the false written report by Herodotus, phi was in the *ratio of the angles*. While from the alleged story the priests told him, it was in the *ratio of the sides*. It's everywhere! It's everywhere! But since neither was the real pyramid, enough of this phi foolishness.

It may have been the first of the Seven Wonders of the World and the only one standing today; it may be five millennia old; it may be built from two million plus blocks, but it's only a pyramid — a childish structure, a jumbled pile of rocks. Despite its orientation to the compass points, its secret burial chambers, its ventilation shafts, and its Grand Gallery, it isn't as wondrous in construction or detailed in design as the smallest dung beetle resting in the shade of its great bulk. It's grandiose but not grand.

THE ANCIENT GREEKS were a different people, set off from the surrounding barbarians, as they called them, by their singular clear-headedness. Our European civilization had its origins in Athens not Jerusalem.

Much detailed knowledge of Greek society survives — yet much is buried in history's dust. Nevertheless, in one form or another some of their writings have come down to us. In a sole fragmentary manuscript, Pythagoras explicitly mentioned phi concerning the division of a line segment. And Euclid summarized all known geometry in his celebrated textbook *Elements;* Book VI deals with the study of ratios — "Proposition 30" is all about phi.

Consider their most important temple, the Parthenon. Dedicated to the goddess Athena, protector of the city, it was then, as now, the showpiece of Greek architecture. Built in the 5th century BC, this edifice[7] fits into a rectangle of height 1 and width phi (φ). The interior housed a chryselephantine statue of the goddess Athena, with an immense coiled serpent at her feet. The designer was none other than **Phi**dias, from whom the golden ratio gets its other name.

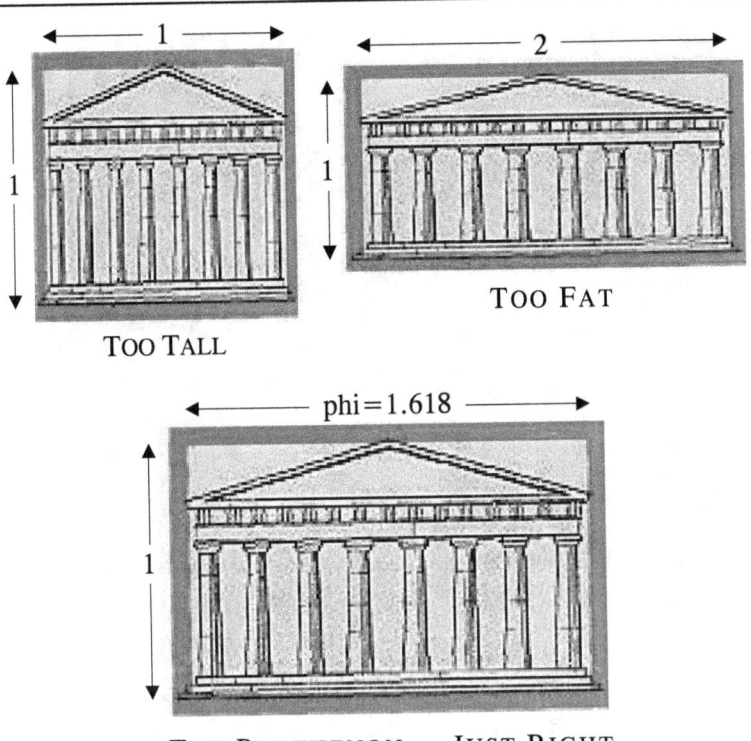

POSSIBLE PARTHENONS

Figure 6.19

Whether the proportions of the Parthenon result from a keen aesthetic sense or are a deliberate application of the golden ratio is debatable. Neither the Greeks nor the Egyptians left any written records considering phi as a most pleasing or beautiful shape. This idea originated in the late 1800s. Nonetheless, any temple built with a 1:1 ratio would be too tall, and a 1:2 would be too fat, while the 1:(phi ratio) is just right.

Of equal importance in their architecture was the theater at Epidaurus (next page) designed by Polyclitus. Still used today for productions of Greek plays, it's a regular tourist attraction. As you can see, a passage (indicated by the arrow) runs directly across the theater. This walkway separates the 55 rows of seats into 34 below and 21 above it. Surprisingly, these three quantities are consecutive terms from the Fibonacci sequence. By dividing, we get an excellent value for phi (34/21 = 1.619). Polyclitus knew his mathematics well.

GREEK THEATER AT EPIDAURUS, ABOUT 300 BC

Figure 6.20

Now that we have outlined phi's march around ancient Greece, we could also follow its tramp through Rome and the Dark Ages, but let's jump ahead to the Renaissance.

PHI, LIKE JANUS, has two faces: one looking forward with reason and compassion, the other staring backward into a dark and occult past. This nocturnal plant was in full bloom during the Renaissance, but Pythagoras planted the seeds. The ancient sage also had two faces. He was the first person to prove anything by the use of deductive logic; he was also the mystic who founded a famous secret society with all its attendant symbols, rituals, and rules. The most important of the latter was the strict injunction *never to eat beans*. Bertrand Russell in his *History of Western Philosophy*[8] said of Pythagoras' bipolar behavior, "The influence of mathematics on philosophy, partly owing to him, has, ever since his time, been both profound and unfortunate."

The Pythagorean society chose the pentagram, or five-pointed star, as the sacred symbol of their order. Drawing this figure in a single unbroken movement was *their sign* of recognition. Every segment of the figure is in golden ratio to the next smaller segment. By joining the intersecting points of the

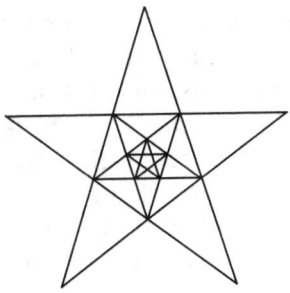

THE PENTAGRAM

smaller interior pentagon a new pentagram is created and the 1:(phi ratio) continues to infinity.

Of course, numbers and figures are neither good nor bad in themselves but only in the attributes we give them. But from this time forward the pentagram and phi had occult, even satanic, associations — *see* the next drawing from *De Occulta de Philosophia* by Agrippa of Nettesheim (1486–1535).

As a species we enjoy discovering connections between things — any things — even if those connections turn out to be illusory. We call the art of distinguishing true from false relationships "science." For thousands of years we have honed and polished this practice, which may have begun with *the hunt*.

GOAT OF MENDES

The kudu passed here at sunrise (because the trail of a morning beetle crosses *over* the antelope's track). The kudu rested here at noon (because the sun being overhead only these trees could provide shade. This kudu is a male (because of the pattern his urine makes with his hoof prints). This male kudu is large (judging by the *depth* of his prints). I have followed this beast for two days; his stride has shortened, and he's very tired (because of his great size, he can't disperse his body heat as I do — the square-cube law at work). This kudu is exhausted and resting in the shade of that acacia tree. . . . This kudu is mine.

The metaphor of the hunt is relentless, but on occasion it can lead to a rotten carcass. Humanity has an overwhelming need to form patterns, even false ones. It's how we adapt and live. It's what makes us human/sapiens.

Some theological arguments are serious, others merely hilarious. One of the latter variety has been disturbing biblical fundamentalists for millennia, to wit, did Adam and Eve have navels? Or, as the gay movement would say, did Adam and Steve? Or Ada and Eve? (If they did have bellybuttons, were they innies or outies?)

On a more serious and historical note, the English writer Sir Thomas Browne said (referring to Adam), "The man without a navel yet lives in me." James Joyce wrote in his novel *Ulysses*, "Heva, naked Eve. She had no navel." Many medieval artists realized the navel problem, so they omitted it from their paintings. Others hedged their bets by allowing the ample fig leaf to wind up and over the navel as well as the pubic area. But from the time Michelangelo painted Adam with a navel on the Sistine Chapel ceiling, artists have followed his lead.

The usual fundamentalist's answer to this inadvertently amusing question about Adam and Eve's aberrant anatomy is laid out in James Harrison's book *The Pattern & The Prophecy: God's Great Code:*

> Since Adam and Eve had no parents, they couldn't have had a navel. If they did, it would imply each was born of a woman. Since this is clearly false, the very presence of a navel would have been a deception, and so couldn't have come from God. Satan deceives; God does not.
>
> Of course, we know directly from the Bible that Adam had no navel. After all, he was created in the image of God. By definition, the Deity has no predecessors, and therefore no navel. But we, the sons and daughters of Adam and Eve, all bear his mark in our nature and on our bodies. The only spiritual difference between humans of today and our first ancestors is original sin. The only physical difference between humans of today and our first ancestors is the navel. *This "birthmark" is just the bodily symbol and reminder of that sin.*
>
> Using man's limbs for units of measurement is an ancient practice. The inch was the length from the thumb's knuckle to its tip; the yard went from the king's nose to his outstretched finger. The biblical cubit was from the elbow to the middle finger, while the foot is obvious. All the prior units are from the body, the body made "in the image of God." However, phi is from a different body [*see* below] and for a different purpose. It derives from the navel, the symbol of man's fall.[9]

Fundamentalists concluded from this that their original, perfect nature was reflected in their smooth, buttonless stomachs, and since the supposed "fall," the navel has been a symbol of original sin. So there you have it — this carcass!

As I have said, not all hunts are successful. Consider the following strange foray attempting to connect the navel, original sin, the pentagram, man, and phi.

The pentagram has five points, the human body has five extensions, and properly placed it fits inside this shape. It's the symbol of man. The figure below is the *Man in the Pentagram,* also from Agrippa's *De Occulta de Philosophia.* Intersecting lines at the genitalia mark the center, but the horizontal line through the navel is more curious. For Agrippa, and some Renaissance artists, the navel divides the body in the ratio of 1 to phi.

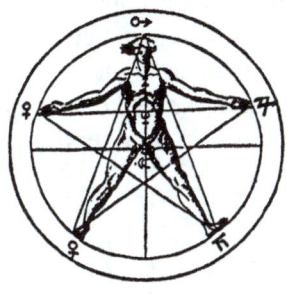

MAN IN A PENTAGRAM

Measure the distance from your head to your navel, then measure the distance from your navel to the floor. Now divide the smaller length into the larger, and you get approximately 1.618.... I urge every reader to confirm this by measuring themselves or their partner. If your personal ratio is larger than 1.618..., consider your body a high-phi model, otherwise you are, unfortunately, low-phi!

Leonardo designed the famous *Vitruvian Man* (*see* below) at the height of Renaissance Humanism. It also shows the navel separating the body in the ratio of 1 to phi. To make this apparent I've written the numbers on the drawing's left side.

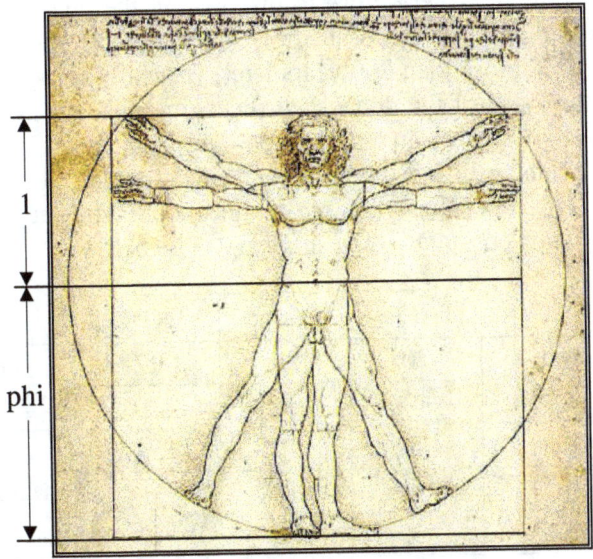

VITRUVIAN MAN BY LEONARDO DA VINCI

Figure 6.21

Because of the strong influence of his friend Pacioli, da Vinci made use of phi in his geometrical research and paintings.

The degree of Pacioli's enchantment with this ratio can be gauged by listing a few of his chapter titles: "The Incomprehensible Effect," "The Excellent Effect," "The Unutterable Effect." To illustrate one of these relationships consider the following: take three golden (phi) rectangles and symmetrically fit them together so they intersect each other at right angles. Then the 12 corners (four per rectangle) mark the 12 vertices of an icosahedron or the middle of each face of a dodecahedron as in these diagrams:

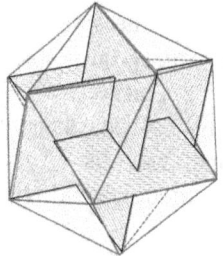 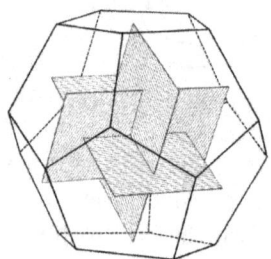

THREE GOLDEN RECTANGLES WITHIN EACH SOLID

Figure 6.22

This "effect" is very close to discovering the duality of the icosahedron and the dodecahedron — historically something that occurred more than 200 years later. Basically it says if you change the faces of the dodecahedron into points and its points into faces, it becomes an icosahedron, and vice versa. The cube and the octahedron are duals of each other, while the tetrahedron is self-dual. These are *real* connections. The number of edges remains constant — *see* summary table below.

	FACES	EDGES	VERTICES	
Icosahedron	20	30	12	Dual figures
Dodecahedron	12	30	20	
Octahedron	8	12	6	Dual figures
Cube	6	12	8	
Tetrahedron	4	6	4	Self-dual

THE PLATONIC SOLIDS

Figure 6.23

A contemporary of Leonardo's, although they never met, Albrecht Dürer (1471–1528) used phi in his canon of proportions but only as a guide, not as an unvarying law. From time to time other artists and architects (such as Le Corbusier) continue to use this ratio in their work, but it has never been a major theme. And often the ratio exists only in the eye of the beholder. Even in *Vitruvian Man* the ratio is actually 1.51, which is far from what it's professed to be — measure it yourself.

In Chapter 2, page 44, I referred to *The Sacrament of the Last Supper* regarding the gigantic dodecahedron and the headless body. But another element exists here. The entire canvas is *supposed* to be a huge golden rectangle — even the placement of other figures is rumored to be done by this ratio. But how can this be? After all, the canvas is 66 by 105 inches, and after division, this yields 1.59 — close but not the golden ring. For an excellent approximation to phi, you can't do better than choose two adjacent terms from the Fibonacci sequence. So, if Dali intended to place his painting within a golden rectangle, why didn't he use 55 and 89 inches or better still, 89 and 144 inches? Both give phi correct to three decimal places.

The Sacrament of the Last Supper by Salvador Dali
National Gallery of Art, Washington, DC

Figure 6.24

PHI HAS TAKEN US on a fantastic voyage from the pyramids of Egypt, to Pythagoras' five-pointed star, through the columns of Athens' Parthenon and the seats of Epidaurus' theater, and Agrippa's *Man in the Pentagram,* to the paintings of today. We have journeyed from Adam's navel to phi's bellybutton. And much of what we've found is pure "fluff."

What can using 1.618 compared to 1.5 possibly add to a painting or a design? Nothing! What extra significance does phi give to a work of art? None! What additional symbolism does the golden ratio bestow? Zilch! As long as the Parthenon was narrower than the too fat ratio (1:2) and shorter than the too tall ratio (1:1), then any number of values would be just right — including 1.5 and phi (*see* Figure 6.19). Furthermore, the supposed "navel ratio" of *Vitruvian Man* and the *Man in a Pentagram* makes no sense. You might just as well find the average length of the distance between your eyes and divide it by the length of your nose. What racial group should you use to achieve this average, Ruandan or Inuit? These are distractions, which when taken to extremes, lead to a kind of hi-phi foolishness that only humans are capable of.

A marvel of the publishing world, Dan Brown's *The Da Vinci Code* cleverly exploits this pseudo marriage between phi and the art world. The plot begins with an albino member of the Roman Catholic sect Opus Dei executing a prominent art curator in the Louvre's Grand Gallery. Left dying, the curator pens a scrambled version of the Fibonacci sequence (i.e. 13-3-2-21-1-1-8-5), strips off his clothes, paints a pentagram around his navel, and arranges himself in the form of Leonardo's *Vitruvian Man* (*see* Figure 6.21). All this is to conceal a conspiracy covering-up the "true" origins of Christianity. From here this book rockets to *oc*-cult status.

But the divine proportion belongs in the field of dreams among the compound flowers, the branching of trees, and the spiraling of pine cones. To tease out his secrets, he's also welcomed in the mathematician's study — this is his true home. Somehow, sometimes, Phidias stumbles into the artist's workshop or party, where, while he does little harm, he brings no gifts. As art spectators, our search for patterns has led us to see designs that are either entirely in our heads or the artiste didn't intend. Or if Phi was invited, the result is no better or worse than if he had been excluded from the guest list. Although some may imagine they saw him, Mr. Phidias Golden has either left the building or he's mute.

The Texture of Infinity

Man is equally incapable of seeing the nothingness from which he emerges and the infinity in which he is engulfed.
Blaise Pascal (1623–62)

IN THE ALLEGORY OF EDEN there were two trees: one of the knowledge of good and evil and the other of eternal life. Had Adam and Eve discovered the second tree the entire human drama would never have unfolded. By analogy, in "Forms and Fractals" we visited the Garden of Simple Complexity and learned about the almond tree of Mandelbrot. And like our ancient mythical parents, we had to exit the garden before reaching the second tree, the fig of Feigenbaum. But unlike Adam and Eve we can still, figuratively, get there.

Mathematicians label the fruit of this second tree — the last in our trio of nature's numbers — δ (delta); we'll call it the *Feigenbaum constant*. Both π (pi) and φ (phi) have been known for millennia, on the other hand, Mitchell Feigenbaum discovered δ in 1976 — in the lifetime of many of us. Here's a number intimately related to the parabola, a curve only slightly more complex that a straight line, yet it shielded a great mystery. Why so late an entrance on the world stage? Why didn't someone discover it centuries earlier? The answer in a word is *computers*. Until high-speed calculation and graphing were practicable, this constant and its related images were as close as the next minute but as distant as the Stone Age.

We have previously met the Feigenbaum fig tree, although at that time we didn't pick any of its fruit. Now it's time to do just that. Recall Figure 3.12, Bifurcation Diagram Two, with its period-doubling cascade to chaos. For convenience, I've reproduced it (*over*) clearly, marking the "m" points for 2, 4, 8, and 16 bifurcations: from limbs to branches to twigs to twiglets, . . . to chaos. The "m" I'm speaking of is the control-knob parameter for behavior in $x_{t+1} = mx_t(1 - x_t)$, our friend the logistic function. Recall that x is between 0 and 1 (vertically), and m can vary from 0 to 4 (horizontally).

Take successive distances, d_i, between bifurcations (branching points) and divide the larger number by the smaller. Surprisingly, these quotients are not random but *approach* a special ratio we call the Feigenbaum constant. That is, the further along the trail of bifurcations you go, the more precise the ratio becomes. So, as i goes to infinity, d_i/d_{i+1} goes to delta:

δ (delta) = 4.66920 16091 02990 67185 32038...

Self-similarity is the trademark of the fig tree. Yet as you can see in Figure 6.25, this repetition is not exact. For example, the two crosshatched sections aren't precisely similar to the section in d_1 — the lower is too fat and the upper too thin. But as you progress along the cascading twiglets, the progeny become closer and closer to the progenitor. In the ideal case each twiglet is the same as the one before but reduced by a factor of δ (delta). At this limit, d_i/d_{i+1} no longer goes to the Feigenbaum constant, it's already there.

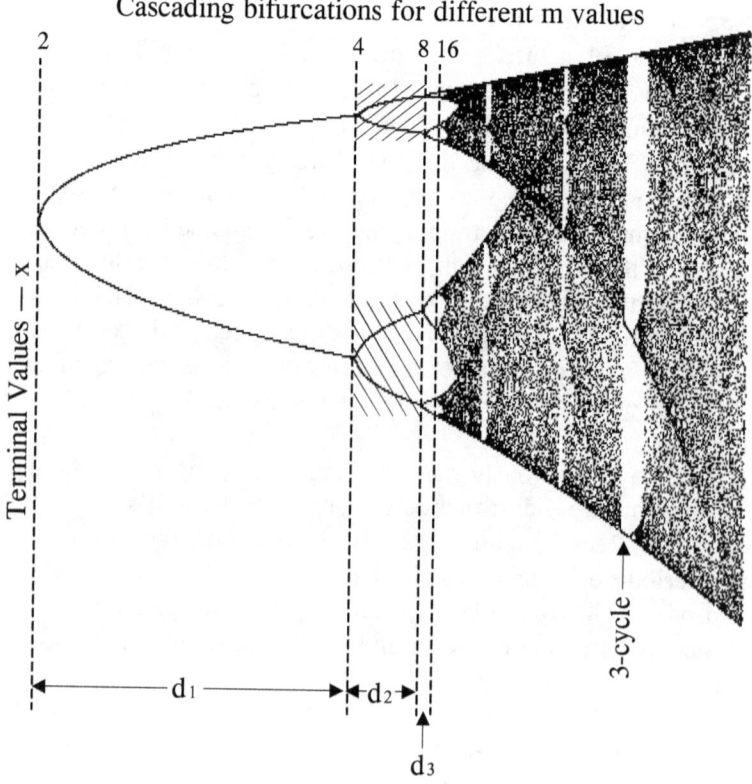

THE FIG TREE AND THE FEIGENBAUM CONSTANT

Figure 6.25

Beyond this first point of accumulation is the field of chaos, though a kind of "ordered chaos" — an oxymoron for our age. In this bedlam are sudden patches of regularity: 3-cycles, 5-cycles, 7-cycles, and so on (marked on the above figure). But the tangle commences with simple period doubling. And this doubling scales to perfection at the Feigenbaum constant.

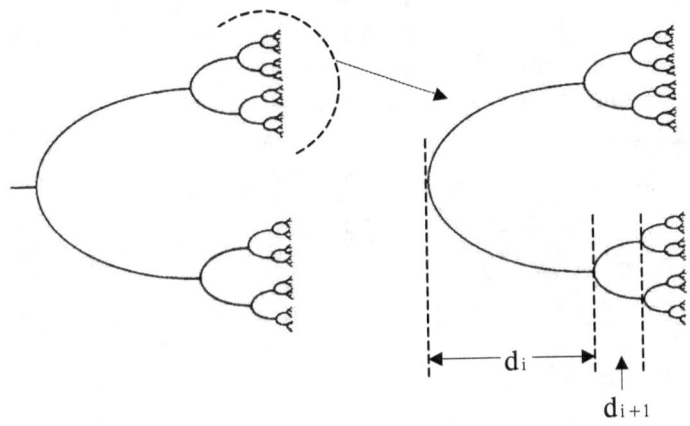

SELF-SIMILARITY IN THE IDEAL FIG TREE

Figure 6.26

This irrational number 4.66920. . . is a universal constant. We have already investigated two other such ratios, pi and phi — there are many more. On page 99, I wrote about certain universalities in Feigenbaum's fig tree: the bifurcation graphs of the logistic and sine functions were identical. In brief, pendulums, streams, weather, oscillators, and other systems all have remarkably different equations, but all involve period doubling on the road to chaos. As a matter of record, all smooth graphs that rise initially and fall off later have identical bifurcation graphs. And now from this graph we conclude they all have the same scaling factor — the Feigenbaum constant. *Delta is a constant for all functions approaching chaos via period doubling.* And by transcending any particular equation, its power is unmistakable.

IN A BROAD SENSE human activities can be considered as art. The hunter on the track of a lion may have previously drawn its image by torchlight on the walls of a cave. The artist is well known for what he *omits,* attempting to capture the essence of his subject. Painting isn't close-up photography. The hunter also is notable for what he *leaves out* in trying to understand and interpret the important elements of the animal's trail. Tracking isn't large-scale cartography. Any scene contains innumerable data points most of which are irrelevant to the artist and hunter. Both are minimalists.

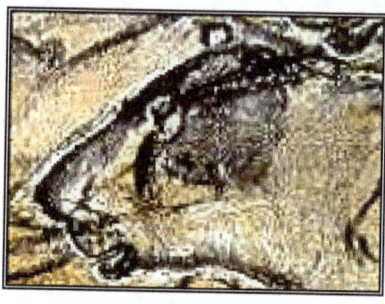

CAVE PAINTING OF A LION

Mankind's oldest known artwork — radiocarbon-dated at 30,000 BC — covers the walls of the recently discovered Paleolithic cave at Chauvet in southeast France. From its magnificent galleries consider the charcoal and red-ocher drawing to the left. These were the times when lions roamed Europe. Note the strength of the neck and the suggestion of movement and power. More to the point, note the omission of background and the absence of any detail in the hair. This is the way of imaginative artists and scientists — to see beyond the particular and the present to the universal and the future. These cave murals aren't messy daubs, but rather "canvases" by fur-clad Leonardos worthy of the Louvre.

The old bison saw them first. They were spread out in a line working their way up the valley on both sides of the stream. He instinctively knew they were after him and not smaller prey. They carried deadly atlatls (throwing sticks), bows, and quivers of arrows on their backs. Finely chiseled arrowheads decorated the ends of their weapons. It wasn't just their weapons he feared but their tracking talents. The leader was following the old bison's spoor, finding a hoof print here, some loose hair on a bush there, and trampled twigs elsewhere.

In his youth, the bison had eluded or outrun the bulky hunters that are gone now — the ones with the low brows and broad shoulders. They had been unable to follow his trail with the skill of these tall swift newcomers.

Later that same day the Paleolithic hunters carried the gutted bison to their encampment where it was roasted and devoured. They rested in their finely sown skins and furs made with pockets to hold berries for snacks, or rawhide and arrowheads for repairs on long hunts.

THE RAGING BISON, painted 28,340 BC

For two days they rested, and on the evening of the third day they go to the caves. Torches light their way and drive out the immense cave bears that sometimes settled there. Deep in one chamber a hunter mounts scaffolding and begins to sketch the old bison he had speared. This hunter-artist had kept the image in his head waiting for this moment. Now he transfers it to the cave wall to inspire everyone for a new hunt. With a deft hand and a few colors, he outlines the body of the beast and catches its strength and bravery when in the last seconds before death it turned to face its adversaries.

In blackness unlike anything nighttime can bring, these caves extend almost 1¼ miles (2 kilometers). They contain a connecting gallery, three vestibules, and three main chambers. These are covered with masterworks breathtaking in their vitality and strength, stunning in their use of perspective and shading — techniques art historians thought weren't discovered for millennia. And eons before Georges Seurat, these Stone Age artists were using pointillism. By taking finely ground red ocher into their mouths and steadily expelling it, certain objects — particularly the human hand — could be outlined. The caves of Chauvet demolished the theory of gradual growth in artistic skill.

The hunters' practical creations were no match for the grand cave paintings. But art and science are both human conceptions requiring individuality and creativity. The fields through which the hunters walked up the valley were fields of Fibonacci flowers with phi hidden in their petals and spirals. The disc of the sun overhead, the shaft of the spears in their hands, and the meandering stream at their feet all concealed the number pi. The controlled chaotic stream and the gathering storm harbored the last in our trinity of famous numbers, delta. But all this glory lay in the future. This would be the paradise created by science and mathematics, a second Eden from which no one would drive humanity out. But the hunters knew none of this, for theirs would be an incredibly long journey. Art and science arise from the same human ability — the gift to see the future in the present. To anticipate the next day, to prepare for it, to work by delaying today's pleasure for tomorrow's glory.

The days of glory for art have long been with us — though some days are larks and others are crows. With a single leap these Paleolithic hunters mounted the stallion of great art and rode off to create their own Sistine Chapels. And they did it with pride and a sense of individuality that we can feel to this day. As Jacob Bronowski writes in the *Ascent of Man*[10], "All over these caves the print of the hand says: 'This is my mark. This is man.'"

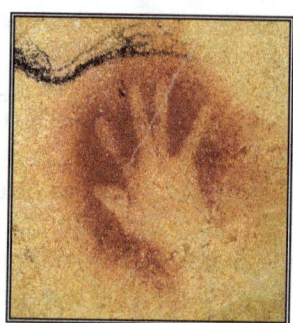

THE HAND OF MAN

Chapter — 7

LIFE ON THE EDGE

Nature does nothing in vain, and more is in vain when less will serve; for Nature is pleased with simplicity, and affects not the pomp of superfluous causes.
Isaac Newton (1642–1727), *Principia*

WINTERTIME and the living is hard. Reynard is on the prowl. Because of the deep snow, he hasn't hunted in two days. It's −40°F/°C; the red glow of dawn fills the east as he meanders through a field of wisteria bushes and poplar saplings. Reynard has no destination, he's going nowhere, and everywhere — this is the hunt. It's the prey that's important, not his location; he follows his ears and nose. Since the snow is deep, he treads with great care *conserving energy* however possible. His track pattern is unusual for a four-legged animal: he leaves only half the expected paw prints. His smaller hind paw lands in the impression just made by the front paw on the same side. That is, he "double prints," as we do through deep snow when following a friend because it's less arduous. Reynard dreams of mice and voles or perhaps a plump partridge sleeping in its overnight snow cave. He hasn't eaten in two days.

To conserve energy in snow, many other animals such as wolves and coyotes also double print. Unnatural selection and a full food bowl have rendered dogs incapable of doing so — they commonly make a four-print pattern. Through a hundred thousand years, natural selection pressures have favored energy-conserving behaviors. In a million-year march, a minuscule pebble in a boot can be crippling.

FOX DOUBLE PRINTING

These energy-saving behaviors, these maximizing and minimizing activities, are found across the entire spectrum of life. Often they're subtle, usually they're overlooked, but they're always necessary for long-term survival. Consider the humble bee. Bernd Heinrich in his lucid and lively book *Bumblebee Economics* reveals their innumerable adaptive behaviors. On a summer's day, in the same field of wisteria where Reynard conducted his winter hunt you can see bumblebees going about their tasks. Here's one survival adaptation Heinrich noted:

> The wider resources are scattered, the less efficient it is to recruit and defend specific items, and the more difficult it is to patrol and defend an area. Competitors then appear to work peacefully (without contact) side by side, but they may still compete relentlessly by trying to remove resources faster than the next individual. Aggressive encounters then become a liability, for even the winners lose — they have only expended time and energy that could have been used for foraging. The nonaggressors, which do not interrupt their foraging, reap more food energy and are competitively superior. Such competition, called scramble or exploitation competition, generally results in the depletion of resources to the very minimum of economic profitability. In turn, it selects for energy economy and foraging efficiency in the contestants.[1]

This is *natural selection* at work, at all times, in all places. As the greatest evolutionary geneticist of our time, Theodosius Dobzhansky wrote, "Nothing in biology makes sense except in the light of evolution."

Maximums and minimums can be found throughout the universe. Mathematicians have sought them in geometry for millennia. After the death of Archimedes, Marcellus (*see* page 50), who had built up such respect for his long-successful adversary, buried him with elaborate ceremony. Archimedes had been so fond of his own treatise *On the Sphere and the Cylinder* that he told friends he wished to have them engraved on his tombstone — with the latter enclosing the former. This Marcellus did.

Of all three-dimensional objects, the sphere has the greatest volume for a *fixed* surface area. Hence, as we saw in Chapter 1, it retains heat best. But what shape of cylinder has the maximum volume for a *fixed* surface area? Should it be tall and thin like the tube for tennis balls, or short and flat like a shoe-polish container, or perhaps some intermediate form? This is an important question for companies manufacturing cans. Interestingly, it can be proven that the best shape has its height exactly equal to its diameter. Such a cylinder viewed side-on from a distance, would appear to be a square: in the case of the tombstone, a square enclosing a circle (*see* diagram at the right below).

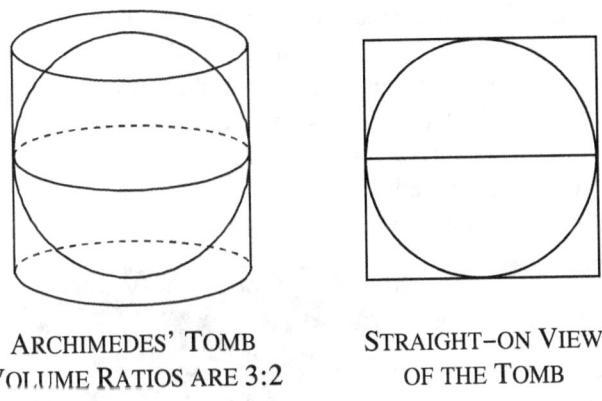

ARCHIMEDES' TOMB STRAIGHT-ON VIEW
VOLUME RATIOS ARE 3:2 OF THE TOMB

Figure 7.1

The Sage of Syracuse made many great discoveries, including the integral calculus — this almost two millennia before Newton. So, why have such a simple grave marker? Well it turns out that the exact ratio of this cylinder's volume to the sphere's volume is 3:2, and remarkably their areas also have the same ratio.

I maintain that for Archimedes this ratio had the force of Pythagorean philosophy behind it — that power of number over the flux still haunting scholars today. After Pythagoras had proven his famous theorem, he sacrificed a hundred oxen to the gods for the inspiration. This is the exaltation every scientist feels when he does the experiment and the numbers come out right and say, "This is part of the structure of nature herself."

The tomb triggered a curious ancient sequel. The Roman statesman and philosopher Cicero (106–43 BC) not only recorded this but initiated it:

> When I [Cicero] was questor in Sicily [in 75 BC, 137 years after the death of Archimedes] I managed to track down his grave. The Syracusians knew nothing about it, and indeed denied any such thing existed. But there it was, completely

surrounded and hidden by bushes of brambles and thorns. I remembered having heard of some simple lines of verse which had been inscribed on his tomb, referring to a sphere and cylinder modelled in stone on top of the grave. And so I took a good look round all the numerous tombs that stand beside the Agrigentine Gate. Finally I noted a little column just visible above the scrub: it was surmounted by a sphere and a cylinder.[2]

Although Cicero cleaned the site and repaired the tomb, the following centuries swallowed it up again. His generous act was perhaps the most memorable contribution of any Roman to mathematics! Many artists have painted this scene but (necessarily) with the sphere on top of the cylinder; unfortunately, this obscures Archimedes' discovery of the simple volume and area relationships between these objects.

Cicero Discovering the Tomb of Archimedes
by Benjamin West, 1797, in a private collection

Figure 7.2

We have seen some examples of maximums and minimums in *nature* concerning foxes and bumblebees and in *mathematics* regarding cylinders and spheres. The worlds of both biology and physics maximize certain resources and simplify various processes. To paraphrase St. Paul, a little of this is clear, but much is seen through a glass darkly. Life dances precariously on the edge between maximizing resources and death, and the physical world hears the same drum beat. In this chapter I intend to explore this dance of nature and life.

At First

There's the room you can see through the glass — that's just the same as our drawing room, only the things go the other way.
Lewis Carroll (1732-98), Through the Looking Glass

WE CANNOT SAY "In the beginning . . ." for nature, but with science this might just be possible. Recall the two elderly friends, Epius and Phemius, from earlier chapters who often argued over the merits of science versus art. Every evening in their seniors' residence they play billiards. One night, tiring of their regular game, they decide to do individual point shots requiring a bounce off one cushion.

"Blue ball in the near pocket off the far cushion," said Phemius the poet.

"How do you know where to hit the cushion?" asked Epius the engineer.

"It's the *halfway point* I've marked with the arrowhead," he responded. He shoots! . . . He misses!

Epius set up his shot so that the *angle of incidence equaled the angle of reflection*. He shoots! . . . He scores!

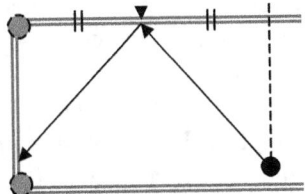 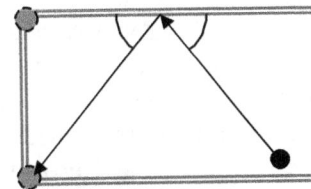

EQUAL DISTANCES   EQUAL ANGLES

The poet mused aloud, "Why was your shot with equal angles so accurate while mine with equal distances wasn't?"

"As soon as I've erased tomorrow's dinner menu from this chalkboard, I'll explain why equal angles work," replied the scientist.

"You can't erase that, the staff will be angry."

"What'll they do? Take my bathroom privileges away? Now that would be a pretty sight. Here, help me find a piece of chalk so I can draw a diagram." And so he continued:

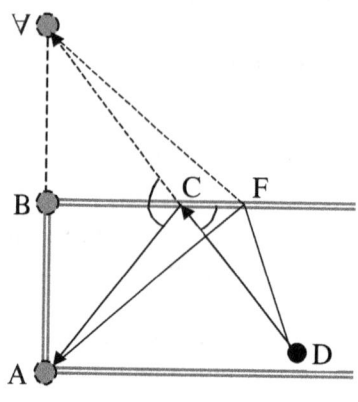

"Go the Other Way"

A straight line on a flat surface is the shortest distance between two points — Euclid knew this. Think of the top cushion on our billiard table as a mirror (line BCF) and reflect pocket A in it. It's a property of mirrors that image (∀) is as far "into" the mirror as the pre-image (A) is "out of" the mirror — Alice knew this (*namely* AB = ∀B). Everything else follows.

Triangles ABC and ∀BC are congruent (math talk for identical), and so AC = ∀C. Then the *straight distance* from D to C to ∀ is precisely the same length as the *reflected distance* D to C to A — and this is as short as it gets. Take any *other* point F on the cushion/mirror and reason similarly, and the crooked distance from D to F to ∀ must be longer than the straight distance from D to C to ∀. *To say it's the reflected distance is to say it's the shortest distance.* With this knowledge, many problems and puzzles are quickly solved. As my logic professor would rattle off, "a bachelor is an unmarried man." By that he meant you were saying the same thing twice — just as we are here with "reflected" and "shortest."

From before the time of Euclid the Greeks knew light reflected from mirrors with equal angles. But it wasn't until 400 years later that Hero of Alexandria showed the reflected distance to be also the shortest distance. This is the first optimum principle of nature.

This "bouncy" behavior isn't an isolated phenomenon of billiard balls and light. This is how basketballs recoil from backboards; how hockey pucks career off rink boards; how baseballs fly off bats, and how your voice echoes from a barn wall.

Back to the billiard table where our two elderly friends are doing some "trick shots." The scientist was determined to impress his friend by a bouncing behavior that generalized the previous one. He challenged the poet with the following question: "If I shoot a ball with no English [i.e. spin] and bounce it off two cushions, what's unique about its path?"

"I can't say," his friend replied.

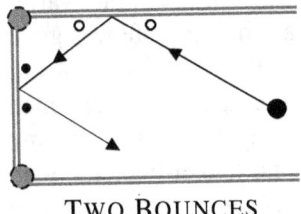

TWO BOUNCES

"Well, as I'm always telling you, take a scientific approach by actually trying a shot." Phemius did and observed that the ball reflected from the second cushion *parallel* to its original path. And a few more shots convinced him this behavior was invariable.

Outside the seniors' residence moonlight sparkled diffusely from freshly fallen snow. By a further generalization of this "bouncing" behavior, the moon was compelled to confess a favorite secret: her exact distance from the earth. Consider how light reflects from three mutually perpendicular walls (*namely* a corner). The diagram below shows how *exit rays are parallel to the entrance rays*. Similarly on a squash court, a ball without spin that strikes three walls comes out of the corner parallel to its entry. (Be careful to view the diagram with the corner pointing away.) A bicycle reflector is just a collection of a hundred or more small "cube-corners," which cause a car's headlights to be reflected directly back to the motorist.

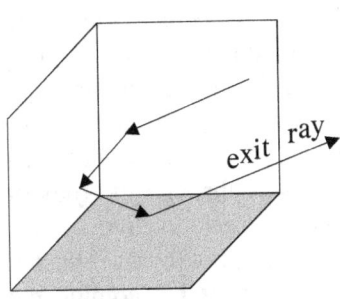

THREE BOUNCES

Four U.S. Apollo missions and one Soviet Lunakhod lunar probe deployed cube-corner reflectors on the moon's surface. When beams of laser light sent from earth bombard these, measurements of the two-way travel time allow extremely accurate calculation of the moon's distance. (The perigee, the shortest distance, is 221,589.648 miles or 356,613.970 km.) All this white magic is possible because cube-corners reflect laser light *straight back* to the sender. Simple, neat, ingenious!

"Epius, where did you learn all these things about reflections?"

"Don't you remember the time we were in Egypt? I was an apprentice in Hero's workshop and you were studying their art."

"I do recall their flat art with its endless symmetry."

"I may have learned my scientific skills from Hero, Phemius, but you taught me all I really needed to know."

"That's not possible. Until now I didn't know any of this."

With a sigh Epius said, "When I left Troy after the war, I went home, married, had children. For a long time life was good — I thought I'd be happy forever. But I was a forceful, uneducated man trained only to bare-knuckle box and assemble rough boards into huts and sheds. Eventually my wife died and my children married and moved away. They no longer bothered with me. Once more I went down to the sea and sailed away, this time to Italy in search of something — I didn't know what. After several years of wandering my mind turned to Odysseus and that fateful night we spent in the Trojan horse together. I wondered if the man who had blessed me was blessed himself with great longevity and health. So I sailed to Ithaca only to be told that he too had become restless in his old age and sailed away. In the words of Tennyson, 'Come, my friends, 'Tis not too late to seek a newer world.' He was never heard of again.

"But I met you! You opened my mind and calmed my personality. Your studiousness, love of poetry and music inspired me to study and deal with ideas. I had found my newer world. You taught me to love learning — all the rest was easy."

In the natural world many nocturnal animals have evolved a "straight-back" concave reflector rather than the cube-corner of science. This reflector — a layer of cells behind the retina called the tapetum lucidum — gives the animal "white-eye" like the wolf and fox below with that caught-in-the-headlights look. Since this cellular layer is absent in humans, we correspondingly get "red-eye" from a network of blood vessels. The purpose of the tapetum is to reflect light back through the rods of the retina a second time thereby giving creatures like wolves and foxes excellent night vision. Each photon is reflected *straight back* to the same photocell that missed it coming the

EASTERN CANADIAN WOLF   RED FOX

Figure 7.3

other way so the image is not distorted. If the photon is missed on this second try, it escapes the eye socket and becomes part of the white-eye phenomenon.

All the previous reflected paths have been off walls or mirrors perpendicular to each other. And as we saw a few pages earlier, reflected paths are also the shortest paths. Let's consider a two-reflection puzzle where the mirrors/walls are *not* at right angles.

A cowboy at point C wishes to feed and water his horse before returning to his tent at point T. Using the diagram below, should he go to the pasture first and then the river, or the river and then the pasture? As an added problem, find the shortest possible distance in which he can do this.

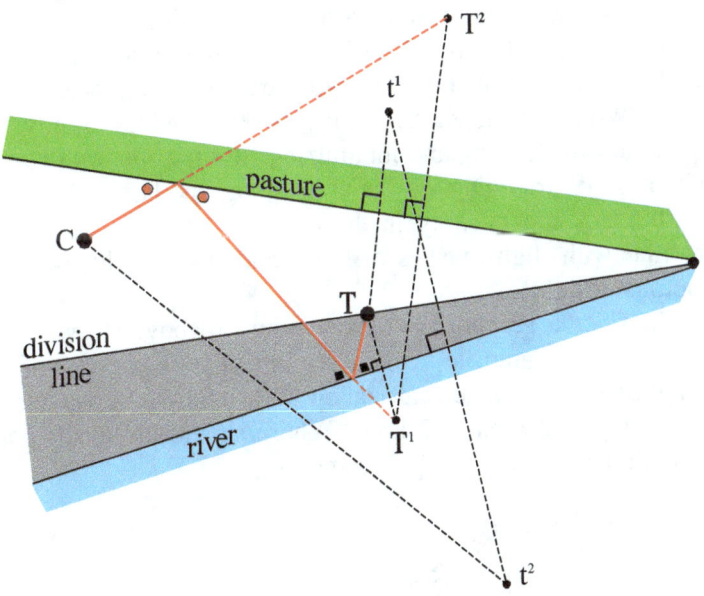

THE COWBOY'S FIRST DILEMMA

Reflect the tent in the river's bank to get T^1, and then reflect this image in the pasture's edge to get T^2. When the order of reflections is reversed, we get images t^1 and t^2. The red zigzag line is the *shortest path;* it's equal to the straight-line distance CT^2. Note the congruent triangles. The other distance (river first and then the pasture) is Ct^2. All his positions can be classified into three groups:

- those nearer to T^2 — white area— (pasture first)
- those nearer to t^2 — gray area — (river first)
- those that are neutral — the division line (either first).

The last group of points forms a straight line between the other two. If the cowboy is in his tent, what he does first is immaterial: the minimum path from T–pasture–river–T is the same distance as T–river–pasture–T; similarly, at the point where the pasture and river intersect. Joining these two points creates the division line of neutral positions.

As with various puzzles, the cowboy's dilemma is solved by the fact that the shortest distance is also the reflected distance — two reflections in this case.

The poor cowboy has an infinite number of longer paths he can follow — the primrose paths — and only one is shortest. But from a broader perspective, this seems an artificial problem involving cowboys, rivers, pastures, reflections, shortest distances, tents, and so on — positively nothing to do with the real world. If you think this way you are mistaken, deeply mistaken. We're enveloped by a quantity that behaves exactly with this kind of reflecting and minimizing. It's the *sine qua non* for life. It's the first thing we see at birth and the last at death. It's the fundamental quantity in the universe, the one we're most familiar with: light, waves of light, photons of light, waves of photons of light.

Consider a second dilemma for our cowboy. He wishes to return to his campsite, which is currently on an island. His steadfast horse Trigger can gallop at 20 mph (32 kph) but can swim only at 2 mph (3.2 kph). Where should he hit the beach to minimize his return trip to camp?

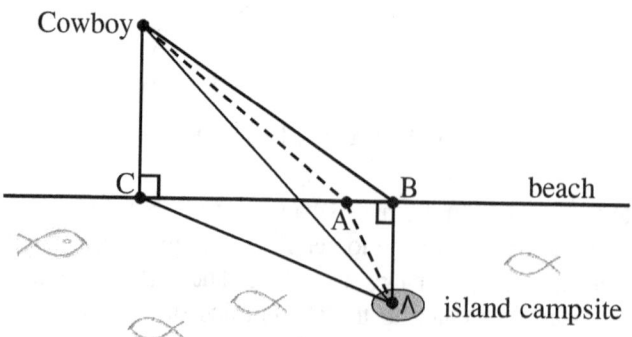

THE COWBOY'S SECOND DILEMMA

The question is ambiguous. Are we to minimize the time or the distance for the return trip? In the above diagram the *least distance* is the straight dotted line. On the other hand the slashed doglegged path, cowboy–beach–island, is the path of *least time*. How is that possible — aren't shortest paths always the fastest paths? No, not always! The worst situation is to go directly to the beach at "C" and have the long swim to the island (path at the left). Going directly to "B" and taking the shortest possible swim to the island is also not the best strategy (path at the right). Why? Because the greater distance and time to "B" over "A" outweighs the *small decrease* in swimming time. By using something called Snell's law[3] and the calculus of variations, the exact point "A" for the minimum time can be located. Accordingly the cowboy has two extremes he can use: the one of least distance and the other of least time. Save for Hollywood and some movie aficionados no one cares what cowboys do. The vital question is which of these two paths does *light* follow?

To me, the answer to this question seems quite unexpected: *light always traverses space in the path of least time*. When the light beam is entirely within a homogeneous medium, then the path is straight. Nevertheless, on every boundary between different mediums — if it doesn't reflect — it obeys Snell's law and refracts (bends). When passing through a series of mediums, the path is especially erratic. Historically, the greatest amateur mathematician of all time, Pierre de Fermat (1601–65), had the initial insight that the common thread among straight, reflected, and refracted paths wasn't the least distance but the least time.

Einstein established that light in a vacuum travels at the maximum speed possible (warp one on *Star Trek*) of about 186,000 mps (300,000 kps). As the density of the medium increases, the speed of light decreases, ultimately going to zero when the medium is opaque. And the greater the difference in density between two mediums, the greater the refraction. This is a commonplace observation. We have all noticed how a spoon appears to bend in a glass of water or a reed in a pool — something that was first investigated by Claudius Ptolemy in AD 140.

All the geometric optics of Hero, Snell, and others are mere skating on a pond and now and then cutting a more important design than ordinary — to paraphrase Newton — while the whole ocean of truth lies under our feet. Immediately below the ice stretches the classical theory of electromagnetic radiation; this in turn is supported at a deeper level by modern quantum mechanics. And Richard Feynman in his book *QED: The Strange Theory of Light and Matter*[4] rests all these strata on the apparent bedrock of quantum electrodynamics.

Forget the ice beneath your feet! Gaze at the starry night arched over your head! The sky both you and the astronomer observe is as unreal as the vision Vincent Van Gogh painted in his most popular work *The Starry Night* (Figure 7.4, below). How can that be? The moonlight entering your eye left the moon's surface 1.2 seconds ago, while the light from our morning star — as Thoreau calls it — takes 8.5 minutes to reach us. And the gentle twinkle from our second-closest star, Proxima Centauri, needs 4.3 light years just to softly touch your retina. Incredibly, that small smudge of light, low in the eastern sky called the Andromeda galaxy, has been journeying at 186,000 miles per second for 2.2 million years. Yet this is only the wading pool of the cosmic ocean. We expect the stars of this mighty galaxy are still burning brightly, but we can't be certain — we haven't *seen it* in over two million years. When we consider the entire luminous universe, our own or Van Gogh's, we're looking forward in space but backward in time. We see the past, never the present. The concept of "now" is unreal, a practical phantom for our everyday convenience.

The Starry Night by Vincent Van Gogh, 1889,
the Museum of Modern Art, New York

Figure 7.4

Something else is here — a refinement on the observation that space and time are interconnected. As we have learned from Fermat, light propagates through the changing mediums of space in the shortest time possible. Thus the starry panorama may be ancient but it's also the *youngest universe we will ever witness*.

Circulating widely in the ancient world, before Hero, before Euclid, was the problem of Queen Dido of Carthage (about 900 BC). Like all legends, it had elements of fact and particles of fiction. She was a Phoenician princess of Tyre driven from home by her evil brother. With a few ships and some supporters, she sailed westward along the coast of North Africa seeking a suitable site for a new city. Eventually arriving at present-day Tunisia, Dido bartered with the local ruler King Jarbas for a parcel of his land. He agreed under one condition: she could only have as much land as could be surrounded with the skin of an ox.

Dido made the most of his challenge. She had her men cut the hide into thin strips and tie these pieces together to form a single rope — possibly 1000 to 2000 yds (914 to 1829 m) long. Now, these Tyrian exiles were people of the sea. (Herodotus reported that some years after Dido, a Carthaginian named Hanno circumnavigated Africa — all this two millennia before Vasco da Gama.) Therefore the territory she was to claim must border on the Mediterranean. So the puzzle was *to maximize* the amount of land you can enclose with the skin of an ox when one side borders on the sea. What is this shape?

Basic mathematical concepts arise from our interactions with the natural world. Most of us have played with string or straw and know that the opening (area) is greatest when circular. As previously mentioned, less complex peoples have traditionally built their huts, teepees, and igloos with a circular base because this maximizes their floor area for a given amount of wall material. A circle of the same perimeter as an equilateral triangle has 65 percent more area.

So it's not a leap of imagination to see Dido instructing her men to stretch the hide in the form of a semicircle along the beach. This is the shape yielding the maximum area from the rawhide rope of some 33 to 132 acres (13 to 53 hectares). Ever after, the Carthaginians lived within the skin of an ox.

Let's return to the cowboy for a related example of this reasoning. Presently our hero is leading a wagon train of homesteaders across the Western plains of America when a war party of Indians appears. What should he do? More by instinct than mathematical ability he maneuvers the covered wagons into a circle. This exposes the least perimeter for the largest area with no weakest point. Recognizing the strength of his tactics, the Indians ride off into the sunset. Other animals also form defensive rings. When attacked by polar bears, wolves, or man, adult musk oxen adopt a protective circle with their calves in the center. Many ancient, medieval, and renaissance cities were circular: more area and less work in building the walls. A few like Cologne, Germany, were actually semicircular.

MEDIEVAL MAP OF COLOGNE ON THE RHINE

Implicit in the previous paragraphs is the assumption that of all figures with a *fixed perimeter*, the circle encompasses the greatest area. (This is also true of Dido's solution because we may picture a mirror image semicircle in the sea.) Although the ancient Greeks tried unsuccessfully to rigorously show this, it wasn't until the 19th century that a proof was found. Mathematicians call this the "isoperimetric problem." Tie a length of string into a loop. Put all your fingers inside the loop to make the largest possible opening. For maximum area of curves with a fixed perimeter, the circle cannot be improved.

But what about figures with straight sides? Consider one such, a quadrilateral (four-sided polygon) with a *fixed perimeter*. Put four fingers inside the loop of string and deform it to have a maximum area with linear sides. What shape do you have? The answer — a square — seems as intuitively obvious as the circle for "curved sides," but considerably easier to prove.

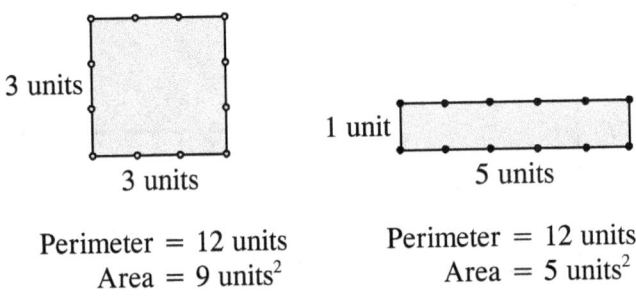

Perimeter = 12 units
Area = 9 units²

Perimeter = 12 units
Area = 5 units²

The above figures have identical perimeters (12 units), but wildly varying areas (9 versus 5 units²). Clearly, if you're building a home with this perimeter, then a square plan provides almost double the floor area for the same quantity of wall material.

Imagine Queen Dido in her quest for land was further restricted to enclosing an area facing the sea with *three straight sides*. By applying the 12 units above, what dimensions should she use to achieve a maximum? The answer isn't immediately obvious. Considering our knowledge of squares, a reasonable first solution might be the following:

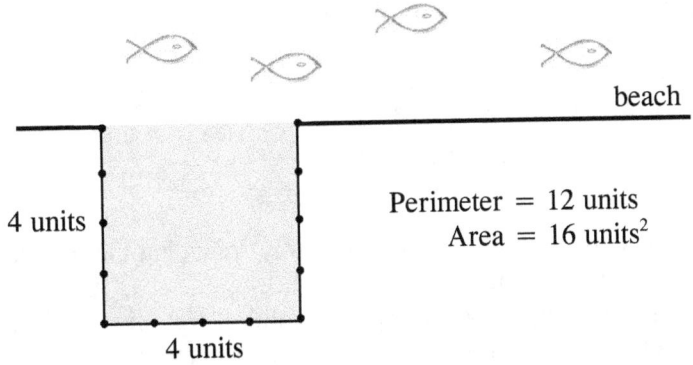

Perimeter = 12 units
Area = 16 units²

Surprisingly, this isn't the maximum area. Here our common-sense intuition fails us. The reader is encouraged to try other sizes for the rectangle, always keeping the perimeter fixed at 12 units. You will almost certainly find a larger area — given that you do, it's still not clear why the above arrangement (4x4 units) isn't optimal. This specific situation is analogous to Dido's semicircle; however, even this approach won't provide a *general* solution.

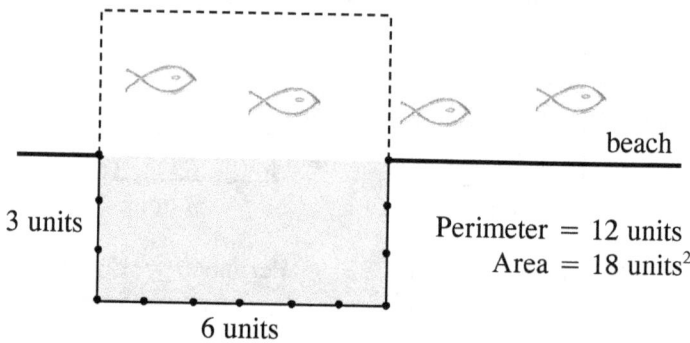

The maximum answer (3x6 or 18 units²) can be rigorously demonstrated with a little calculus. So for the curved and linear figures on the beach, the optimal shape is a semicircle and a hemi-square respectively. Is there anything else unexpected here? Let's look further.

Consider a final example using our 12 units of perimeter (any fixed length would work equally well). After his wagon train adventures, our versatile cowboy has settled down to a farming life; Trigger has become a plow horse. The cowboy/farmer wishes to enclose two bordering fields with his cows in one and his bulls in the other. How should the fields and fencing be arranged for maximum pen size?

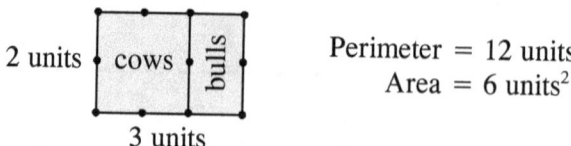

This extra dividing fence was costly, reducing the total area (6 units²) to one third of the enclosure-by-the-sea. Again, this arrangement can be proven with calculus — yet there is a much simpler way. All these examples — the square, the enclosure-by-the-sea, and the two pens — have a common theme, not immediately apparent: *the sum of all the vertical distances equals the sum of all the horizontal distances*. When this is so, the arrangement is optimal, providing a general solution for all objects with straight sides of whatever configuration. And with a little extension, it can be applied to every polygon. Look for yourself. Try a few other possibilities: two, three, or however many dividing fences; two fences in a corner; and so on. As Newton said, "Nature is pleased with simplicity," and so is mathematics.

For surfaces and volumes, extensions can be made into three dimensions. A cube has three pairs of parallel surfaces. Since these are equal to one another in area, the volume they surround is maximal — this situation is analogous to the square. This approach, having sides parallel and/or equal, can be broadened to include all the Platonic solids from Chapter 2. An interesting question arises. If the surface area of each solid is *fixed at 6 units^2*, what are their volumes relative to a single cubic unit? You can see from Figure 7.5 that as the number of faces increases — with the area fixed — the volume also rises. The tetrahedron is 25 percent smaller and the icosahedron, 26 percent larger than the cube. You might have anticipated this since the icosahedron is the most sphere-like of the solids and rolls readily as a die.

Platonic Solid	Number of Faces	Surface Area Units^2	Relative Volume
Tetrahedron	4	6	0.75
Cube	6	6	1.00
Octahedron	8	6	1.07
Dodecahedron	12	6	1.20
Icosahedron	20	6	1.26
⋮	⋮	⋮	⋮
Sphere	infinite	6	1.38

RELATIVE VOLUMES OF THE PLATONIC SOLIDS

Figure 7.5

Consider Archimedes' tomb again with the sphere inscribed in the cylinder (*see* Figure 7.1). Since the cylinder touches the top and bottom as well as the equator of the sphere, its height must equal its width. This implies that of all such containers with a fixed surface area, its volume is, like the sphere's, at maximum. *Generally, the more symmetrical an object, the greater is its volume.* The full Styrofoam cup holding your morning coffee has certain symmetries; by squeezing its top you reduce these and your cup runneth over.

THERE IS AN ECONOMY in nature with everything she does; she gives no second courses, no encores. As Kepler said, "Nature uses as little as possible of anything." Light travels everywhere in the least time even if reflected or refracted. And science has uncovered a wider principle, one that holds Fermat's as a special case — even Newton's laws can be deduced from it.

To gain some sense of this new principle, consider the following. Suppose you have been hiking all afternoon and it's time to go home. If you walk 2 mph for 2 hours, you would carry out twice as much *action* as you would in traveling 2 mph for 1 hour. And this last *action* is double what it takes to walk 1 mph for 1 hour. So, by transference, walking 2 mph for 2 hours requires 4 times as much *action* as walking 1 mph in 1 hour.

Using this common-sense example, *action* is defined as the product of mass, distance, and velocity:

$$action = mass \times velocity \times distance.$$

In our example, the mass is your own. By using the fact that kinetic energy equals ½ times the mass times the velocity squared, the above formula may be rewritten as:

$$action = 2 \times energy \times time.$$

For details, *see* the Chapter Notes.

This formula's power resides in the fact that nature always "chooses" — among an infinite number of possible paths — the path of *least action*. If the time were to increase, the energy would diminish in proportion, or decrease the time at the cost of increasing the energy. On your way home from the hike, you probably followed nature in choosing the path of least resistance, avoiding hills, valleys, and unnecessary detours. Water running downhill follows the path of steepest descent, the quickest way down. Objects in free fall in a gravitational field take the path of minimum *action*. Planets orbiting stars obey the *least-action principle*.

This principle is credited to several scientists and at least three mathematicians, but Pierre-Louis Moreau de Maupertuis (1698–1759) seems to have been the first to express it in the above form. Its use is restricted to objects in motion. Maupertuis said "Nature always minimizes action," and he exploited the generality and economy of the principle to infer the existence of god. Actually, the order was reversed. Because of his devoutness, he looked for such a generality in the assembly of natural laws. With the unifying principle of least action, he believed he had proof of god's handiwork and, hence, existence.

Without some expertise in the area of physics it's difficult to gauge the merit of this idea. The best we can do is turn to an unbiased expert for advice. From Thomas A. Moore in the *Macmillan Encyclopedia of Physics*, we get just that:

> The least-action principle is an assertion about the nature of motion that provides an alternative approach to mechanics completely independent of Newton's laws. Not only does the least-action principle offer a means of formulating classical mechanics that is more flexible and powerful than Newtonian mechanics, variations on the least-action principle have proven useful in general relativity theory, quantum field theory, and particle physics. As a result, this principle lies at the core of much of contemporary physics.[5]

Strong praise for a powerful principle. Keep in mind, however, it *applies best* to the nonliving world in a context of frictionless motion. The lifeless world is a razor's edge that allows no variation — behave this way and no other.

In the world of the living there is an analogous principle, one where *slight variations* are tolerated, while larger ones are punishable by death. In every ecological niche, life hovers around a cluster of particular behaviors and bodily adaptations. The mechanism that brought about these adaptations is the master narrative of the 20th century. The philosopher Daniel Dennett[6] said, "If I were to give an award for the single best idea anyone has ever had, I'd give it to Darwin." He wasn't referring to evolution, which had been discussed since the time of the Greek philosopher Empedocles (492–432 BC). Even Darwin's grandfather, Erasmus, wrote about evolution — an idea as firmly established as heliocentrism. No, Darwin's dangerous idea was the discovery of the *mechanism for evolution* — the force that causes descent with modification. We call it *natural selection*.

All creatures, great and small, dance upon this edge of life. Any variance too far and all is lost. From womb to tomb, this is our passage — no broad primrose path for us. In his lyrical "Ode to the West Wind," Shelley rages against this state compared to the freedom of the wind:

Wild Spirit, which art moving everywhere;
Destroyer and Preserver; hear, oh hear!
. . . .
Oh, lift me as a wave, a leaf, a cloud!
I fall upon the thorns of life! I bleed!
A heavy weight of hours has chain'd and bow'd
One too like thee: tameless, and swift, and proud.

The wolves run on through the evergreen forests in their eternal pursuit of the deer. And for their part, the deer lead the wolves on a deadly chase. Each hones the other to perfection by natural selection. It's not only the weak and the old who falter and fall; it's the inefficient — the ones who stray too far from the edge. To those who do the dance, whether deer or wolf, belongs the day. It's not a good day to die. It never is. And so the wolves run on through the evergreen forests.

EASTERN CANADIAN WOLF, *Canis lupus lycaon*

The Skin on Water and Air

"I AM ESCAPED WITH THE SKIN OF MY TEETH" is an ancient saying, old even when the author of Job 19:20 wrote it down. Job was referring to his wretched body and toothless mouth — meaning he clung to life *by nothing at all*.

In the real world, there are creatures that live less on metaphor, and more on water, something whose "skin" appears to be *nothing at all*. It's a remarkable ecological niche, a one-molecule thick world, existing between sky and water and supporting an entire community of insects. Three of these you're certainly familiar with: water striders, water beetles, and mosquito larvae.

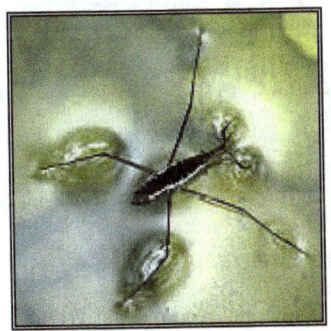
WATER STRIDER

MOSQUITO LARVAE

The first two live *on* and the third lives *under* the "skin." As you can see, the weight of the water strider causes slight indentations in the water's film while the larvae produce shallow dimples.

The English writer Hilaire Belloc whimsically described the water beetle's walking-on-water talents:

> *The water beetle here shall teach*
> *A sermon far beyond your reach:*
> *He flabbergasts the human race*
> *By gliding on the water's face*
> *With ease, celerity, and grace;*
> *But if he ever stopped to think*
> *Of how he did it, he would sink.*

Had the poor water beetle possessed an understanding of molecular physics, it might have reasoned like this: the cohesive forces between water molecules are responsible for this phenomenon of skin on water. The *topmost* molecules don't have other molecules all around them; consequently, they cohere more strongly to those associated with them on the surface and directly underneath. This results in the net inward pull and contraction that we call surface tension.

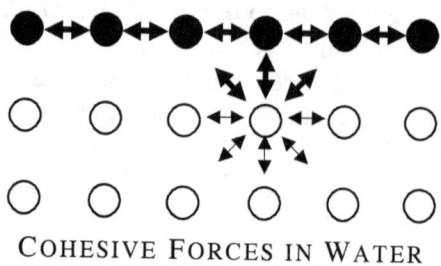

COHESIVE FORCES IN WATER

Knowing this, Belloc could have added the following lines about the whirligig beetle:

> *In circles whirling round and round*
> *The secret never would be found.*
> *But knowledge of atomic structure*
> *Explains the lack of water rupture.*

Surface tension is responsible for a wide variety of effects, some natural, some not. With care it's possible to float a needle on water as well as fill a glass beyond its volume. And the common phenomenon of capillary action is a direct result of this force. Tent materials are rainproof because surface tension bridges the pores of the finely woven material. Touch the tent and break the skin, however, and you'll learn why it's a dog's life in a pup tent. We often wet loose threads, bushy paintbrushes, or frizzy hair, to induce cohesion among the wayward strands. Conversely, soap and detergents lower surface tension so the solution can creep into pores and cracks, thereby removing dirt. By agitating the molecules, heat also lowers the surface tension.

At the same temperature different fluids have widely varying surface tensions — mercury's is seven times that of water, even though for equal volumes the metal weighs almost 14 times as much as water. But because of its great surface tension (*see* Figure 7.6), it maintains a more spherical shape; this form gave rise to its other name, quicksilver — it rolls so fast.

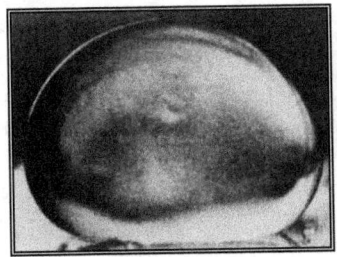 

DROP OF MERCURY   DROP OF WATER

Figure 7.6

For every resting drop, the forces downward (gravity) and inward (surface tension) are in equilibrium. Like the two examples above, stable equilibrium states are defined as having less potential energy than any other nearby state. Reasoning by Maupertuis' principle for these two drops, the process of assuming their present states was accomplished with the least possible *action* (mass x velocity x distance = 2 x energy x time). Furthermore, *the surface area of each is minimal*. If it were otherwise, the forces would be unbalanced and therefore in motion.

"I'M FOREVER BLOWING BUBBLES, pretty bubbles in the air" — are the first two lines of a familiar childhood song. Everyone has blown bubbles, marveled at their brilliant colors and admired their simple forms. I suspect that early humans, bathing in frothy pools, did so millennia ago. Soap film and bubbles have long been beloved by children, mathematicians, and artists. Lest they keep all the fun to themselves, let's see what these *skin on air* objects have to offer.

On the large scale they're simple things. No concatenation and confluence of a thousand forces here, just two: gravity and surface tension. And for small bubbles and film the effects of gravity are negligible. This is a rare situation with just a single, dominant force doing all the work. Furthermore, since the surface tension has the overriding quality of minimizing itself, this property will prove extremely useful — so useful that this nightingale sings a powerful and completely new song cycle, solving previously intractable problems.

A Boy Blowing a Soap Bubble by Jean Chardin, 1739, National Gallery of Art, Washington, DC

At least five prominent artists have employed bubbles as subject material: Chardin, Hamilton, Manet, Millais, and Murillo. The dreamy depiction of *A Boy Blowing a Soap Bubble* by Chardin (left) exemplifies this genre. We, like the peeking child, wait for the bubble to burst at the moment nearest maximum size and the scene to dissolve, but it never does. Mathematicians, as we will discover, see soap film and bubbles differently than artists do. Both views have their merits.

The defining characteristic of science is skepticism: to consider whatever conclusion you have arrived at as tentative and possibly erroneous. Similarly for the conclusions of others be they Newton, Galileo, or Einstein. Scientists have taken as their credo the well-known words of Oliver Cromwell: "I beseech you, in the bowels of Christ, think it possible you may be mistaken." Of all human communities, only the scientific holds this principle paramount. It's the fountain of progress. Virtually everything biologists believed in 1850 has been proven false, yet the company of biologists flourishes.

Contrast this with other human institutions. After the embarrassing trial of Galileo in 1633 for just discussing the possibility of heliocentrism in his *Dialogue on the Great World Systems,* the Catholic Church shrewdly avoided a similar response to Darwin's dangerous idea. But for Galileo, who spent the final years of his life blind and under house arrest, forgiveness was 360 years in coming. And that only grudgingly in 1992 after 13 more years of evaluating his trial and "heresy" — all this for being right, or as right as anything ever is in this world.

Yes, a few scientists have exhibited the same dogmatic behavior as the clergy, but this is not the norm, not the proven path to improvement. Return to the credo of science. How do you discover errors in hypotheses, theories, and laws? You do a well-defined, controlled experiment. So now we wish to test the truth of the statement that *soap bubbles and film always minimize their surface area.*

Bubbles first: Take a solution of water, soap, and a little glycerin, and blow a bubble of air (*namely* a fixed volume). As we know and mathematicians have proven, a sphere has the least surface area for a fixed volume (*see* Figure 7.5). The slightest deformation of the sphere and its volume decreases — this is why an orange squirts when squeezed.

Film second: Take the wire ring on the left below, and dip it into your soap solution. After withdrawing it, observe the flat sheet of solution covering the whole interior. Notice also a thread with a loop of fixed length caught in the film. Now for the white magic! Take a *dry* finger and carefully touch inside the thread's loop. Voilà, you get the picture on the right below. The soap film is pulling on the loop of thread like firemen on a net — outward in all directions. So the hole in the film is as large as possible, thereby allowing the film to be as small in area as possible. Every maximum implies a minimum of another sort, and vice versa.

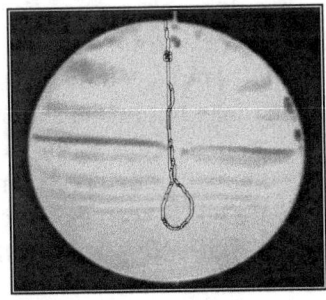 

SOAP FILM   MINIMUM SURFACE

Figure 7.7

The loop of thread is easily pushed around inside the outer wire ring because this doesn't change the area. However, you can't deform the loop; any such action will increase the area of the soap film, and surface tension (or surface energy) will quickly restore the *circular* hole. Here practice supports theory.

The film's iridescent colors result from interference patterns like rainbows from a prism. The film is thinner toward the top, as with ancient cathedral windows, and eventually vanishes as gravity drains the water away. Then the film turns black and will be only millionths of an inch thick — perhaps the thinnest object any of us has ever seen.

Another example to consider involves finding the shortest distance between towns. The least distance between any two towns is straight — the way the Romans built roads. What about three towns? Mathematicians call this, and its generalization to an arbitrary number of towns, the "expressway problem." Here are three possible routes connecting towns A, B, and C.

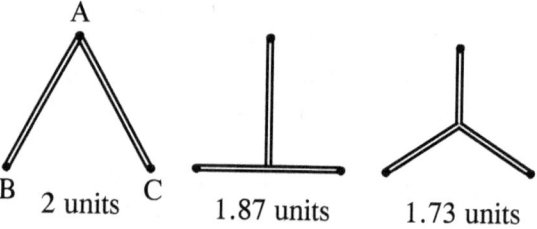

JOINING THREE TOWNS

With a little arithmetic it's possible to calculate the total length of roadway for each, as I've done above in the special case of towns equally spaced one unit (mile or kilometer) apart. In the real world, this simplification would rarely occur. As we shall see, however, our soap solution works for all towns in whatever configurations and separation distances you may wish.

The shortest route above (1.73 units) may not be the shortest possible. With their area-minimizing property, this is an ideal task for soap film to solve. Place three thumbtacks, points down, between two sheets of Plexiglas, and lower the whole apparatus into our soapy mixture. The film will adhere to the tacks (towns) and join them with the *least possible film* — the answer you want. Withdraw from the mixture! *Look! See!*

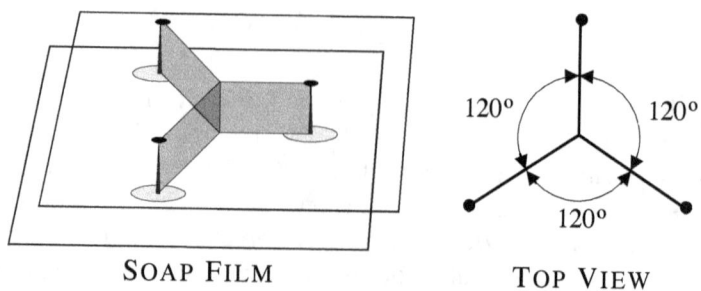

SOAP FILM   TOP VIEW

THE THREE-TOWN SOLUTION

Figure 7.8

The top (cross-sectional) view shows 120° angles. The lines are straight; the angles, exact; the results, surprising. Mother Nature is seldom so clear with her plans, or this revealing with her clothing. Nevertheless, it's in her character that three-way junctions *always meet* at 120°. This is the angle that minimizes — a favorite of hers. Whether in the bubbles of babbling brooks or is the froth of ocean foams, this is how all triple junctions meet under the influence of surface tension.

Moreover this pattern isn't restricted to the world of water. Consider homogeneous elastic materials like mud and rock! Under stress they shatter suddenly at triple nodes and open up at 120° angles as in the photographs below. This allows them to form the minimum cracking. Plainly, the forces here aren't surface tension, but the outcomes are identical. Elasticity in mud is influenced by different drying rates and water levels. And when substances are neither homogeneous nor elastic they fracture in other ways.

CRACKS IN MUD   CRACKS IN STONE

Inelastic materials like pottery glazes (next page) crack sequentially and at right angles to each other — as does mud when various parts dry at different rates. The cracks release all the tension in a single direction parallel to themselves, and so any new stress must run in the opposite direction — perpendicularly. Nonetheless, within this matrix, areas isolated by stress lines at right angles still show mostly 120° cracks. A near perfect example of this (also on the next page) is found in the wings of dragonflies — one of earth's most ancient insects. Note the short veins that meet the straight rib lines perpendicularly while at all other vein junctions the angles are 120°.

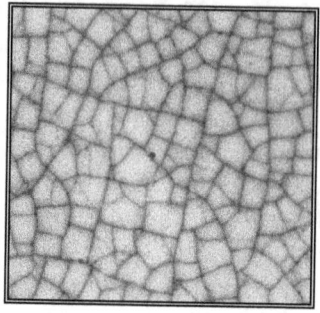

CRACKED GLAZE
ON POTTERY

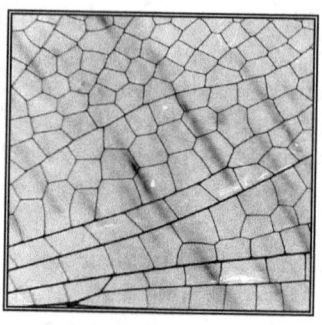

DETAIL OF A
DRAGONFLY WING

Let's expand our previous expressway puzzle (also called Steiner's problem after Jacob Steiner, a 19th century Swiss geometer) by introducing a fourth town. Arrange these on the vertices of a square, one unit per side. Again, the challenge is to join all four, A, B, C, and D, with the minimum length of roadway. We have more choices than before. Here are a few — perhaps the reader can think of others. Beneath each, I've shown their relative lengths. Note that the pair on the right differ by only 3½ percent.

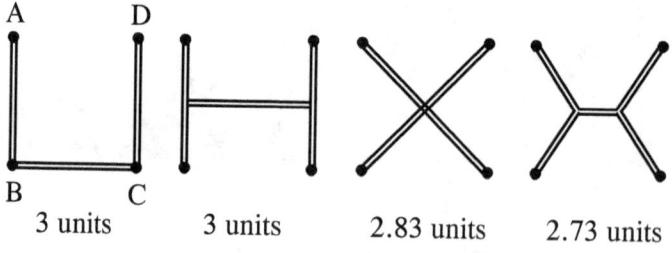

JOINING FOUR TOWNS

Again, the minimal solution noted above (2.73 units) may not be the best possible. With some calculus, the minimum answer can be found, but for us there is a better and quicker way: our magic elixir. Introduce a fourth thumbtack into our soapy sandwich and arrange each at a vertex of a square — although *any* quadrilateral will suffice. Immerse the whole apparatus into the mixture. The film will adhere to the tacks (towns) and join them with the least possible film — the minimum roadway. Look and see!

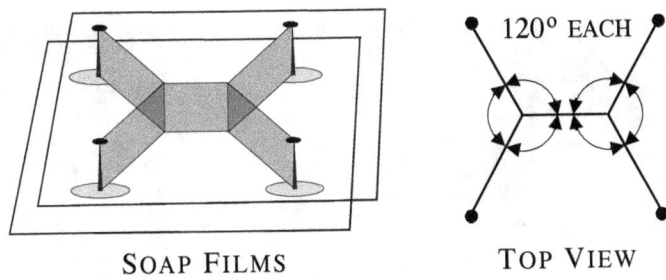

SOAP FILMS   TOP VIEW

THE FOUR-TOWN SOLUTION

Figure 7.9

The minimum solution is as unexpected as it is familiar and consistent with the three-town answer. You could lay a protractor on the Plexiglas and check the angles; they're exact, and the soap film is straight. Since the tacks are at the vertices of a square, the entire network could have formed at right angles to its present orientation without affecting the film's area. More significant, hidden here is a larger pattern, and as we shall see, a well-known one, almost an old friend. A brief look at the five-town pattern will begin to make this clear.

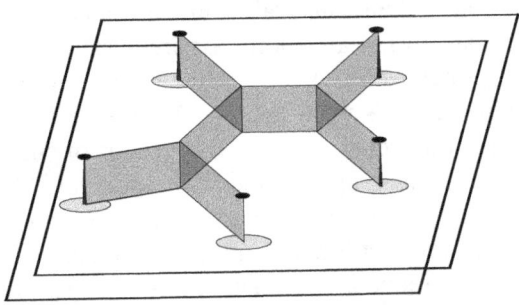

THE FIVE-TOWN SOLUTION

Figure 7.10

The five tacks are joined by the smallest possible film surface, creating three new points. All the angles around these new points are precisely 120°. Initially, upon removing this tack sandwich from the soap mixture, the configuration — due to gravity — may be distorted. But as the solution drains away, the above *minimum* arrangement appears. To see the whole pattern, rather than more pieces and parts, turn the page.

HONEYCOMB

Behold the humble honeycomb! To make beeswax is "expensive" and time-consuming. Over the millennia those bees who have used it *most efficiently* have had an evolutionary advantage, hence their survival. Other advantages occur with this design that are not as readily apparent as economy. Structurally, hexagonal shaped cells with their shared walls are more rigid than circular cells with their stand-alone design.

This formation occurs widely when groups of similar objects pack together as close as possible. Consider your dinner plate spread with a large number of warm peas. Press them together and the honeycomb pattern with its 120° angles emerges. This design appears frequently in nature. We see it in the seed arrangement of sunflowers, the hexagonal packing of royal tern nests, the rods in retinas, and the compound eyes of insects.

 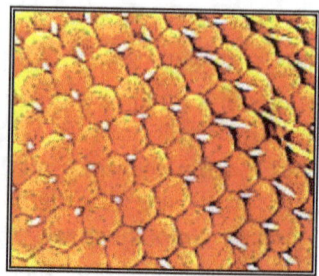

PEAS ON YOUR PLATE   INSECT COMPOUND EYE

Humans have always been intrigued by the geometry of the honeycomb. Many have attributed its design directly to a deity. Consider a traditional European instance from the writings of Sir Thomas Browne (1605–82):

> ... indeed what reason may not go to Schoole to the wisdome of Bees, Aunts, and Spiders? what wise hand teacheth them to doe what reason cannot teach us? ... but in these narrow Engines there is more curious Mathematicks, and the civilitie of these little Citizens more neatly sets forth the wisedome of their Maker.

Numerous arguments have been put forward "to prove" the existence of god. Amusingly, the Catholic Church laid down a dogma declaring that existence of the Deity can be proved by pure reason — it would have been more to the point just to do it! Nonetheless, there's a favorite among these arguments: prod any theist and he'll soon trot out *the argument from design in nature* — Sir Thomas Browne's assumption.

The Anglican clergyman William Paley (1743–1805) set forth this classic argument for the existence of god in his book *Natural Theology*. When Charles Darwin was a young man attending the University of Edinburgh, Paley's works were required reading. And scientists much admired the cleric's careful reasoning. Paley begins his book with a famous passage:

> In crossing a heath, suppose I pitched my foot against a *stone,* and were asked how the stone came to be there; I might possibly answer, that, for anything I knew to the contrary, it had lain there for ever: nor would it perhaps be very easy to show the absurdity of this answer. But suppose I had found a *watch* upon the ground, and it should be inquired how the watch happened to be in that place; I should hardly think of the answer which I had before given, that for anything I knew, the watch might have always been there.[7]

Using this opening statement, Paley reasons by analogy: unlike a stone, watches are designed for a specific purpose — to tell the time. No part of the mechanism may be changed without affecting the whole, whereas any part of the stone could be removed without harm. Visibly, the watch shows every element of careful, thoughtful purpose, and design. Hence, the watch implies a watchmaker or makers.

The human eye is many orders of magnitude more complicated and wondrous than any watch. Therefore the eye must imply an eyemaker. (To avoid impinging on his monotheism Paley shuns the plural term "eyemakers.") And we call this eyemaker, god. Q.E.D.

Paley published *Natural Theology* in 1802. For its time, and in the absence of cogent alternative explanations of how the human eye — or any other eye — came into existence, his was a believable argument. Darwin himself fretted long over the eye's evolution; even his wife Emma questioned him on it.

But after 1859 and the publication of *The Origin of Species* and *The Descent of Man,* it was no longer possible for any thoughtful human to accept Paley's reasoning. The eye evolved from other, usually simpler, eyes. In the valley of the blind the man with a thousandth of an eye is king. And in the arc light of Darwin's dangerous idea — natural selection — the argument from design was dead. A real watchmaker has foresight, planning for things to come. Natural selection has no purpose, no foresight, no blueprint, no sight at all. None whatsoever! Everything is done for the here and now. As Richard Dawkins[8] wrote, "If it can be said to play the role of watchmaker in nature, it's the *blind* watchmaker." Only those modern oxymora, scientific creationists, cling to such fragile hand straps as the argument from design while their train plunges into a darkening tunnel. Darwin has led us on a long night's journey into day where we can dance in the light to celebrate every living creature's evolution. We are neither ascended from apes nor descended from angels. But ours is a heroic past!

AFTER THIS DETOUR, let's go back to the bubbles. To this point, we have encountered two maximum/minimum ideas of importance:

- the isoperimetric problem — the circle enclosing the maximum area with the minimum perimeter (Figure 7.7)
- the expressway problem for three points — the soap film joining these with the minimum length of roadway (Figure 7.8).

Everyone enjoys discovering connections between apparently unrelated phenomena. In literature, we call these connections parallelisms; in music, themes; in painting, designs; in psychology, archetypes; and in science, theories. The question is, are the isoperimetric and expressway problems related? The answer is yes, and the connection is straightforward:

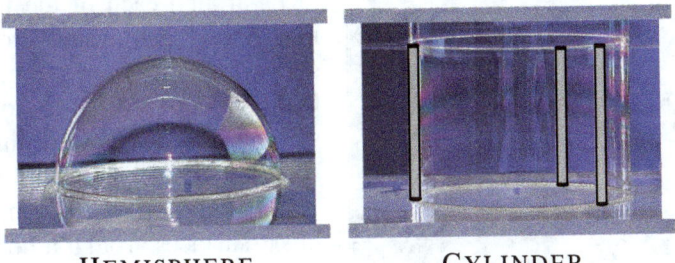

HEMISPHERE   CYLINDER

Blow a soap bubble on a moist glass plate; note the perfect hemisphere formed — the kind you see around your kitchen sink and bathtub. As you blow more air into the bubble, the half-sphere configuration is maintained. A second wetted plate lies parallel to the first and above it. At the instant the ever-enlarging hemisphere touches the second plate, it morphs into a cylinder meeting both plates perpendicularly. Notice also that the three pegs separating the parallel plates are *in* the cylinder's wall. Viewed from above or below, the pegs are *in* the circle's circumference. And as we know, the circle is the solution to the isoperimetric problem.

With a straw, slowly extract all the air out of the bubble — viewed from above, you appear to be removing the circle's area. This disc will continuously deform through a series of outlines until all the air is gone and it becomes the solution to the expressway problem. This is their connection.

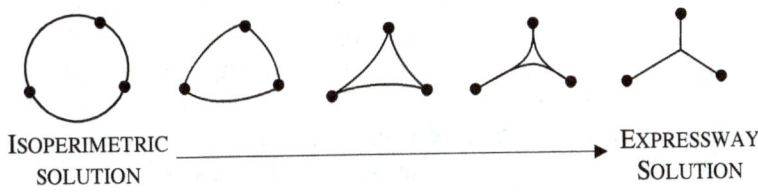

ISOPERIMETRIC SOLUTION   ⟶   EXPRESSWAY SOLUTION

TOP VIEW OF THE CYLINDER

In this book's second chapter "The Perfect Five" we explored many — but not all — the fascinations of the Platonic solids. Take the simplest of these, the tetrahedron with its six edges, and immerse it in the soap solution. Slowly withdraw the wire frame now connected by shifting iridescent surfaces, which are themselves joined by four straight lines in the middle of the tetrahedron, making angles of 109.5° all around. Whither our 120° angles you ask? They're here between the faces around each of the four new lines. Nature isn't prodigal with these soapy angles. Consider her short inventory: she displays 120° when three surfaces meet around a line; 109.5° when four lines meet in six surfaces around a point; and 90° when a soap film touches a hard surface (*see previous page*). In every pool, in all babbling brooks, in your beer stein, only these three angles exist. Lewis Carroll's statement about the five regular solids applies with greater force to these ephemeral angles: they are "provokingly few in number."

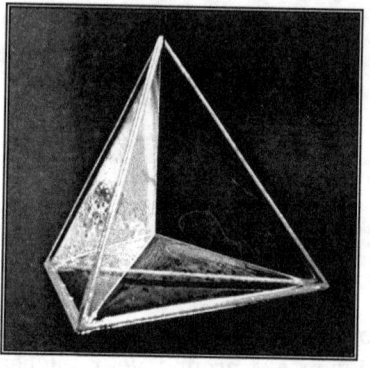

TETRAHEDRAL SOAP FILM

The angles may be few, but they're ubiquitous in nature's workshop. Consider the methane molecule, simplest of the alkanes, a series of organic compounds of H and C (hydrogen and carbon). Ethane, propane, and octane are other, progressively more complex family members. The shape of the methane molecule is exactly that of the above soap film in a tetrahedron, right down to the 109.5° angles between the atomic bonds. The electrons on the hydrogen atoms try to stay away from each other, thus maintaining the lowest possible energy level. Sounds familiar?

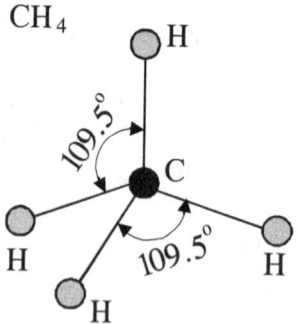

METHANE MOLECULE

Methane, commonly called marsh gas, can be seen bubbling up from decomposing organic material on the bottom of ponds and swamps. It's lighter than air, and floats in a pale eerie haze until blown away by a strong wind. When ignited by man or nature, it burns with a mysterious blue flame. This is probably the source of unexplained lights in swampy areas, strange apparitions, and many UFO reports.

Of course any wire frame arrangement can be immersed in our soap solution. The iridescent configurations seen after withdrawal are always beautiful and often striking. The straight lines, the exact angles, the sculpted structure all clash with our previous experiences, like Chardin's dreamy painting of *A Boy Blowing a Soap Bubble*.

As well as exploring the tetrahedron in the second chapter, we investigated the tesseract, the four-dimensional cube. Analogy and unfolding were our guides, but there is a second path to understanding the hypercube: analogy and shadow.

Using shadows to derive inferences is an ancient tactic. Aristotle (384–322 BC) proposed a spherical earth on the basis that the shadow of the earth on the moon during a lunar eclipse is round, since only a sphere *always casts* such an outline. By drawing an object you reduce the dimensions from three to two and in a sense depict its shadow. Early rock artists did exactly that:

Shadow Art of the San People, Kalahari Desert

With analogy as our guide, we will climb from two, to three, to four dimensions. First let's project the shadow of a wire-frame cube on a plane. This is an alternative way for Flatlanders to get an understanding of the properties of a three-dimensional cube. It has 8 vertices, 12 edges, and 6 faces, and so does its shadow although with much distortion. Our friend Mr. T. Square is amazed to see this just when he was beginning to comprehend the folded-down cube. The volume is missing in the projection, and we see two squares one inside the other connected vertex to vertex (*see* Figure 7.11), and with the remaining "squares" lodged between as four trapezoids.

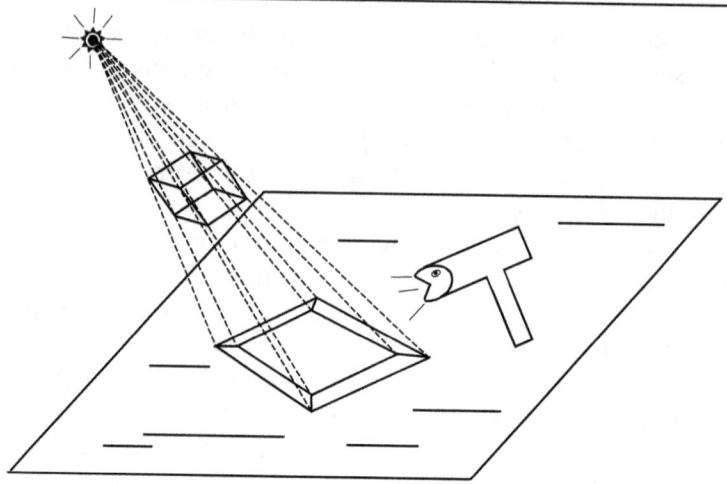

SHADOW OF A CUBE PROJECTED ON A PLANE

Figure 7.11

If three-dimensional objects cast two-dimensional shadows, then it's reasonable to infer that four-dimensional figures cast three-dimensional "shadows." What would the shadow of a hypercube look like? What are its various parts? The latter question is answered by checking the last row of the following table; note that similar to the cube, the hypervolume is missing. Yet the answer to the first question — how these many parts fit together — is far from obvious. We're in the analogous position to Mr. T. Square.

	No. of Points	No. of Edges	No. of Faces	No. of Cubes	No. of Hyper-cubes
Cube	8	12	6	1	0
Cube's Projection	8	12	6	0	0
Hypercube	16	32	24	8	1
Hypercube's Projection	16	32	24	8	0

PROJECTION PARTS

Figure 7.12

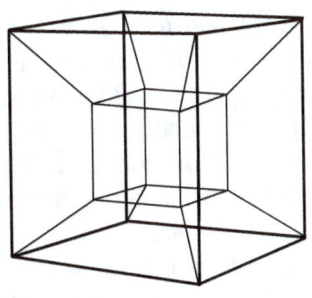 

SHADOW OF A TESSERACT   SOAP FILM IN A CUBE

Figure 7.13

Who knows what complications lurk in the hypercube? The Shadow knows. He knows how to assemble all the bits and pieces to construct the exact projection. His image (above left) has two obvious cubes joined at the vertices: one exterior, one interior. Let's count the others! Jutting inward from each large face of the outer cube to each small face of the inner are the remaining six "cubes." Granted, each is deformed into a frustum (truncated pyramid) just as the squares in Figure 7.11 were deformed into trapezoids. All the points, edges, and faces are easily enumerated and correct — you can check this with the table on the previous page. We have met these eight cubes before in the folded-down version of the tesseract and in Salvador Dali's *Corpus Hypercubus*.

The Shadow doesn't know the soap film figure (above right) with its shimmering surfaces, straight edges, and remarkable central cube that fascinates most people. Nature has such beauty, such peculiarity in mundane objects; wonders lie in the small things of creation. As Thoreau wrote, "Only that day dawns to which we are awake."

Take a wire-framed cube, submerge it in our ever-present soap solution, then carefully withdraw. If a bubble of air has been trapped, the above figure results. If not, a minimum surface will connect all the edges with a square at the center. Now take a straw and blow a bubble on this square — magically the cubical bubble appears.

Some of nature's smallest and most exquisite creations are the radiolaria we investigated in Chapter 2. The next drawing is from Ernst Haeckel's *Monograph of the Challenger Radiolaria*,

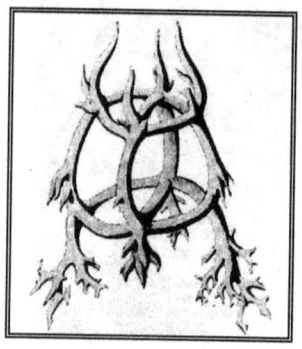

Lithocubus geometricus

and it shows the microscopic skeleton of *L. geometricus*. In life it was a small mass of protoplasm surrounded by a soap-like froth of cells. In the interfaces of the froth silica-bearing fluid tended to aggregate resulting in a figure bearing a striking resemblance to the soap film in a cube of Figure 7.13. If a bubble were blown into the center of the soap film in a tetrahedron, a related radiolarian skeleton could be found. At the bottom of the ocean Haeckel discovered an enormous radiolarian ossuary with innumerable forms of the fallen.

A PUZZLE AND A PARADOX exist here. The puzzle first: why does the central "cube" in Figure 7.13 have *curved edges*? Recall that nature allows only *two angles* between films and lines and *one* between film and hard surfaces. At the corners on the interior cube, four curved edges meet in six surfaces — count them. Hence, all the angles around that point must be 109.5°. (To determine an angle when an edge or edges are curved, draw the tangent at the point on the rounded side.) The curving is to adjust the "cube" so that its angles are 109.5° rather than 90°. Nature is a martinet!

Now the paradox: although I earlier wrote that *only* gravity and surface tension influence bubbles, that's not quite the whole story. Within every bubble is a third force, *air pressure*. We all know that if you blow up a balloon and then release its neck, it will career around the room like a drunken housefly. This results from the escaping air. The question is, which would squeeze the air out more quickly, a large or a small balloon (bubble)? We'll do something very Western, very scientific — an experiment to determine the correct answer.

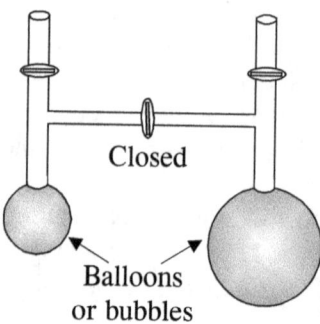

Closed

Balloons or bubbles

Consider the apparatus at the left: two vertical pipes each with a valve and a crossover pipe with a valve. Through the vertical pipes blow up balloons or bubbles of different sizes and shut all the valves, as in the diagram. We are to anticipate what will occur when the central valve is opened and the air is allowed to flow. Four outcomes are possible:

- All the air will flow from the larger bubble to the smaller.
- All the air will flow from the smaller into the larger.
- They'll equalize in size.
- Nothing will occur.

Open the crossover valve. Voilà, the small bubble empties into the large bubble! So the second choice is the correct one. Therefore the pressure must be greater in a small bubble than in a large one. Most people are surprised by this outcome; some find it paradoxical.

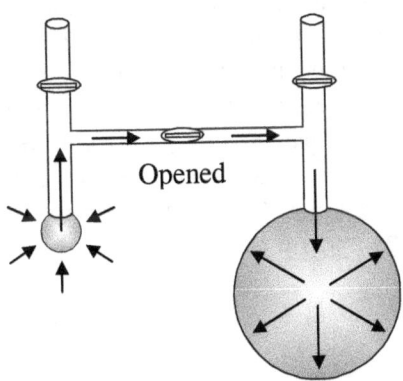
Opened

The child in each of us feels that the bigger of two similar objects is the more powerful, the more grand, and the more significant. Every language has this propensity built right into the meaning of its words. For example, "grand" implies both "large" and "powerful." Animals have always known this. Confront a barred owl in the north woods and it spreads its broad wings and raises them vertically while fluffing out its feathers to appear more than twice its normal size. Hikers can use this strategy to their advantage when unexpectedly meeting a black bear. By opening their jackets and spreading their arms, they appear larger and more menacing to the bear. This is the psychological reason why we feel the outcome with the bubbles is paradoxical. But humans and animals can be mistaken: the smallest creatures among us are often the most deadly, for example, viruses. It's not the meek who shall inherit the earth; it's the microscopic.

Let's look at the pressure more closely so that paradox is converted to understanding. When a bubble's radius is doubled, its pressure is halved. Since the soap film is always stretched with the same force — whatever the bubble's dimensions — the pressure must depend exclusively on the *curvature of the bubble*. The larger the bubble, the less it's curved. The earth is so large that it took humanity eons to discover that it even had a curvature. Think about a trampoline. The greater its deformation from flat, the greater its rebounding force (pressure). And if you wish an arrow to go further, you pull the cord of the bow back further and so increase its curvature; similarly for a slingshot. Mathematicians say the pressure (P) in a soap bubble is inversely proportional to its radius (R), meaning as one rises the other falls, and vice versa. Symbolically we have:

$$P \text{ is proportional to } \frac{1}{R}$$

A single bubble floats iridescent and brilliant — as short-lived as a shooting star. Two bubbles caught by a zephyr gently touch. They don't form a single volume; rather they share a wall or partition, which is part of a third much larger sphere (L). Nonetheless the surface area is minimized. The smaller bubble (S), because of its higher pressure, pushes the partition *into* the medium sphere (M) until they're in *equilibrium*. By using the formula above, examining the figure below, and realizing that the pressure to the left balances the pressure to the right, we reach the following general formula:

$$P_S = P_M + P_L$$
$$\frac{1}{S} = \frac{1}{M} + \frac{1}{L}$$

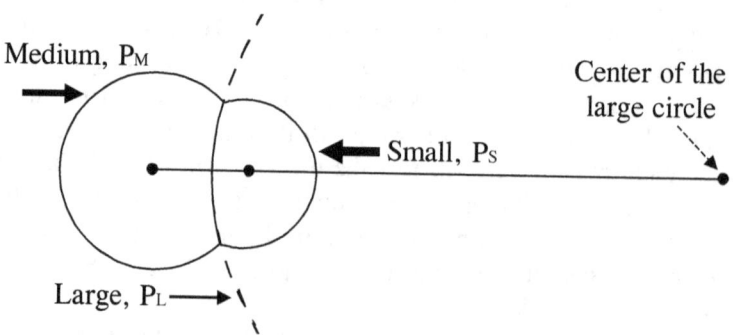

TWO TOUCHING BUBBLES AND A GHOSTLY THIRD

Figure 7.14

(a) Two Unequal Bubbles

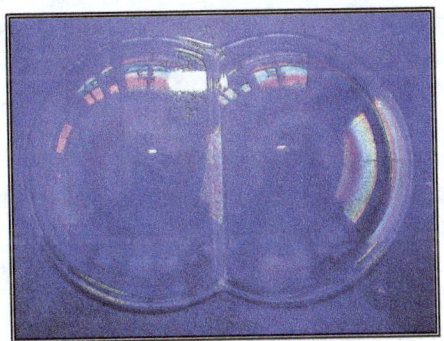

(b) Two Equal Bubbles

(c) Three Bubbles Forming
the 120° Junction

BUBBLE–GALLERY TRIPTYCH

Figure 7.15

A little glycerin added to the soap solution will help seal minor holes, and when protected from air currents and particles, a bubble can last for months. The water seeps from between the inner and outer film; the iridescent colors fade; the sphere becomes black and theoretically only two molecules thick. If a hole starts, usually at the top, the bubble doesn't explode like a balloon or a bomb; rather the opening progresses as a circle over the entire sphere, winking out at the point antipodal to its start.

An entire geometry resides in soap bubbles, one that Euclid would have enjoyed, and an algebra that Descartes would have celebrated. Most of it was discovered by the Belgian scientist Joseph Plateau whose great work *Statique Expérimentale et Théorique des Liquides* was published 30 years after he became permanently blind. He did the geometry of the bubbles with the help of sighted assistants, but he discovered the algebra himself. Consider the following example.

Draw three dotted tangents at the top point where the bubbles intersect. Every tangent makes a right angle to its related radius, as well as forming 120° angles to each other. From this it's possible to deduce that the angles marked with the tiny solid triangles are both 60°. You could draw the following diagram on a piece of Plexiglas and by judiciously blowing the right-size bubbles observe that theory and reality harmonize flawlessly.

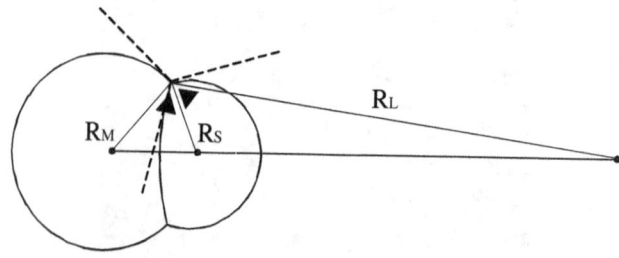

TWO TOUCHING BUBBLES IMPLY TWO EQUAL ANGLES

Figure 7.16

We're standing in the anteroom to Plateau's mansion, but we have no more time to visit here. There are new houses and lands to explore. One backward glance and we'll be gone. Any close inspection of a soap film or bubble reveals a world in disarray with molecules in pandemonium. Yet out of this develop the most incredible and beautiful forms, ubiquitous in time and space. Peter S. Stevens in his wonderful book *Patterns in Nature* summarizes the mathematics of soap bubbles whether in your bowl, bathtub, or beer:

> You can thus see algebraically, as well as geometrically, that nothing is left to chance in the joining of the bubbles. Every junction, every curvature, and the length of every partition is determined absolutely by the [seemingly] random motions of the molecules elbowing their way into the centers of the film.[9]

THERE IS ONE ROAD in the natural world, but two travelers: the lifeless that paves the path and the living that dances along the way. They're not unrelated. The lifeless world doesn't exist *on* an edge — *it is the edge* — straight and narrow. In this world everything is done with the greatest economy to create the most parsimonious universe, whether it's light traveling in the shortest time or, more generally, Maupertuis' least-action principle. As Newton wrote, "Nature is pleased with simplicity," the simplicity of doing with less. We discovered this economy in planetary paths, cracking rocks, molecular angles, and the froth and foam of running brooks. Strikingly, in *As You Like It* (Act II, Scene 1), Shakespeare wrote almost the same thing:

> *And this our life exempt from public haunt*
> *Finds tongues in trees, books in the running brooks,*
> *Sermons in stones, and good in everything.*

All creatures great and small of the living world must practice an economy, whose purpose is to stay alive and pass on genes. We have seen this in Reynard's double printing, the bumblebee's dedication to the work ethic, and the reflective layer on the retinas of nocturnal mammals. At a deep level, chemistry and physics make the rules that life dances to. If your niche is the water's skin, you dare not gain too much weight, lose too many legs, or sharpen your nails. Charles Darwin wrote the book on the innumerable minor rules of survival and the one supreme principle, *natural selection*.

REYNARD ON THE HUNT

IT'S LATER NOW, the sun is up, but a light snow is beginning to fall while Reynard hunts on. As one of his many survival strategies, he had previously cached a red squirrel; it made a nice snack and has fueled his body. Like the lemmings that his arctic cousins devour in great numbers, the meadow voles in this field of wisteria go through high and low cycles, and this is a low year. Nevertheless, Reynard hears a subnivean vole seven body lengths away; he creeps closer, but not so close as to alert the ever-vigilant prey. The fox and the vole have been playing their evolutionary game for eons. Each has helped to shape what the other has become, and they are still engaged in this evolutionary contest, a tournament worthy of the mettle of both.

Here's a dilemma for Reynard. He must eat voles to live, yet he can't approach them because the least sound — the snap of a twig, the crackle of a leaf, or the crunch of compressed snow — sends the creature scurrying away into its network of tunnels. The solution: do a standing long jump from as far away as possible. He must launch himself like a rocket, bullet, or football; the physics is the same, discovered by Galileo and explained in his famous *Dialogue on the Great World Systems*. Just five factors determine the range (distance) of any projectile whatsoever, and on this planet, we have no control over one of these — gravity. So the remaining four decide how far foxes can jump, and within these, Reynard practices great art and science:

- The weight — foxes are the lightest of the canids.
- The length of time this force is applied — foxes have the longest hind legs of related dog-like carnivores.
- The muscular force exerted on the ground before flight.
- The angle — the fox must find the one that maximizes his range.

At any given time, however, he is unable to significantly alter the first three of these factors: his weight and the *power* of the exerted force — the latter is determined by the length and strength of his hind legs. So these four conditions reduce to one, the angle of launch. This was a classic problem of early ballistics. The angle required for the greatest range is the same for all projectiles be they living or dead, pigskin or rock, frog or fox. It's 45° — *see* the diagram above. For Reynard every day is another Olympic event — jumps must be swift, high, strong.

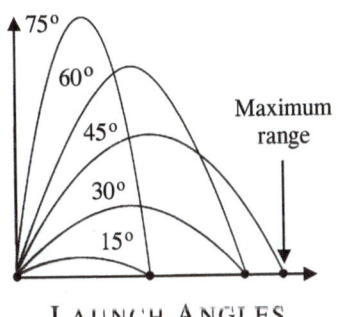

LAUNCH ANGLES

Observation of foxes proves they do use an angle very close to 45°. If the snow cover is firm and hence requires a greater downward force to trap the vole, then the angle is higher, lower if the snow is soft. Through practice, he knows that any two angles whose sum is 90° have the same range.

Why does Reynard lunge? To catch voles of course. He's the ultimate mousetrap, but there is more to it than that. Animal behaviorists have been in a long retreat from earlier denials of all human feelings to other creatures, and attributing their every behavior to instinct. I believe Reynard also lunges for the thrill of it! His ember-colored body soars in an exciting graceful parabola to land exactly on his supper. The hedgehog knows only one thing; the fox knows many.

THE FOX

Chapter — 8

SHATTERER OF WORLDS

In Eden I

Madam, I'm Adam.
Eve. (She replies.)
Even in Eden, I win Eden in Eve.
Mad Adam! (Eve)
Tut tut. (Snake)
— a cautionary tale by Anonymous

IN THE BEGINNING everything was beautiful, healthy, and *symmetrical*. Hence, Adam was uncertain whether he should speak left to right or right to left so he spoke in palindromic sentences; Eve and Snake responded in kind. And thus the first humans "deified" this wordplay that we have been burdened with ever since. Some zealots of the craft extend this to words with reversible meanings, declaring that when Genesis 1:1 says "heaven and earth" we are meant to interpret this as a three-word list with "and" as "DNA." After all, they state, Hebrew does read from right to left. (Perhaps I had best end this paragraph before my embarrassment makes my face even "redder".)

On a serious note, mathematician Hermann Weyl in his classic little book *Symmetry* points out that this word is used in everyday language with two meanings. In one sense it denotes being well proportioned or balanced, beautiful. This first meaning is general and a little vague, but the second is particular and exact: it defines symmetry as "bilateral symmetry" — implying its essence is purely geometrical. In the natural world, this left-right equality is found in insects, birds, animals, and flowers. If a slide transparency or a negative of the natural world is flipped horizontally (bilaterally reflected), it's usually impossible to know unless you're familiar with the scene. This second sense has wide application in art and nature. It's one of the birds in the aviary of Heraclitus that has been singing for millennia; let's listen to its song. It sings as it flies.

In the first sense of this word, we *unconsciously* equate health with symmetry. A mother immediately checks her newborn for signs of asymmetry: hands and fingers, feet and toes — similarly for birthmarks. Once the infant has passed this visual examination, it's judged healthy. Any animal in the wild that lacks left–right equality is sick, be it a limp, a closed eye, a tattered ear, or whatever. A wolf with a game leg will do its best to hide the limitation: it's never to the animal's advantage to reveal a weakness. Your dog will do the same. Every healthy animal grooms itself symmetrically.

As we age, symmetry trickles away, carrying our health along with it to be replaced by a cane, an arthritic hip, a mastectomy, a sling, a knee brace, and such. Although there is life after death in the form of mold, all our bodily symmetry will have vanished. First we decay, and when our bodies rise again they will be wildflowers, then rabbits, and then foxes and wolves. This is the cycle of life — as well as of symmetry, something we will investigate in depth. Cycles, like symmetry, pervade our world. The poet Shelley celebrated the mundane water cycle in his poem "The Cloud."

> *I am the daughter of Earth and Water,*
> *And the nursling of the Sky;*
> *I pass through the pores of the oceans and shores:*
> *I change, but I cannot die.*

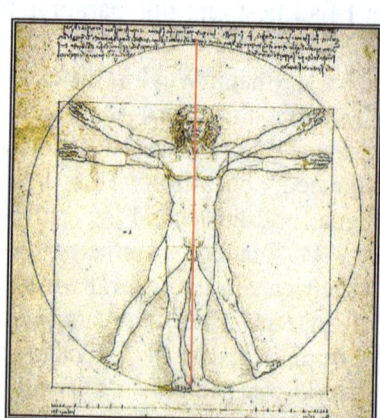

Symmetries of *Vitruvian Man* by Leonardo da Vinci

If all aspects of the first meaning are now evident, the second was always clear. Consider the bilateral symmetry of Leonardo's most well known drawing, *Vitruvian Man*. A centered vertical line divides it (at least the top section) into two mirror image halves — they're equal to each other but, as Alice says, "only the things go the other way." With bilateral symmetry, half a loaf is as good as the whole.

Beauty is *not always* culturally determined despite a common argument citing the full-figured women of Ruben's paintings as weighty proof. Bilateral symmetry is a biological constant transcending country, culture, ethnicity, and advertising campaigns. Why this transcendence? Perhaps we're wired that way. Many mothers realize their infants prefer looking at symmetric rather than asymmetric patterns. Or perhaps it's a form of imprinting; after all, the loving face that babies *first* focus on has bilateral symmetry, providing a reference for security and survival. As Blaise Pascal noted in one of his many observations, "Our notion of symmetry is derived from the human face. Hence, we demand symmetry horizontally and in breadth only, not vertically nor in depth."

"PHEMIUS, COME AWAY from those women and join me here where we have the sun to warm our backs." A tall, handsome man broke loose from a throng of lace and perfume and walked toward his old friend.

Seating himself he said, "Thanks for helping me." Phemius knew from long experience that Epius wanted to talk.

"Over these many years I've wondered why women find you *so* handsome. Do you remember the great banquet to celebrate the completion of the Colossus of Rhodes?"

"Of course. I sang and recited my poems for half the evening, to great applause as I recall. And since Chares had died and you had become the master builder of the Colossus, you were honored as well," Phemius reminisced.

"Those were great times. But what I recall was you surrounded, mauled almost, by a bevy of beautiful women. Now I think I understand why they chased you so — it's all here in this science magazine article I've been reading. In a word, *symmetry* is what women and men find attractive."

Phemius smiled and asked just what that had to do with him.

"Everything! Look at your face. It's perfectly symmetrical about a vertical plane. You even part your hair in the center. And you look the same — except for a few gray hairs — as you did millennia ago. Match this with your singing and poetry and it's a wonder you get any sleep at all. Since the women in this place outnumber the men seven to one, you're finished, you're toast."

"A hero like yourself must have done well with the ladies?" enquired Phemius.

"Good god man, look at me. I'm as large as an ox and my face is mangled from years of boxing in my youth. The right side of my mouth droops, my left ear looks like a vegetable, several of my teeth are missing, and my forehead and jaw are matted with scars. My face is so rough it can hold two days of rain. You're Apollo, I'm Hephaestus the gargoyle — symmetry next to rubble." In his frustration, Epius was almost raging.

"My friend, my friend, I recall that you won the boxing competition at the funeral games for Patroclus — Homer wrote of your unforgettable victory; I have sung songs about it. Legend says that with your *great heart* you went over and helped your opponents after they fell. Among all the Achaeans, you were the only one to show such generosity; you were the champion. That's the man I remember. If you say I have beauty on the outside, then you have it in your heart." With this, Epius' anger subsided as quickly as it arose.

A group of the remaining women viewed this implausible pair from a distance. One of them said their friendship was special, almost sacred, and shouldn't be disturbed. So the two old friends sat together and talked the sun down the sky.

MAKING A NEW DISCOVERY is exhilarating; making a new discovery about something commonplace is remarkable. We take things for granted. We must get on with our daily lives; yet, some things deserve a second look.

In Chapter 2, we investigated "mirror" images in different dimensions and discovered that if you're one dimension higher than the object observed, it has no handedness for you. Consider reflections again, in particular that disheveled person staring back at you each morning from your bathroom mirror. Do you understand everything about your mirror image? Really? I thought I did before reading Martin Gardner's *The Ambidextrous Universe*. Early in that wonderful book, he poses a most intriguing question about reflections:

> Let me try to confuse you with a simple question. Why does a mirror reverse only the left and right sides of things, not up and down? Think this over carefully. The mirror's surface is perfectly smooth and flat. Its left and right sides do not differ in any way from its top and bottom portions. If it is capable of transposing the left side of your body to the right, and the right to the left, why doesn't it also switch your head and feet? . . .

Since the mirror exchanges left and right, what happens if we give the mirror a quarter turn clockwise? Will it turn the image of our face upside down? We know, of course, that no such thing will happen. Then why this spooky, persistent preference for left and right? Why does a mirror reverse the room horizontally but fail to turn it topsy-turvy?[1]

Outwardly, mammals and butterflies each have one plane of symmetry; inwardly they have none. The Great Pyramid at Giza has four planes; a cylinder has an infinite number of vertical planes yet only one horizontal (*see* below). Can an object exist that has no planes of symmetry? Yes, and this is the key to Gardner's conundrum. Such objects are everywhere — your hands are the most common example (Figure 8.2). Each is referred to as the *enantiomorph* of the other, implying they cannot be forced to match by any movement whatsoever. Siamese twins are exact enantiomorphs. If one is right-handed, the other will be left-handed, and unbelievably, even their viscera are transposed.

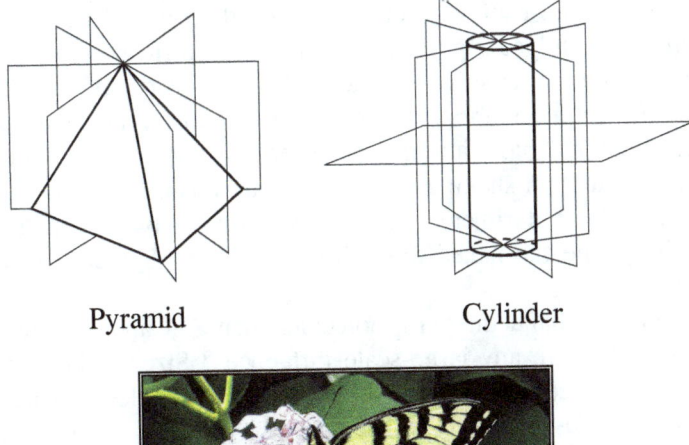

Pyramid   Cylinder

Tiger Swallowtail Butterfly

PLANES OF SYMMETRY

Figure 8.1

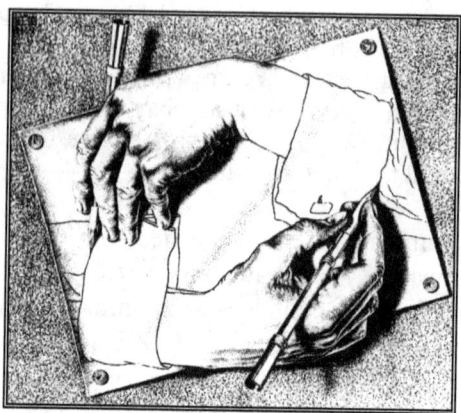

Drawing Hands by Maurits Escher

Figure 8.2

The French scientist Louis Pasteur broke new ground with his discovery of left- and right-handed molecules — perhaps the smallest of all the enantiomorphs. A lesser person — one who had lost a sense of wonder at the world — would never have noticed or judged significant such minute differences. Apparently Pasteur was so overcome by his discovery and its implications that he embraced one of his chemistry assistants and exclaimed, "I have just made a great discovery . . . I am so happy that I am shaking all over and am unable to set my eyes again to the polarimeter!" Archimedes must have felt the same elation as he ran naked through the streets of Syracuse yelling "Eureka!"

Does the handedness of molecules matter? Can anything this minuscule have any large-scale influences? Size, or lack of it, has never been a measure of importance. Whales can perish from a virus; bacteria ruled the earth for a billion years. Opposing handedness has profound effects on biochemical interactions, and this has far-reaching consequences for humans. The drug thalidomide occurs equally in two forms: one treats morning sickness; its enantiomorph causes horrific birth defects. The devil is in the details.

Every DNA helix is right-handed, implying it's non-superimposable on its mirror image. (When a spiral or helix is right-handed, it means nothing more than that it coils clockwise going away from you when viewed from either end.) Nonetheless, a few books still illustrate DNA as a left-handed helix.

But some molecules — especially those from the living world — occur in both forms. Objects of opposite handedness can be abundant in one form and scarce in the other. In many of nature's creations, however, both enantiomorphs occur with equal frequency.

Conus marmoreus

Most snails have right-handed shells, but occasionally a mirror image "sport" is found and becomes a collector's gem. Even Rembrandt (1606–69) drew a seashell with the wrong handedness. This is a postcard of a *signed* etching by the artist of *Conus marmoreus*. Since it coils counterclockwise as it goes away, the shell must be left-handed rather than right. Note the lip of the cone should turn the other way. It's very unlikely the Dutch master had a sport specimen.

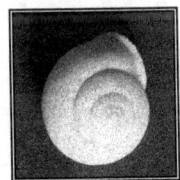
Right-handed Shell

Left-handed Sport

ENANTIOMORPHIC TWINS

NOW WE CAN ANSWER Gardner's question: "Why do mirrors reverse images left to right and not top to bottom?" Superficially, you have one plane of symmetry dividing your body into left and right halves — the bilateral symmetry of art and biology. Pretend you are standing in front of a large mirror covering an entire wall. You see your image and all the contents of the room. You imagine yourself standing behind the glass, facing the other way so that you have rotated 180° and reversed your left and right sides. Aye, there's the rub! When your body generally has bilateral symmetry, it's easy to think of yourself being half-turned to match your mirror image. We have done the puzzle on ourselves — we were never reversed left to right. *This is not what reflections do.*

Let's look deeply into mirrors and see what they really do. In Chapter 7, we noted that light and billiard balls both reflect so that the angle of incidence equals the angle of reflection. Viewing your image straight on at 90° makes these angles equal to each other — but we need something more, something new. That something is simple and obvious: your image is as far *into* the mirror as you are *out* of it. So, the mirror hasn't reversed left and right, but *front to back*, perpendicular to the mirror as in this diagram. If the mirror were on the ceiling, your image would be on its head and the direction of the reversal apparent.

                             MIRROR
Real World   Reflected World

So, that's the answer to the puzzle. To summarize: because of a *rough* bilateral symmetry, we see ourselves in the mirror rotated 180°. Our active imagination deludes us. Mirrors don't reverse left to right; they do reverse front to back.

One dilemma remains, who is this reflection in the mirror? When you wink your right eye, your image winks its left. A scar on your left arm would be on its right. The *image is your enantiomorph,* one of those things that "go the other way" — your Siamese twin that never was. You are on the inside looking out; you don't see yourself as others do.

Previously, *enantiomorphs* were defined for objects with no planes of symmetry. Haven't I been saying all along that the human body has a bilateral plane? Not exactly! At every instance its use has been qualified with words like *rough* and *superficial.* Any close inspection of a face puts a lie to that idea because the said symmetry is shattered by subtleties far more profound than hair parting and scars.

With a computer and a scanner or digital camera, you can test the symmetry of your own face. Digitize a photograph of yourself, cut away one half, make a duplicate of the remaining part; now flip (reflect) this piece horizontally and join it to the other half. You have created a kind of doppelgänger; someone you recognize but pray isn't you. Do the same on the previously discarded half and create a triptych of photographs, two of which will be caricatures. To demonstrate this I've done some cutting and pasting on Albert Einstein's familiar face.

| Two Left Halves | Normal | Two Right Halves |

REFLECTIONS ON EINSTEIN

Figure 8.3

The two outside photographs, particularly the one on the right, appear bizarre. The amount of deviation from the "normal" image is some measure of Einstein's asymmetry — all this without a part in his hair. This clearly establishes that humans do not possess true bilateral symmetry, and so we do have an enantiomorph — the specter in the mirror.

IT IS OFTEN STATED that bilateral symmetry is common in the art of primitive, less complex cultures, but we see none of this in the caves of Lascaux, Altamira, and Chauvet. Nonetheless, the early Egyptians and Assyrians used it extensively in their art and construction, and it had a revival in the religious paintings of the Dark Ages. Consider the following mural of a threshing scene from Thebes, in Upper Egypt, which shows strong elements of bilateral symmetry:

FROM THE TOMB OF MENNA, 18TH DYNASTY

Children are delighted with bilateral symmetry. Put a dab of paint in the crease of a sheet of paper, fold, press, and open it. They'll squeal in pleasure at the design; adults soon tire of this too obvious, too simple pattern. Bilateral symmetry is everywhere in our society from cups to birdbaths, bicycles to cars. It's part of the background; we hardly notice it. Historically, Japanese art and culture have attempted to avoid the banality of a too noticeable symmetry. In their ritualistic tea service, the cups and urn are placed to prohibit such an effect.

Gravity, the force that molds planets into spheres and ponds into smooth glassy sheets, also sculpts living bodies with a horizontal symmetry. And the need for movement to find food gave birth to front-back and hence to right-left orientations. Stationary life forms have the first feature (horizontal symmetry) but lack the other two (front-back and right-left). Together, gravity and movement explain the origin of bilateral symmetry, and it seems reasonable that they have shaped life everywhere on this planet and beyond.

In Western art, bilateral symmetry is always associated with an axis perpendicular to the earth. From the faces of our parents to the profiles of trees in the field, this is what we have come to expect. As Pascal said, "We demand symmetry horizontally and in breadth only, not vertically nor in depth." Can the reader recall a single painting or drawing with any elements of vertical symmetry? Maybe, but they're rarer than a bluebird in springtime.

Take a slide or negative of any natural scene. Other than placing it in a new position with the same orientation, you can make three significant transformations:

- Flip (reflect) it horizontally.
- Flip (reflect) it vertically.
- Rotate it 180°.

The first change won't be evident unless you know the landscape — a natural scene might just as well "go the other way." The second and third options, however, will appear strange. Occasionally you can use rotation to your advantage. Say you have a painting on your wall that you have grown to dislike. Hanging it upside down offers something entirely new, though it may be a distorted image in familiar colors.

| Scheherazade | Sherlock Holmes | Dog |

A few rare "artists" are capable of drawing reversible figures as above. The English artist Rex Whistler as part of a 1930s Shell Oil advertisement drew the two figures on the left, Scheherazade–Prince and Holmes–Hood. The Dog–Cat figure is from American illustrator Peter Newell who published two books of reversible drawings in *Topsys & Turvys* around the year 1900. Everyone finds the middle picture impossible to discern without inverting the page.

To the prohibition on vertical symmetry one common exception exists, where a scene of trees, rocks, or stars is reflected in a lake or river. A famous example is van Gogh's *Starry Night on the Rhone* (Figure 8.4) although in this case the stars seem to emerge from the water. In an act of infrequent realism, van Gogh unmistakably represents the Big Dipper stretched across the night sky. Folk legend says this must be fall since the dipper is opening up to catch the bounty of the harvest season. And his letters to his brother Theo tell us he did paint this scene in Arles shortly before midnight in October 1888.

The effect that van Gogh painted, Matthew Arnold described in the final two lines of his almost forgotten poem *Sohrab and Rustum*. The story is about the brave and gentle Sohrab slain by his father Rustum who doesn't know him. The depiction of the Oxus River in the poem's concluding lines — a metaphor for life, death, and possibly resurrection — is no mere afterthought on Arnold's part, but is artistically right and inherently beautiful:

*A foil'd circuitous wanderer — till at last
The long'd-for dash of waves is heard, and wide
His luminous home of waters opens, bright
And tranquil, from whose floor the new-bathed stars
Emerge, and shine upon the Aral Sea.*

Starry Night on the Rhone
by van Gogh, 1888, Musee d'Orsay

Figure 8.4

HISTORY IS FULL OF CHARACTERS renowned for their particular qualities. Heroes, artists, mathematicians, sculptors, and despots — Odysseus, da Vinci, Archimedes, Michelangelo, and Hitler come to mind. But very few are noted solely for their mathematical brilliance. One such that hardly anyone recalls is the transcendent French genius Évariste Galois (1811-32). He didn't begin the study of mathematics until he was 16 when he devoured all the books of Euclid's *Elements* in a few days. His originality was so profound that no one could follow or share in his genius, yet legend says his temper was so great that during his entrance examination to the École Polytechnique he threw an eraser at the supervisor's head. Apparently he considered the questions insultingly easy; no surprise to say, he was not admitted. In an act of incredible daring touching on arrogance, he tackled a problem that had baffled the world's best mathematicians for hundreds of years. More incredibly, he solved it so completely that he gave scientists a concept to work on for the next two centuries. Not to stretch the credulity of the reader too far, Galois scribbled out 60 pages of mathematics and letters in the two nights before he died pointlessly in a duel over a woman who didn't love him. Leopold Infeld wrote a poignant biography of this tragic youth titled *Whom the Gods Love*[3]. During the reign of Louis XVIII, Galois's outspoken republican sympathies often landed him in trouble or jail. He desired perfection from an imperfect world, within a flawed social system, and in difficult times. Ultimately he asked too much. The full title of Infeld's book comes from the ancient Greek poet Menander who wrote, "He whom the gods love dies young."

What was the problem Galois unraveled? It's one every schoolboy and girl knows something of, equation solving. This is an ancient topic: by 1600 BC the Babylonians had discovered the general quadratic formula — the bane of modern high school students. They even tackled specific cubic and quartic equations with some success. It's amazing the speed with which advanced mathematical concepts developed in early human cultures. *Homo sapiens* have a profound need, both practical and aesthetic, for mathematics. The Greeks, however, with their emphasis on geometry rather than algebra didn't greatly advance equation solving. That honor fell to a collection of colorful Renaissance characters, one of whom we've met previously in Chapter 2, Gerolamo Cardano.

A cubic equation involves the third power of the unknown in addition to its square and first power. As an example, consider $x^3 - 3x^2 - 6x^1 = -8$. The roots (numbers) that make a cubic equation true are normally difficult to find, but since I've concocted this one, I know they're 1, -2, and 4. Test this by substituting 1, -2, or 4 for x and you'll see they make the equation *true*, meaning the left hand side will equal the right. The unknown with the highest degree determines the number of roots; here that's three. The general formula to solve every cubic is credited to Scipio del Ferro and Niccolo Fontana, nicknamed Tartaglia, "The Stammerer." Fontana's jaw and palate were split open by a sword when he was a boy, and it left him with a permanent speech problem. Nonetheless, he was a brilliant mathematician and readily answered to his nickname. Cardano persuaded him to share his secret for solving cubics but promised never to reveal it. Tartaglia did, Cardano didn't. After a decade of silence, Signor Gerolamo published his famous *Ars Magna* of 1545 informing everyone about the method — but with full credit given to Fontana.

The next mountain to climb was the general quartic, the equation of the fourth degree. Enter Cardano again. He hired a sometime street urchin called Ludovico Ferrari for menial chores around his home, but realizing the boy could read and write he made him his secretary. The youngster quickly absorbed much mathematics and it became apparent this boy was extraordinarily gifted. To Cardano's credit he educated him, and Ludovico in turn became the master's lifelong friend and most ardent defender. And then something extraordinary happened: Ferrari discovered a general method for solving *all quartics*. This too was published in the *Ars Magna*.

The mountains get ever higher. Now it was the turn of the quintic, or general equation of the fifth degree, to yield to human genius. But it didn't happen, at least not right away and not in the manner anyone expected. The techniques that had worked before — the ropes, pulleys, and wedges — were of little use on this higher peak. But progress was made in the foothills. Joseph-Louis Lagrange showed that the formulas for solving quadratics, cubics, and quartics all required the same procedure, and that ploy failed on the quintic. Yet there could be other paths up the mountain that no one had found. Even that hope was cut off when Niels Hendrik Abel proved the peak was unscalable by any approach. But Abel still hadn't penetrated to the heart of the mystery. However, because of one man, a boy really, all that was about to change.

Galois's technique was all about symmetries, and this was odd because the subject at hand was algebra not geometry. Obviously our concept of symmetry needs to be expanded. The young genius concentrated on permutations, or as we say, rearrangements, and these are actually symmetries. By working with permutations of the solutions of equations, and focusing on those preserving *all algebraic relations* between these roots, he scaled the quintic mountain. In his analysis, Galois discovered an unexpected property of any two consecutive permutations: you always get another permutation of the same system. He was the first mathematician to use the word *group* with respect to these permutations. From this lofty peak he realized equations solvable by a formula are of a special type, and the quintic are not of that type. This ended more than three millennia of investigation!

It's beyond the scope of this book to delve into the complexities of Galois's permutation groups. Yet this chapter is partly about symmetry breaking and this will be most easily understood by exploring groups — *simple* symmetry groups.

BOTH THE ARTIST and the scientist love symmetry, and for the same reasons: the sense of beauty and order it brings to the world. The artist may prefer to explore all the possibilities of a particular type of symmetry while the mathematician prefers to explore all possible types of symmetry. These approaches are compatible, for beauty is the driving force for artists and all the greatest scientists, and mathematicians.

The time has come to expand our concept of symmetry beyond the bilateral and horizontal; this is what mathematicians do best — generalize. Some say they generalize to such a degree that their ideas are about nothing at all. But that's a lie. The larger the net, the more fish it captures; the bigger the sail, the faster the ship. Ian Stewart and Martin Golubitsky in *Fearful Symmetry*: *Is God a Geometer?* comment on the defining and expansion of the concept of symmetry:

> It took humanity roughly two and a half thousand years to attain a precise formulation of the concept of symmetry, counting from the time when the Greek geometers made the first serious mathematical discoveries about the concept, notably the proof that there exist exactly five regular solids. Only *after* that lengthy period of gestation was the concept of symmetry something that scientists and mathematicians could *use* rather than just admire.[2]

What is this precise formulation? All good definitions should contain the original idea (here it's bilateral symmetry) as a special case, and this new definition does that. Symmetry in Hermann Weyl's first sense is an aesthetic quality, a value judgment, meaning it's subjective and almost impossible to formalize. We won't touch this. It's his second sense — bilateral symmetry — that we'll concentrate on. And the envelope please. The definition: *symmetry is a verb not a noun by which we mean it is a movement, a transformation, not a thing*. There are an infinite number of movements, but we want only those that leave *unchanged* the overall form of the object to which they are applied. On certain objects, bilateral or horizontal reflections (movements) can do that. What are the others?

Well, they vary depending on what you're transforming. Let's pick an easy object and find *all* its symmetries.

The equilateral triangle is the simplest of the polygons. So that the reader and the writer can tell it has been transformed, I've distinguished its angles by the symbols O, X, and V. These *movements* come in two varieties, rotations around a fixed point and reflections about a fixed mirror line (L). Apply each in turn to the identity (I). Here's the complete list:

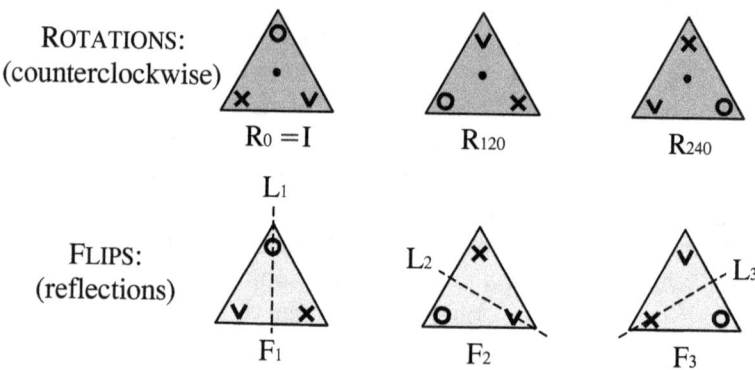

THE SYMMETRIES OF THE EQUILATERAL TRIANGLE

Figure 8.5

These are the six symmetries for the equilateral triangle. Take note that the first of the flips is our old friend bilateral symmetry. Without the markings O, X, and V, we wouldn't be aware anything had moved, and a rotation of 360° is the same as no rotation (R_0), also called the identity (I). A square has eight symmetries, four rotations, and four reflections. A starfish has 10, five of each type.

With six operations, like the six numbers on a die, there are 36 products, those light gray, medium gray, and darker gray entries in the table below. Interpret *juxtaposition* to mean *is followed by*. For example, $F_3 R_{120}$ = F_1 means you *flip first* about L_3 and then you rotate 120° ccw. This table contains the essence of what it is to be a group — Galois's great idea. Every mathematical group must have at least four properties, which the following acronym will help you remember: *cl a n i* (rhymes with Danny).

- *cl* — *closure,* any application of a flip and/or a rotation is another flip and/or rotation.
- *a* — *associative,* i.e. $(R_{120} F_2)F_1 = R_{120} (F_2 F_1)$.
- *n* — *neutral element,* a transformation that does nothing. Here that's R_0, also called the identity I.
- *i* — *inverse,* every transformation must have one,
- i.e. $R_{120} R_{240} = I$ or $F_1 F_1 = I$.

	I	R_{120}	R_{240}	F_1	F_2	F_3
I	I	R_{120}	R_{240}	F_1	F_2	F_3
R_{120}	R_{120}	R_{240}	I	F_3	F_1	F_2
R_{240}	R_{240}	I	R_{120}	F_2	F_3	F_1
F_1	F_1	F_2	F_3	I	R_{120}	R_{240}
F_2	F_2	F_3	F_1	R_{240}	I	R_{120}
F_3	F_3	F_1	F_2	R_{120}	R_{240}	I

D_3: SYMMETRY GROUP OF THE
EQUILATERAL TRIANGLE — CLOSED SET

Figure 8.6

Informally, we see that a set is closed when you can't transform your way out of it. It's like the Eagles' pop song "Hotel California" which says, "You can check out any time you like but you can *never leave.*" A set consisting of only F_1 and F_2, say, isn't closed because $F_1 F_2$ = R_{120}, and this product isn't in the set (at the left). The associative property simply means that if you have three transformations you can do the first two and then the third or the last two and then the first; it's irrelevant to the answer. Remember that brackets mean "do us first." Now for a more subtle point: it's possible to be associative and not commutative, meaning you can't *generally change* the order of the transformations. Putting on your shoes and then lacing them

	F_1	F_2
F_1	I	R_{120}
F_2	R_{240}	I

OPEN SET

up isn't the same as lacing them up and then putting them on. As an example, consider:

$$F_1F_2 = R_{120}$$
$$F_2F_1 = R_{240}$$
In general, $F_1F_2 \neq F_2F_1$.

The neutral element is obvious — it does nothing. That's analogous to "0" when you add, or "1" when you multiply. Next, every transformation of the group must have an inverse, and elements may be their own inverses. When adding, the inverse of 4 is −4; when multiplying, it's 1/4. That is, if combining two transformations produces the identity, they're considered inverses of each other. From the table for D_3, it's easy to find all the inverses of this group. A random set is unstructured, but a set with "clani" is a group.

For our purposes the salient feature of Figure 8.6 is the *light-gray subgroup* of rotations in the upper left portion of the table. Check that it has *clani*. What about the identical dark gray set in the lower right? Isn't it also a subgroup? No, because it lacks the first property, closure — the product of any two reflections always yields a rotation. Since the triangle has been flipped twice, it's the same way up so this must be a rotation as the table shows. We call the entire 6x6 array the *di*hedral group of order three or D_3, since it's composed of two parts. D_4 is the dihedral group of the square with four rotations and four reflections. D_n — the dihedral group of order n — has 2n elements, half of each type.

Is it possible to have an object with no line of reflection yet with a point of rotation? Yes, consider the Isle of Man's running-legs symbol or the clockwise Nazi swastika to the left.

SWASTIKA

Any rotation through 0°, 90°, 180°, or 270° leaves the latter design unchanged; hence, each movement is a symmetry. Mathematicians use the symbol Z_4 for this cyclic group with four elements, and Z_3 for the cyclic subgroup in Figure 8.6 generated by three rotations. This can be generalized to Z_n, the cyclic group of n elements — as with a regular n-sided polygon. So far, all these groups have been finite.

Hermann Weyl in his little gem *Symmetry* recalls that during a lecture in Vienna just before Hitler's henchmen stormed into

Austria he mentioned the swastika as a symbol of terror far more terrible than the snake-girdled Medusa's head:

> And a pandemonium of applause and booing broke loose in the audience. It seems that the origin of the magic power ascribed to these patterns lies in their startling incomplete symmetry — rotations without reflections.[4]

Z_N AND D_N are the only finite groups on the plane. Enumerating all possible cases is one of the things mathematicians do. Entire university departments are devoted to the craft, an ancient pursuit. Theaetetus' inventory of all five regular solids was a momentous beginning (*see* page 32). Let's continue his work with a list of all their symmetry groups.

Consider the cube, the most familiar of the five. Every such solid has eight vertices (corners), with three edges meeting at each vertex. Let A be a fixed point in space. Now all its corners may be rotated to rest *at* A, and having done so, the cube can be cycled into three new positions *about* A. So this solid has 8 x 3 = 24 rotational symmetries. In addition to *every* rotation there must be a unique reflectional symmetry. Therefore, there are 24 of each, yielding an aggregate of 48 symmetries of the cube. Similar reasoning can be applied to the remaining Platonic solids, and by doing this we uncover the following table of symmetries:

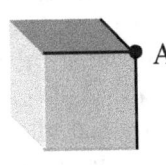
CUBE

REGULAR SOLID	NUMBER OF ROTATIONS	NUMBER OF REFLECTIONS	TOTAL NUMBER OF SYMMETRIES
Tetrahedron	12	12	24
Cube	24	24	48
Octahedron	24	24	48
Dodecahedron	60	60	120
Icosahedron	60	60	120

COMPLETE SYMMETRY GROUPS OF THE PLATONIC SOLIDS

Figure 8.7

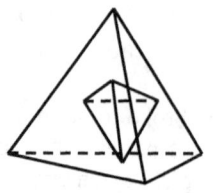 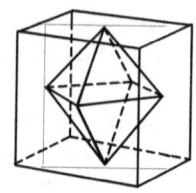 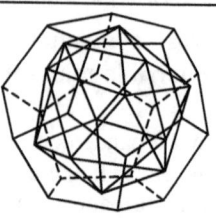

DUALS OF THE PLATONIC SOLIDS

Figure 8.8

Some inspection of the above table is worthwhile. Note the cube and the octahedron have the same number of symmetries, as do the icosahedron and dodecahedron. What significance, if any, does this have? The identical rows of symmetries — emphasized by the colors — are the result of *mathematical duality*. Place a dot in the middle of each of the six faces of a cube; these dots will be the vertices of an octahedron, its *dual*. Because of this, every symmetry of the cube must also be a symmetry of the octahedron, that is, they have the same groups. By placing dots in the middle of each of the 12 faces of the dodecahedron, you mark the vertices of its dual, the icosahedron; hence, they too have the identical groups. What about the tetrahedron? It has nothing left to be the dual of — nothing but itself. This simplest of the solids has four faces and four vertices implying it is self-dual. *See* the above figure. Unexpectedly, the Platonic solids give us three groups, not five.

Duality could have been introduced in Chapter 2 *without* reference to groups just by noting that the vertices of one dual are the faces of the other (and the reverse); the edges stay unchanged. Alternately, cutting the corners off a Platonic solid (i.e. changing points to faces) does the same thing — it creates the dual of that solid. *See* the Chapter Notes or page 234 for a table of all their points, edges, and faces.

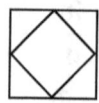 In two dimensions, the number of sides equals the number of vertices, so every regular polygon is self-dual. To see this, place dots in the center of each side and join. Voilà, you have its twin.

This property of duality continues into higher dimensions. You may recall that in the fourth, six regular polyhedrons exist. Of these, two are self-dual. And the hypercube is dual with the crosspolytope, and one hypervolume is bounded by 120 dodecahedrons while its dual has an astonishing 600 tetrahedrons. Zounds! Fortunately, we won't go there.

Amazing, isn't it? These products of the study hall can walk out the door into the fields, skies, and oceans to find their own "duals" in the real world. Viruses shaped as dodecahedrons; the once terrible poliovirus as a minuscule icosahedron; the ubiquitous methane molecule as a tetrahedron. The radiolaria have an entire Platonic gallery, and Kepler was honored with a molecule mirroring his most majestic failure. Truncated icosahedrons drift through interstellar space as Bucky balls. A remarkable alliance exists between the firstlings of our minds and the firstlings of nature's workshop.

NOTHING YOU PERCEIVE is perfectly symmetrical. Nevertheless, we can imagine things that are — that's really what geometry is all about, seeking perfection visually as well as logically. The circle *in the mind* has perfect symmetry: it has infinitely many rotations about its center and infinitely many reflections in axes passing through that center. And as before, any two reflections yield a rotation. This two-part group is labeled $O(2)$ — the orthogonal group in two dimensions. Its rotational subgroup is $SO(2)$ — the special orthogonal group in two dimensions. As n goes to infinity, D_n goes to $O(2)$ and Z_n to $SO(2)$.

Let's take everything up a notch by considering the sphere. All its infinite rotations and reflections constitute $O(3)$ — the orthogonal group in three dimensions. By themselves the rotations form an infinite subgroup $SO(3)$ — the special orthogonal group in three dimensions. Because these groups of symmetries on circles and spheres are infinite, that would seem to be large enough so we'll stop here.

With respect to the *number* of symmetries, rotation group Z_n and dihedral group D_n are finite. These forms are common themes in art and architecture. Exquisite examples of each are found in the harem of the Alhambra, Spain, a virtual textbook of such groups. Maurits Escher studied them and gathered much of his inspiration there (*see* below). But why would a palace devoted to erotic pleasure have no representations of the human form? Historically, Islam took the Second Commandment's prohibition against graven images *literally:* "You shall not make for yourself an idol in the form of anything in heaven above or on the earth beneath or the waters below." Otherwise, the harem with its fussing eunuchs, blind musicians, and naked women would have had its walls covered with figures reminiscent of *Playboy*.

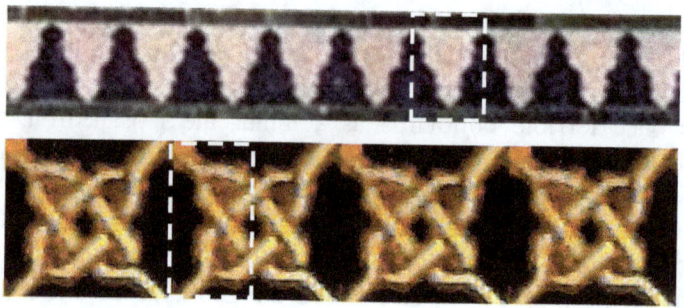

TWO FRIEZE PATTERNS FROM THE ALHAMBRA

Figure 8.9

Repeating patterns add a new operation to rotation and reflection: translation, moving something in a straight line. The novel patterns created by a repeated motif are termed *space groups* or frieze patterns, the kind of designs we see on borders around the top of a room. Maurits Escher's art has influenced a whole generation of mathematicians and scientists with its eerie amalgam of the commonplace and the bizarre. As can be seen in the next figure, this modern Dutch master brings something unique to old art forms like strip patterns.

Horsemen, on a Möebius Frieze by Maurits Escher

Figure 8.10

Without the join in the middle, where the horsemen cross, this wouldn't be a Möebius strip — technically perhaps it still isn't. Nevertheless, if you follow any horseman motif around this band you cover both sides. Why? Remarkably, this frieze has only one side and one edge!

Whether for temples, tombs, palaces, or homes, artists have been designing and artisans building borders for millennia. All of these can be classified into seven different types — the number of completeness for the superstitious (*see* Figure 8.11). Examples of each are found in the decorative tiles of the Alhambra, the woven baskets of Mozambique, and the ornamental art and needlework of old Hungary. I've adopted a method of cataloging based on bodily movements, a product of the fertile mind of British mathematician and puzzle expert John Horton Conway. My list of occurrences isn't exhaustive; Conway's catalog of types is.

JOHN HORTON CONWAY'S CLASSIFICATION

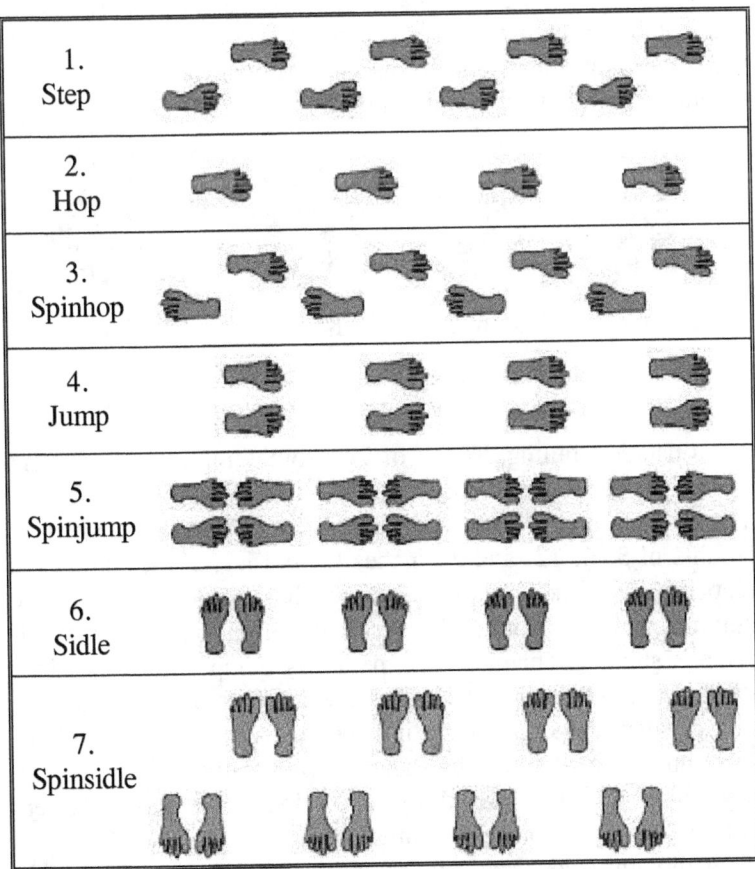

THE SEVEN FRIEZE PATTERNS

Figure 8.11

Crystallographers classify by translation, reflection, half-turn, and glide reflection (a translation followed by a reflection or the reverse), but this more involved system need not concern us. Both schemes begin with a motif: a small part of the frieze needed to generate the whole by judicious movements. In the two frieze patterns from the Alhambra (Figure 8.9), each motif is outlined by a dotted-white rectangle. In Conway's classification, the top is number two, a hop; the bottom is number five, a spinjump. With Escher's woodcut horseman motif, it's mostly number two. Look around the room you are in. Does it have a border? Can you identify which one?

GROUPS HANG OUT in odd places. The explosion is one of nature's basic templates; it and some others will entertain us in the following chapter. Groups and explosions often occur together in ways you would not dream of.

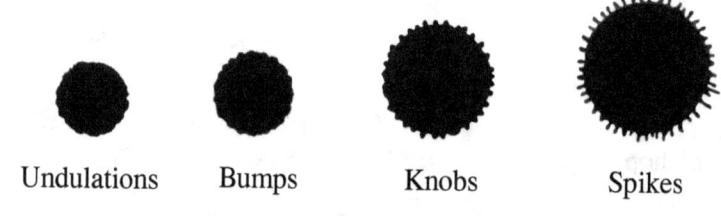

Undulations   Bumps   Knobs   Spikes

THE INKBLOTS

Consider a falling droplet of ink (something in this ballpoint era only older readers will recall). Running across this page are a series of impacts resulting from ink droplets released at heights of 3, 6, 12, and 24 inches (7.6, 15.2, 30.5, and 61 cm respectively). The areas increase but they do not double as the height doubles. Examine the perimeter of each blot: the smallest has slight undulations; the next has bumps; the following, knobs; and the last, spikes. These variations are responses to the greater velocities resulting from increasing the height. Continuing to double the height will result in longer but discontinuous rays like those emerging from moon craters (Figure 9.1). Velocity determines form.

Although never exact, the regularity of the spikes is striking and unexpected. The largest ink spot is the dihedral group D_{48} (by my count) with 96 elements composed of rotations and reflections. When an undulation starts to grow, it sucks ink from surrounding areas, thereby preventing their growth. That is, the growth of a spike inhibits growth *around* the spike. By this response to the rising height, the ink spots produce the regularity we see in the crown of black spikes, above right.

Analogously, higher temperatures promote the growth of longer limbs, fingers, ears, tails, gills, and so on, rather than the growth of body bulk. And this in turn increases the body's surface area, which allows for the quicker dissipation of heat. This reasoning may seem familiar — recall "The Measure of All Things" and *see* Figure 1.7 on the Inuit and the Ruandans.

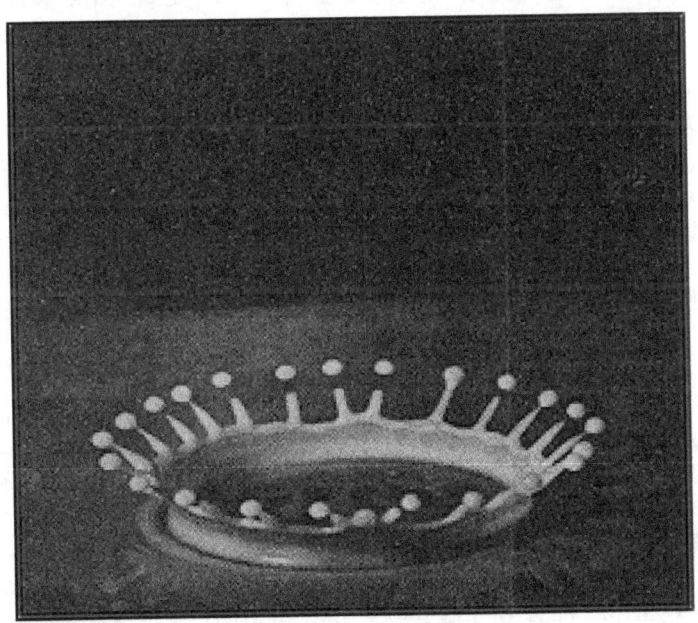

THE DROP-OF-MILK CROWN

Figure 8.12

I cannot leave this point without remembering an example that has adorned numerous books. It first appeared as the frontispiece of Sir D'Arcy Thompson's *On Growth and Form,* the pioneering work on the mathematical features of biological design.

This splash was photographed at $\frac{1}{10,000}$ of a second above a plate covered with a thin layer of milk. The impact produced a smooth circular ring curving upward with 24 spikes — each pinching off a droplet of milk by surface tension. As with milk, so it is with ink, raindrop, and meteor impacts; such examples form a universal template, the explosion. Note that the symmetry isn't perfect. The falling droplet of milk — the cause — will have had minute imperfections in its shape: perhaps it had a wobble, or maybe the angle of impact wasn't perpendicular. The crown — the result — displays these variations, but that's nature's way. If all the preconditions had been perfect, the resulting 24-point crown would be perfect.

A paradox lingers in the spikes of the ink and the crown of the milk. The problem is more than academic; it's the way of the world. The falling drop of milk, being nearly spherical, has *infinite symmetry* with group $O(3)$, or more likely $O(2)$; the symmetry of the bowl and the milk are infinite as well with group $O(2)$. These are the preconditions just before the drop of milk impacts the surface creating the crown with *finite symmetry* of at most 48, 24 rotations and 24 reflections, that's dihedral group D_{24}. Its regularity is that of a 24-sided polygon. As a result, the symmetry has been broken (reduced) from infinite to finite. *Where did it go?*

IT'S DARK. Although the first gray mists of morning fill the east, sunrise seems a long way off as you huddle in the cold rocks with your fellow hunter-gatherers. From the plain below the night roars of the predators and the cries of their prey fill your mind with terror. This is the second night on the same hillock. The previous morning a pride of lions killed a wildebeest at the base of the rocks; they feasted on it all morning. After that, the jackals and vultures took what was left as the lions rested under a nearby acacia tree. The quickly rising sun was about to break over the eastern ridge while the roar of lions could still be heard in the distance. One old male returned to the bare bones of the beast in the hope of finding a forgotten fragment. Finally the moment has come, and with the sun shining full on your face, you and your friends break like a whirlwind from the rocks in an all-out running assault on the male lion. Your numbers, and the stones and sticks you carry, chase him from the standing rib cage. You don't slacken your pace until you reach the worthless prize. Even though you have driven the lion away, it seems you're a day late. But one of the older men takes a large wildebeest femur, places it on a flat rock, and smashes it with another rock. It splits revealing the rich marrow within — the prize you have been waiting for. Only wolverines, hyenas, and some wolves can crush bones with their jaws and reach this precious food source. Man does it with his brain. Welcome to the world of ideas.

Ideas help us survive and live longer and happier lives. Yet they do more than that. Humanity has a need to understand, truly understand, the world. It's our purpose in life, a purpose we have given ourselves, but just as deeply felt as any commandment from a deity. We began this chapter determined to listen to the nightingale ascending; broken symmetry is its main song.

SOME REMEMBER PIERRE CURIE only as the husband of the great physicist Marie. Perhaps they forget that he too won a Nobel Prize in 1903 for his work on radioactivity. Yet his future reputation may rest secure in his far-reaching observations on the role of *symmetry* in physical interactions. At first sight, Curie's Principle seems to be just common sense written large; he stated it in two logically equivalent forms:

- The symmetries in the cause reappear in the effect.
- Asymmetries in the effect imply asymmetries in the cause.

Let's test it. What is the shape of falling drops of rain? Are they like teardrops as cartoonists draw them, or are they spherical as physicists prefer? Most of the time they are neither.

FALLING RAINDROP

Their form depends on three factors: size, surface tension, and the length of time they have been falling. They begin their descent as tiny perfect spheres, quickly gathering other drops to enlarge their volume. As they approach their terminal velocity, the resistance of the air pressure molds them into a hamburger-bun shape. However, the upward force of this air may also shatter them into a spray of small spheres and start the whole process over. Eventually the raindrop will fall to earth, perchance to hit a solid surface. If a *slowly* falling droplet is nearly — but not quite — spherical, the impact takes on the layered appearance shown in the high-speed photograph below. Now the raindrop had symmetry group O(2) and the collapsing raindrop also has group O(2) — each the same as the cylinder.

COLLAPSING RAINDROP

Hence, the symmetry of the effect is identical to the cause, and we are not bothered about where it went — it's still all here.

Curie's common sense principle seems to pass one trial and fail another. First, the splash of milk went from, at least, O(2) to D_{24}; second, the splat of the raindrop maintained its symmetry at O(2). But don't scientists junk all hypotheses and mathematicians trash every theorem that falls short on even a single test? Yes, but perhaps we haven't interpreted Curie's Principle either fully or correctly. Let's give it another look.

THE KEY CONCEPT to understanding this principle, the one Curie himself probably didn't grasp, is *stability* or the lack of it. We say an object is *stable* if small disturbances don't much change its rest or motion. Consider your breakfast cereal bowl with a marble in the bottom of it. If the marble is disturbed, it will roll about for a few seconds and then settle down at the bottom again — we say it's stable. Conversely, invert the rounded cereal bowl, and carefully balance the marble at the top. In this setup, even the slightest disturbance will roll the cat's-eye away, and that's another marble we've lost. We say it's unstable.

On the vast scales of time and space, simple laws of motion and gravity can sometimes lead to stability. Our existence depends on it. These laws and forces are the panning; our planet is the golden nugget in the bottom of the pan. Carl Sagan and Ann Druyan eloquently make this point in their book *Shadows of Forgotten Ancestors*:

> You start with a chaotic, irregular [unstable] cloud of gas and dust, tumbling and contracting in the interstellar night. You ended with an elegant jewel-like solar system, brightly illuminated, the individual planets neatly spaced out from one another, everything running like clockwork [stable]. The planets are nicely separated, you realize, because those that aren't are gone.[5]

In the long term the solar system will disintegrate to instability, and the planets will end as they began, in interstellar dust. But unlike the young man at the astronomy lecture who was troubled because he thought the speaker had said the sun would become a red giant in one billion years, this disintegration need not worry us. If by some miracle the earth survived this enveloping expansion in the sun's corona, it would long afterward be reduced to circling a white dwarf with less than one percent of the present luminosity. Nothing lasts forever. The exceedingly distant prospect is clear: we will end as we began in chaos and instability. Robert Frost wrote, "Some say the world will end in fire, some say in ice." Make your choice.

Instability can descend from the stars and possess your body. Too many glasses of wine or spirits can bring your personal stability into question — you begin to wobble and weave. To counter this unsteadiness you collapse to the ultimate stable position, the floor. Those who are on an even higher "spiritual" plane may feel they're going to fall off the floor itself. That's the mark of true instability.

Twice before in this chapter we've seen this phenomenon: first, in the minor undulations of the initial inkblot — the precursor for all the bumps, knobs, and spikes to follow; then in the Drop-of-Milk Crown (Figure 8.12). The rising *ring* of milk initially has circular symmetry but becomes unstable, allowing the spikes to appear higher up. In a previous chapter on chaos theory the bifurcation diagram of the logistic function (Figure 3.11) clearly shows that its instability occurs when the growth factor m is higher than 3.

Everyday examples of stability can be found on your desk. The classic instance is balancing a pencil on its sharpened lead point. Now this is theoretically possible, but practically impossible — too many minuscule disturbances, such as air currents and desk vibrations. Like the inebriated man, the pencil assumes the most stable position — flat on the desk. Hold a plastic ruler horizontally between your palms and press lightly. Nothing happens, but a little more pressure and the ruler will bow up or down and so assume a more stable position. Which way it goes depends on minute qualities in the ruler and the pressure of your hands. Instability is omnipresent in the universe, whether out there or down here, and has been so since the beginning of time — literally.

What does instability have to do with symmetry? Everything actually. Without it, symmetry would never be broken, and we might still be a cosmic egg, however large, in the eternal night of space. In the early universe, instabilities with the density distribution of the cosmic soup put the galaxies on the road to existence and all else followed. Instabilities are the parents; broken symmetries are their children. If a symmetric state grows unstable, the entire system will be transformed into something *usually* less symmetric.

I have yet to answer the question about where the symmetry hides when, say, a nearly spherical drop of milk produces only a 24-point crown. In this case the symmetry has been broken from, at least, $O(2)$ to D_{24} — infinite to finite. With the concept of instability in our repertoire of ideas, this question can now be answered. If a second drop were to fall, we would get a different 24-point crown related to the first by rotational symmetry. How does nature choose? By instabilities in the entire setup: the velocity of the drop, wobbles in the drop, tiny waves in the plate of milk, and so on. Theory predicts an infinite number of states; practice chooses one. Consider a coin with two symmetrically related sides. Hypothetically, it can land heads or tails, but a toss produces only one of these. Or run the film of life on earth a second time, and you wouldn't be reading this book because I wouldn't have written it. More to the point, neither you nor I would exist!

In a valid sense, symmetries aren't broken or lost — they're shared. It's the difference between the potential and the actual, the possible and the real. They're the debutantes shuffling at the edge of the ballroom, waiting for the dance of life. Echoes of parallel universes of science fiction fame can be heard here: we live in one such world where I'm writing and you're reading. We have seen that Curie's Principle — the symmetries of the cause reappear in the symmetries of the effect — is not wholly correct. Usually, but not always, the symmetry is broken or better still *shared*. Yet the aggregate of all possible solutions remains symmetrically equal. The symmetry is shared over an infinite number of solutions whose particular origin is an instability in the cause. Ian Stewart in *Fearful Symmetry* restates the chemist's tenet in the following accurate form: *Extended Curie Principle* means physically realizable states of a symmetric system come in clusters, related to each other by symmetry.

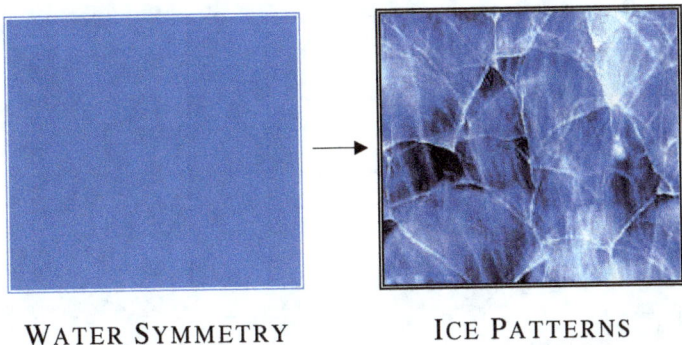

WATER SYMMETRY   ICE PATTERNS

Figure 8.13

Confined in its square container, the water has symmetry group D4: four rotations and four reflections. Theoretically, water has infinite symmetry since the container may be imagined as large as you please. The molecules are in complete random motion, and it's this randomness that imparts the symmetry. Since no direction is different from another, it is isotropic as physicists say. The water molecules in ice are fixed in certain directions with *limited* symmetry at the lattice level, so its pattern is most striking. Clearly, symmetry and pattern are different concepts. We have a precise definition for *symmetry*: it's a transformation that leaves an object appearing unchanged. Scientists define *Pattern* as something formed by the process of symmetry breaking. So, water has symmetry and ice has a pattern. More symmetry implies less pattern, and conversely. An amorphous patternless cloud of water particles is isotropic and hence has infinite symmetries: rotations, reflections, and translations. From this chaos, a snowflake crystallizes with its beautiful pattern of group D6. The symmetry has been broken, and a pattern born.

In the last two decades scientists studying interstellar spaces have found spectral traces of molecules previously thought only to exist on earth: formaldehyde, hydrogen cyanide, ammonia. Apparently the crystal lattice of ice aligns and concentrates the atoms facilitating the formation of complex molecules. Even adenine, one of the four building blocks of life, forms out of a frozen solution of hydrogen cyanide and ammonia. On the other hand, most biologists believe that life formed in warm seas from melted meteors in the days when the earth was young and very hot. Some say life began in fire, some say ice.

Bilateral Symmetry ⟶ Broken Symmetry

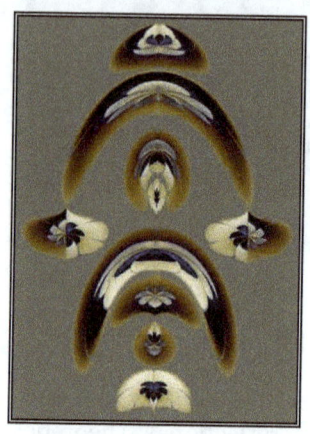
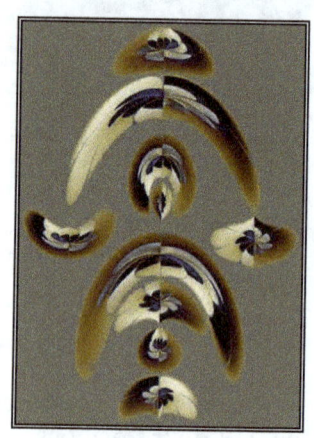

My Variation on
Gold Fountain III

Gold Fountain III

DIGITAL ART BY LEO BLEICHER (right)

This distinction between symmetry and pattern as opposite ends of a spectrum is a significant concept for artists. Intuitively most painters, especially in the orient, have avoided bilateral symmetry. The human brain correctly interprets too much symmetry as a lack of pattern. This is seen in the two contrasting pictures above. After a while, an abundance of regularity is dull, like the black surface of a calm pond before a pebble falls, or a blank canvas before the first brush stroke.

WE HAVE GATHERED a number of eggs in our basket of ideas. Initially we learned that in a broad sense physical symmetry is related to health and beauty — at least as applied to the mammalian body. Show me an animal with a large asymmetry, and I'll show you a sick animal. Much of the cosmetic industry and the work of plastic surgeons attempt to retain or restore bilateral symmetry. On the other hand, *small* exceptions can be intriguing — witness the mole above the upper left lip on supermodel Cindy Crawford.

One strange egg posed the puzzle of why mirrors apparently reverse images left and right but not up and down. The path out of this paradox was the realization that mirrors do neither; rather they reflect front to back. An object reversed front to back and with no plane of symmetry is an enantiomorph of the original.

Put your right hand on a mirror; the image in the mirror is a left hand. But it's *not* your left hand because the ring, moles, and freckles are missing. Pasteur discovered some molecules come in enantiomorphic pairs, as do rare seashells and rarer Siamese twins.

Another egg, very pear-shaped, refuses to lie flat, but chooses to stand on its larger end. Early Egyptian and Assyrian art and architecture specialized in bilateral symmetry. In all ages, vertical symmetry has been uncommon in art except for reflections in water. So very habituated are we to seeing only bilaterally, that upside down drawings are incredibly hard to comprehend.

The largest egg in our basket is spherical, white, and exceptionally beautiful. It's a gift from a young Frenchman Évariste Galois. Since the egg's rotations and reflections obey the clan rules, they form the infinite group of transformations $O(3)$. Groups extend the concept of symmetry to actions that leave the object to which they're applied apparently unmoved. As well, they're marvelously useful in solving subtle mathematical problems, such as proving the insolvability of general equations of the fifth degree.

Five perfect eggs (Chapter 3) huddle together as if for warmth in a corner of the basket. These are the golden orbs of history, legend, and the natural world — and the topic of the final book of Euclid's *Elements*. We discovered that these compose only three "groups" because of the concept of duals.

A bigger clutch of seven eggs lines the edge of the basket. Here we have all the possible frieze patterns generated by rotations, reflections, and translations on a single motif. Mathematicians, in an automatic reflex, imagine the frieze to be infinitely long in both directions. This allows the entire strip to be moved without detection; hence, these transformations — including the translations — form symmetry groups.

A troublesome egg refuses to rest but rolls about with an off-center balance point. This instability gets things moving, causes disasters, breaks eggs, destroys symmetry, and thereby creates patterns. Pierre Curie deposited a soft-shelled turtle's egg into our basket. Since eggs that are hidden in cavities or the soil have no need of camouflage, they're chalk white. Additionally, because there's no danger of falling from cliffs or open nests, they're spherical. Now the troublesome egg in its travels happened to strike the reptile's egg on one end causing

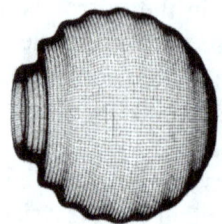

COLLAPSING EGG

its leather shell to collapse. This reduced its symmetry from O(3) to O(2), and although both groups are infinite, the latter is nonetheless a subset of the former. Yet no symmetry has been lost or broken; rather it is shared around through the possibilities — meaning the egg could have been hit anywhere.

Two unrelated eggs (below) are still in our basket. The owl's has symmetry group O(3); the grouse's — considering the color — has none but the identity (I). At first glance the sphere's shining white surface is mildly interesting, but your attention promptly shifts to the other's intriguing lacework patterns. Symmetry makes us sleepy; patterns captivate.

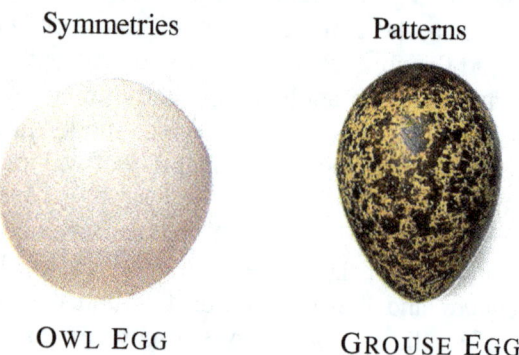

Symmetries   Patterns

OWL EGG   GROUSE EGG

With our basketful of eggs, it is time we whipped up an omelet to get a sense of their taste and texture. In the last few pages, we've had some appetizers; now it is time for the full-course meal.

AS JADED INDIVIDUALS we seek extravagant events. Our entertainments become increasingly violent with more explosions, more shootings, more deaths. Our music grows to painful levels and our sex to multi-orgasmic planes; we miss the depth that contrast brings. No subtleties for us. We want our daily bread stuffed with jalapeno peppers. We ignore nature's fine texture, its small threads slip through our fingers. On this point in the last phrase of his book *A Year In the Maine Woods,* Bernd Heinrich writes, "... the subtle matters, and the spectacular distracts."

We'll start with the subtle but end with the most spectacular event in all creation. Better yet, we'll start with what we know: Figure 8.12, the drop-of-milk crown. Each of the 24 spikes is a column of milk, but instabilities (surface tension and waves) result in each spike pinching off at least one droplet. When the crown falls, it produces outward and *inward* sets of concentric circles. The smaller ones converge at a central node sending up a column of milk, the "Raleigh jet," which breaks up into a series of equal-sized drops because of unseen waves of instability. This behavior characterizes thin columns of fluid. The photographs to the right are of the same jet at slight time delays.

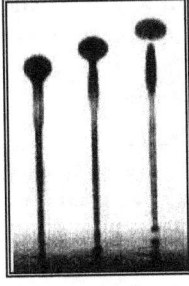

RALEIGH JETS

A fall morning sparkles with silver cobwebs — omnipresent but unseen until dew adorned. We give their geometry a passing glance and walk on by. But as Mae West joked, "It is better to be looked over than overlooked." Let's take Miss West's sage advice and linger awhile. These silver beads are not purposefully geometrical at all; their form arises from simple physical laws — the spider is an inadvertent architect. In this way the web differs from the bees' honeycomb in which the cells are arranged to use the minimum amount of wax possible while retaining strength and rigidity. This isn't physics on the bees' part; it's evolutionary programming.

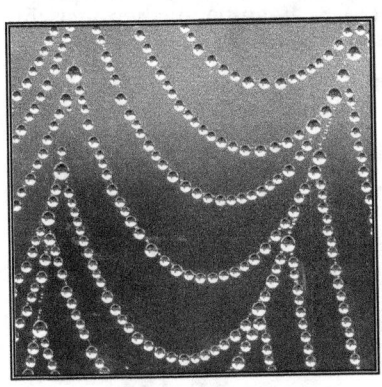

BROKEN SYMMETRY

Figure 8.14

Now look at the evenly spaced dewdrops on the spider's web shown to the left. Does the regular pattern of drops evoke any wonder in you? It should. The dew coats the gossamer threads in a sheath of water — a hollow cylinder coating a solid one. Or this is what we imagine should happen, but it doesn't, at least not for long.

Let's play mathematician, meaning let's generalize. Assume we have an endless, straight gossamer that's coated with a sheath of water. Choose *any point* on the gossamer. An infinite number of translations and reflections may be achieved because all points on the thread are equivalent from the standpoint of these transformations. But as we know, this is a temporary state since the sheath is unstable — surface tension compresses it in the direction of the thread, and there are waves, always waves. Instability does the only thing it can — seek a more stable configuration, and this is what you see in Figure 8.14. This new state has less symmetry: you can only translate and reflect from the *center* of a drop or at a point *exactly between* two drops. The symmetry has been broken and this beautiful pattern has replaced it.

WHEN I WAS A TADPOLE, I foolishly debated a "scientific" creationist. And almost before the debate began, he veered from evolution to the origin of order out of chaos. I can still hear him demanding, "How can patterns emerge from total randomness?" Paradoxically, they do just that! *The most highly symmetrical systems are always the most random.* And it's by the breaking of these symmetries that we create many of our most beautiful patterns.

The antepenultimate example of *spontaneously* broken symmetry — and a lead up to our final illustration — is the Heisenberg ferromagnet. Every magnet consists of oodles of little magnetic domains which we'll *represent* by a gazillion little compasses placed on a round table. Imagine further that this table is shielded from the earth's magnetic field. Each compass is now free to move subject only to the magnetic fields produced by its neighbors.

At the start the needles point in random directions, and the overall field produced is zero — their individual fields subtract as often as they add. What is the total symmetry? Since no preferred direction exists, we are free to rotate or reflect the table (really the compasses) about its center or any line through that center. So, the symmetry is infinite.

Now orient a handful of compasses in a small section of the table so their needles point in the same direction. You can do this by using a strong, external magnet on that section of the table and then switching it off. This newly created magnetic domain will spontaneously cause all the remaining needles to follow suit and align themselves in the same direction. The result has only two axes of symmetry — one in the direction of the poles and the other perpendicular to them. Clearly, the symmetry has been shattered from infinite to just two. In addition, this new setup is stable. Move any needle from north and it quickly springs back to attention. The original configuration was highly symmetrical but unstable; the new arrangement has little symmetry but it is stable.

Curiously, creating symmetry in a magnet is easier than breaking it. If you heat a magnetized piece of iron beyond 1,418°F (770°C), the so-called Curie point, its alignment vanishes — the heat has restored the symmetry. As we shall see, restoring symmetry in this manner is a primary concept of modern physics.

My attentive readers will find it easy to notice broken symmetries. Here's a set of penultimate examples. The mud flaps behind the rear wheels on a large truck (when its speed is suitable) swing alternately and break the symmetry of swinging together. Airplanes *appear* bilaterally symmetrical, but they fly with their nose slightly to the left or right of the flight path (switching every few minutes) and so break the symmetry. Why? The nose-first position is unstable and could cause the craft to shudder and possibly shatter. Desert sand dunes, ripples on a beach, the walk of animals: these are examples of symmetry breaking — not that it all vanishes. Occasionally what appears to be broken symmetry isn't. Consider the chimney swifts of the air, best described as cigars with wings. Because of their extremely erratic flight, ornithologists long thought that they flapped their wings alternately. Stroboscopic photography dispelled this myth, but the illusion is persistent.

ALL GOOD THINGS MUST END. Our ultimate example is as big as it gets: the universe. In the beginning — the true beginning — the world was without form (pattern) but had perfect symmetry. As far as we know, it was born along with time in the Big Bang. Before that, there was nothing — like you before conception. The early universe — the first nanosecond — was incredibly hot and had the utmost simplicity in addition to its symmetry. As it expanded, it cooled. And here's the key point: *it cooled at varying rates,* as shown so clearly in the picture below of the Big Bang's background microwave radiation.

The red spots are hot, blue are cold. Be aware that the heat mirrors the density distribution: the red spots are denser as well as hotter. We have this wonderful picture of instability thanks to the satellite known as the Wilkinson Microwave Anisotropy Probe (WMAP) launched in July 2001. This picture captures the oldest light in the universe! It's from 379,000 years after the Big Bang — more than 13 billion years ago. That's equivalent to taking a photograph of an 80-year-old person on the day of their birth.

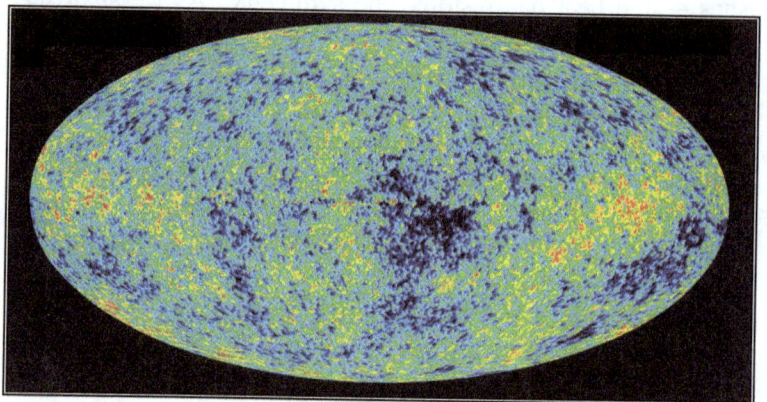

COSMIC FINGERPRINT: ECHO OF THE BIG BANG

Figure 8.15

Heinz Pagels in his elegant book *Perfect Symmetry: The Search for the Beginning of Time* summarizes what we have come to know about symmetry breaking and the Big Bang:

> Here, for the first time, we see a remarkable feature of the modern theory of the origin of the universe: the further back in time we go, the hotter the universe becomes, and broken symmetries are restored. The universe and all its particle interactions are becoming more and more symmetrical as we descend deeper into the big bang. This feature holds out the hope that the universe becomes simpler, more symmetrical and more manageable in its very early history, a hope to which physicists cling in their model building.
>
> Conversely, were we to progress forward in time, we would see that as the temperature falls, those perfect symmetries are broken.[6]

We are puny creatures! We are parasites devouring the third planet of an ordinary star, one of 300 billion, in a small arm of an out-of-the-way spiral galaxy, a minor part of the Local Group of Galaxies, a minute fraction of the Virgo Super-Cluster, one of innumerable such clusters — a mote in the eye of the universe. Yet, we are privileged to have a vision of creation *built on evidence.* We are splendid creatures! You and I are rolled out of stardust, baked in the furnace of broken symmetry. "What a piece of work is a man! how noble in reason! . . . in action how like an angel! in apprehension how like a god!" (*Hamlet,* Act II, Scene 2.)

Sometimes gods can be strange creatures. One such was Robert Oppenheimer, the scientist in charge of the Manhattan Project to create the atomic bomb. His mastery of physics was intimidating to lesser mortals; his knowledge and recall of poetry was legendary. He learned Italian just to read Dante in the original — he was both scientist and poet, Epius and Phemius.

At the Trinity test site of the bomb, on the Alamogordo Bombing Range, Oppenheimer recalled everyone's reaction to the successful detonation of the world's first nuclear device:

> A few people laughed, a few cried, most people were silent.... There floated though my mind a line from the *Bhagavad-Gita* in which Krishna is trying to persuade the Prince that he should do his duty: "I am become death, the shatterer of worlds."[7]

This is but one face of Krishna-Shiva. The complete Hindu god is the destroyer of symmetries and the creator of patterns; the lord of the cremation ground yet a symbol of regeneration and the wild dancer of change.

CHAPTER — 9

RIDERS IN THE WHIRLWIND

The basis of the growth of modern invention is science, and science is almost wholly the outgrowth of pleasurable intellectual curiosity.
Alfred North Whitehead, *The Viking Book of Aphorism*

SUNDAY MORNING and all is well and quiet and dull. The odor of lavender and old lace permeates the halls and common room of the seniors' residence where Epius and Phemius share an apartment. Life seems lifeless. Sunday afternoon is the time the visitors come: sons and daughters with their children performing their ritual duty. The residents sit and talk about the same stories, the same tall tales, with the usual complaints. All's well. But the children are different. They don't know the rules: the rule to be quiet, the rule to act proper, and the supreme rule to never talk about anything interesting. The children are special.

Epius and Phemius have never had a visitor — all their friends, wives, sons, and daughters, have been dead for a very long time. Instead, they play with the children who have escaped to the common room. They don't play from a distance and at a height, but on the floor at the children's level. Epius gives them puzzles to try, some mental and some tactile, while Phemius plays word games and shows them how to draw faces. Sometimes the youngsters bring puzzles for the two oldest children at heart to attempt. One young girl took Phemius' drawing pad and outlined the following two figures. She asked him to redraw each figure without removing his pencil from the paper or going over any line twice.

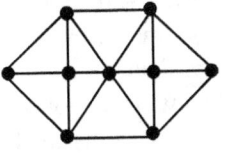 

WINGED HOUSE   RECTANGLE

With remarkable concentration, the ancient poet worked on the puzzle for 20 minutes. Eventually he retraced the winged house without going over any line twice, but the rectangle — the simpler figure — proved impossible. After the mothers had collected their children and left, he gave the problem to his roommate who prided himself on being a solver of puzzles. Reasoning together, they noted that the vertices (nodes) were either odd (o) or even (e) according to the number of lines that

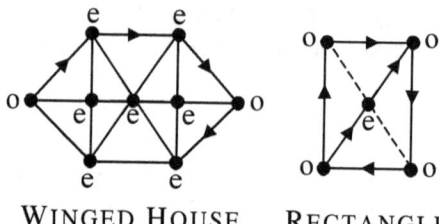

WINGED HOUSE   RECTANGLE

ran into each. The winged house had two odd vertices but the rectangle had four. To solve the first problem, start at either odd node and you will finish at the other (partial solution indicated above). Why? By passing through a node, you *take away two roads* leaving it either odd or even — as it was. The only way to remove an odd vertex is to start or finish there. If you must keep your pencil on the paper, you can do this *only once*. So, if like the rectangle, it has four (or more) odd nodes, the puzzle cannot be done.

After supper Phemius and Epius discovered this was an old puzzle dating back to 1736 and the seven bridges of Königsberg. At that time mathematician Leonhard Euler's analysis of the nodes as even or odd provided a general solution to this and all such problems:

- If all vertices are even, you may start anywhere and finish at the same vertex.
- If two nodes are odd, you start at one and finish at the other.
- More than two odd nodes make the problem unsolvable.

Late into the evening the two blessed men poured over increasingly complex figures. Sometime after midnight Phemius noticed that their drawings always had an *even number of odd vertices (0, 2, 4, 6, . . .)* — the winged house has 2 and the rectangle has 4 odd vertices. Epius with his mathematical skills managed a general proof, and interested readers will find this in the Chapter Notes.

Although the reader will likely never be called upon to solve a retracing problem, yet the latter point about the number of odd vertices tells us something noteworthy about the world. When Leonardo drew and van Gogh sketched, they didn't consciously abide by this law — they had no choice. Every circuit board, every blade of grass, every bush, every tree of whatever complexity must conform to this pattern in nature, this twist in the fabric of space: *the number of odd nodes is always even.*

WE LIVE IN A WORLD OF PATTERNS, and mathematics is their science. Where do they come from? As we have seen, from many places: chaos, fractals, natural selection, the florets of flowers, soap bubbles, broken symmetry, and the structure of space. The specificity of each place and time creates an incredible variety of patterns, but we will seek their common features — the grains in the stone. Here are three: spirals, meanders, and explosions. And, of course, all these have an even number of odd nodes.

Humans have been drawing these three patterns for as long as our collective memory allows. Even prehistory attests to this. And the Hubble Space Telescope gives a gallery of *nature's* celestial love affair with these forms. Certain individuals interpret vast designs like the Whirlpool Galaxy (*see* below) as a commandment from god to fall on their knees. We call these uninformed folk creationists. Yet the form found in galaxies is the same form we see in fiddleheads the size of your thumb (again, *see* below). A similar reduction in size takes us to the microcosmic world of spirochetes and spiral viruses.

THE WHIRLPOOL GALAXY

FIDDLEHEAD

AFRICAN KUDU

Spirals are a familiar pattern. We can see them in toilet paper rolls, in the horns of antelope and sheep, in the curving shell of the chambered nautilus from the depths of the Pacific Ocean, and in the decaying orbits of satellites from the heavens above. We see them in hurricanes and tornadoes; we observe them swirling down our bathroom sinks and toilets. (No hard evidence exists that they flow clockwise in Australia.) What unique properties do they possess? Why are they a major template in nature's toolbox? We will answer such questions.

Eddies and whirlpools are eccentrics living in rivers and streams: they form environments insulated from the rest of the watery world around them. By draining energy from the downhill flow, their whirligig travels may even move them *up river*. In this behavior, they're analogous to living organisms, which also make the environment pay for their existence. By digestion and decay, organisms break down the molecules they draw from other living things. Disorder follows in their wake — entropy increases, as physicists say. Yet for a little while, at least on this planet, they live like eddies and grow by flowing against the current.

Horns are the stuff of legend and history, art and literature, powerful in its symbolism. Even today the slang word *horny* implies sexual energy, and some warriors, like Vikings, still decorate their helmets with them. The Bible uses this symbolism frequently, especially in Daniel and Revelation.

MICHELANGELO'S MOSES

ALEXANDER THE GREAT

When St. Jerome rendered the Scriptures from Hebrew to Latin (4th century AD), he mistakenly translated "rays of light" as "horns" — in Hebrew the words are spelled the same. This was in reference to the radiant glow around the prophet's head as he descended from Sinai with the tablets (Exodus 34:29). Now Michelangelo found his description of Moses in this Vulgate translation, hence the horns on top of his head (or so the legend goes). However, he may have known about the mistranslation and still decided to follow this decorative tradition of medieval art. As for the Greek coin shown above (minted at ancient Pergamum) depicting Alexander's horned head, this was a standard image for a powerful leader.

The meander is brother to the spiral and was named after a river with many windings in Turkey. The silhouette of the kudu's horns (top of the previous page) is a meander as is its trail across the Kalahari plateau. In Chapter 6 we encountered this form as a winding stream in connection with pi, and Einstein's model for its twisting as a battle between the terrain and inertia under the overseeing force of gravity.

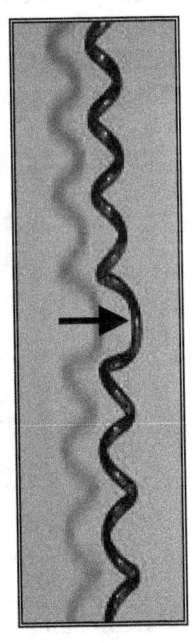

TELEPHONE CORD

The three-dimensional spiral at the right has a meandering shadow. Curiously, squash and cucumbers have spiral tendrils to secure them to the earth identical with this — even to the neutral section indicated by the arrow. The handedness of the spiral reverses on opposite sides of this section. To remove a telephone cord's *neutral zone,* you must work it toward the dangling, twirling receiver and reverse that entire part of the spiral. Note the meandering shadow has no handedness.

Horns and antlers are very different: the first spirals, the second branches. Furthermore, the horns of sheep, goats, and cows are permanent while the headdresses of deer are shed in early winter. Space-filling horns spiral because one or two sides grow more quickly than the others. It's identical, as they say, to having only one oar in the water — you spiral about. If, however, the growth rate alternates periodically, you meander; this is identical to rowing with your oars out of phase. Of course, rowing synchronously creates a straight wake.

Both pictures below show meanders. The windings on the Martian ice cap, like the channels of decades past, may be optical illusions. Scientists think they could be layered hills, buttes, and mesas. As to the processes that created them, that's problematic. But the lowly brain coral we understand: the poor polyps, living creatures with tiny waving tentacles, line the trenches of the coral throwing their excrement on the raised ridges between the dugouts like soldiers tossing sandbags. All their life is spent in a competitive shoving match with the troops in the neighboring trenches. Their fate makes the punishment of Sisyphus in Tartarus seem blessed. This pushing and jostling results in equilibrium — a balance of close-packed, space-filling meanders. Clearly, the processes forming each meander (snow cap and coral) are different, but the visual effect is the same.

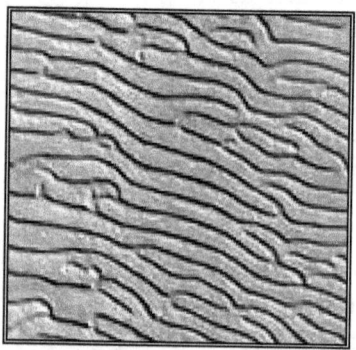
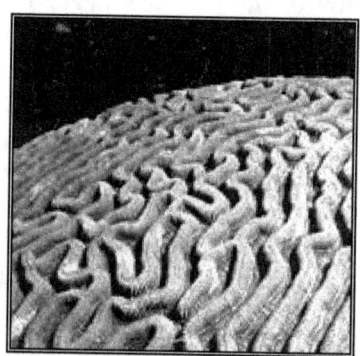

SOUTHERN POLAR ICE CAP ON MARS

BRAIN CORAL

Explosions, the last in our triptych of basic forms, are commonplace and not closely related to either spirals or meanders. As you can see from Figure 9.1, a direct path leads from the center to every outlying point. Although, as with light and sound, the intensity or density of the explosion fills space, yet it falls off with the distance from the center. Mathematically it declines as the inverse square of the distance. That is, move 2 units from the center and the density reduces to 1/4; move 3 units and the density diminishes to 1/9. This falloff is evident in the rays produced by debris thrown out from the famous lunar crater Tycho shown below.

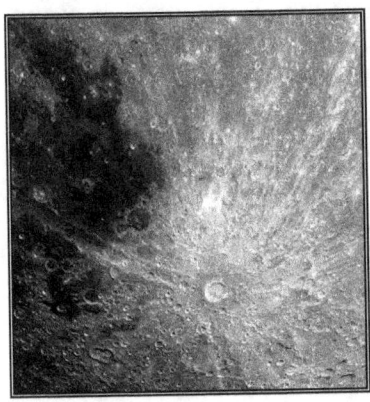

LUNAR RAYS, CRATER TYCHO

Figure 9.1

Explosions were mentioned in Chapter 8 with respect to the loss of symmetry when a drop impacts a pond or dish of milk (*see* Figure 8.12). As the velocity of the drop increases, so does the length of the spikes formed. Similarly, as the average yearly temperature in a region rises, so does the length of the limbs, fingers, toes, and tails of its inhabitants. You may recall this as Allen's rule. With a decline in temperature, protuberances shorten.

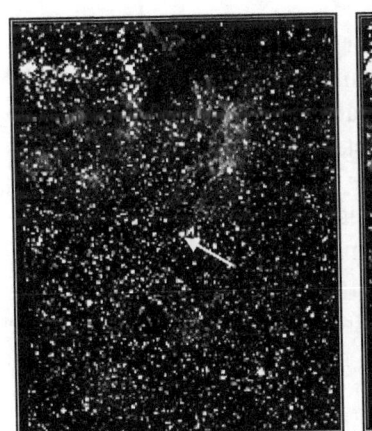 

Before   After

SUPERNOVA 1987A

Figure 9.2

Death and destruction are inevitably associated with explosions, yet it was not always so. The above two photographs were taken of the same region of space on different nights — before and after shots. They show the birth of a supernova from the star indicated by the arrow — one of the most awesomely powerful events in our universe. Actions of this magnitude evade mere words; so we fall back on the quantitative descriptions of mathematics: this supernova had the equivalent power of a 10^{28}-megaton bomb (i.e. a few octillion nuclear warheads).

It's the nature of explosions that they distribute their sound, light, or atoms in all directions as quickly as possible. Stars distinctly larger than our sun collapse catastrophically and then *explode* as supernovas (*see* Figure 9.2). By so doing they seed the universe with elements like iron and carbon created earlier in their nuclear hell from hydrogen and helium. Writing on this point, the late Carl Sagan eloquently described our intimate connection with this stardust of elements:

> The matter out of which each of us is made is intimately tied to the processes that occurred immense intervals of time and enormous distances in space away from us. Our Sun is a second- or third-generation star. All of the rocky and metallic material we stand on, the iron in our blood, the calcium in our teeth, the carbon in our genes were produced billions of years ago in the interior of a red giant star. We are made of star-stuff.[1]

The necessary elements for life would never have come together in a suitable environment for evolution to work on without explosions. Paradoxically, these synonyms for death are the fountains of our existence. The universe is full of wonders, but parts of it, maybe the whole, are not so complex that humans can't understand it. In the last decades we have discovered something truly wonderful — the distant origins of life. We know where we come from: the stars.

With spirals we went from galaxies to fiddleheads; now let's make a similar descent from supernovas to flowers. To get their florets and seeds directly to the sunlight and the wind, many flowers use an explosive design. Here are two:

 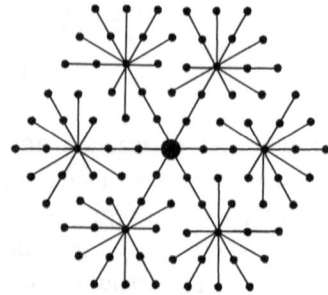

QUEEN ANNE'S LACE   COMPOUND EXPLOSION

Figure 9.3

Queen Anne's lace (wild carrot) is a common roadside plant typifying the compound explosion. The stem terminates in a burst of florets the way streamers on a firecracker end in a barrage of smaller explosions. A popular subject for painters, the center of the lace surrounds a tiny deep-purple floret (colored black here) often overlooked by casual observers. In fall, the many stems fold up to form a mock bird's nest.

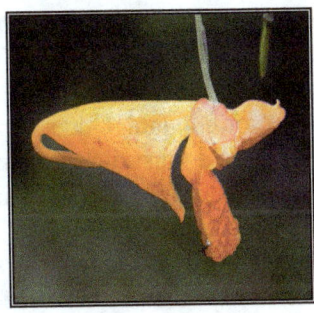
JEWELWEED

POD, SEEDS, AND CASING

Our second flower is the exquisite jewelweed with its orange, trumpet-shaped flowers. It's an annual plant so yearly seed production is a necessity. Unlike Queen Anne's lace, jewelweed has no visible explosive design, but then, neither does dynamite. The pods perform the best known of all plant tricks. A mere touch or gust of wind will burst the ripened pods sending their seeds as far as 7 ft (2 m). In the photograph at the above right, the pod with the arrow has burst leaving its shattered shell casing behind. This hair-trigger reaction is the source of its other common name *touch-me-not*. By October the parent plant has shot its seeds in a wide circle. Then these tokens of eternity sleep away the winter under a blanket of snow to be reborn the next summer. The touch-me-nots are another example of explosions leading to new life.

> *Euclid alone has looked on Beauty bare.*
> Edna St. Vincent Millay

MUCH OF WHAT FOLLOWS was inspired by Peter S. Stevens illuminating book *Patterns in Nature* (New York: Atlantic-Little, Brown Books, 1974), pages 37-48.

Why does nature prefer one pattern over another? In the broadest sense all things evolve to fit their environment whether they be moose, mouse, or mountain — living or dead. We realize that blind luck plays a role in life and death in an improbable number of ways. We realize that things evolve to a state of least energy, least action, and least motion. We realize that the surviving forms are those most likely to survive. Moreover, we should also realize that the previous sentence is a tautology as empty of *empirical meaning* as saying, "A bachelor is an unmarried man" or "Whatever will be will be." Why are certain forms optimal, the most likely to survive, the best of all possible designs? Mere survival is not enough! Understanding doesn't progress by tautologies alone. We must dig deep into the heart of a form's special properties — as Euclid did — and so examine the essence of the thing-in-itself.

Let's focus on three forms, the ones we know: spiral, meander, and explosion. In the last few pages we have seen many examples on earth and in the heavens. The question arises, "Is nature so bereft of imagination that these three forms account for a large part of her design?" Yes, she is, but what wonderful creations come from such a limited repertoire of templates.

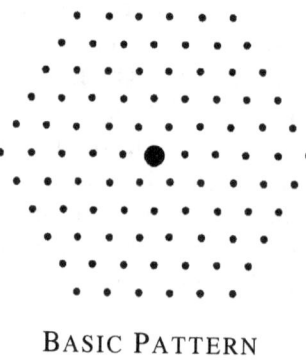

BASIC PATTERN

Consider the array of dots at the left. Every interior point is equidistant from each of its surrounding six points, and the lines joining them make angles of 60° and 120° — we have seen the latter angle before when three soap bubbles meet. The large central dot should be considered the source of growth as with a flower. Like children doing a dot-to-dot puzzle, we'll connect them subject to the following two rules:

- All dots are linked to the large central dot.
- Any two dots connect along only one path.

As an aside, consider this: to maintain the same distance between neighbors on a flat surface, six points must surround each interior point. To do the same on a sphere, you have the liberty of surrounding each point by three, four, or five others. This takes us back to Chapter 2 and the Platonic solids — *see* Leonardo's powerful drawings of these in Figure 2.1. You can easily imagine that each regular solid has its corners on a sphere and so three, four, or five corners — never six or more — encircle every corner.

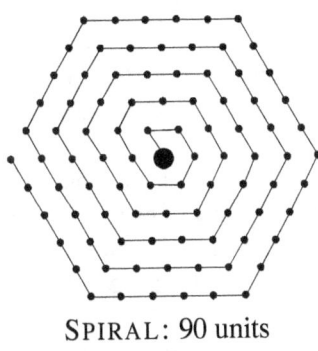

SPIRAL: 90 units

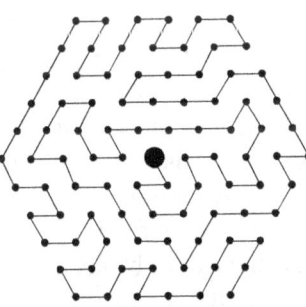

MEANDER: 90 units

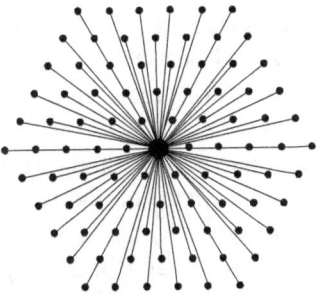

EXPLOSION: 233.1 units

Figure 9.4

In the spiral (left), let's begin at the black dot and move outward to connect all 91 dots.

In the meander (*middle* left), we can leisurely wander this way and that until also connecting all 91 points. If you judiciously set the distance between adjacent dots at one unit, then these two designs have an identical length of 90 units.

Think of the path from the center point to the second point, the path from the center point *through* the second to the third point, the path from the center *through* the second and the third points to the fourth, and so on, as distinct paths. Then the average of all such paths for both the spiral and meander is 45.5 units. If you're a plant, trying to supply food to your extremities, this number is of great significance.

But the explosion is a design with different features. Every dot is directly connected to the central black dot making the total distance 233.1 units — much longer than the 90 units of the spiral and the meander. The average path to all outlying points, however, is 3.37 units — much shorter than the 45.5 units for the other two.

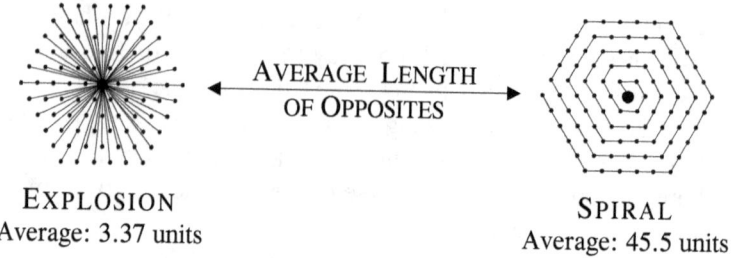

EXPLOSION
Average: 3.37 units

AVERAGE LENGTH
OF OPPOSITES

SPIRAL
Average: 45.5 units

Figure 9.5

Variations of the explosion are favorite designs for flowers since they allow them to *directly* feed their petals, leaves, and seeds, but spirals are rare.

There are other ways to link our 91 dots — for example, the branching designs. We learned previously that Queen Anne's lace has a compound-explosive design (Figure 9.3) with a total length of 107.6 units — less than half that of the explosion. Intriguingly, the average path from the center is only 4.25 units, considerably less than either the spiral or meander's 45.5 units. This branching pattern is an excellent compromise between the extremes shown in the diagram above (Figure 9.5): less building material, but still with quick access to all the extremities.

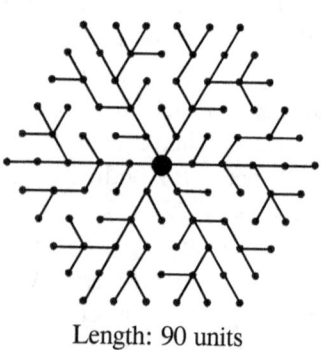

Length: 90 units
Average: 3.73

The outline to the left has six main branches and many more secondary, generating a total length of 90 units, the same as the spiral and meander. Nonetheless, the length of the average path is only 3.73 units — quite near the explosion's 3.37. If this pattern were generalized to three dimensions, it would become an excellent model for sugar and red maple trees.

All branching patterns are a compromise between the direct explosion and the circuitous spiral — somewhere on the spectrum of Figure 9.5. With only a little indirectness, it is possible to achieve a shorter length than 90 units, and branching allows this winning strategy: *see* the next figure. The *total length* of this unbeatable design is 77.9 units — 13 percent below the spiral and meander's 90. And the *average length* is just 4.2 units — 25 percent above the explosion's 3.37 units. That average is more than 1,000 percent lower than the spiral's.

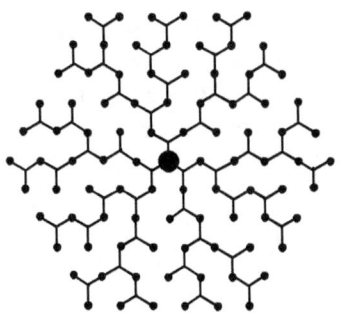

Length: 77.9 units
Average: 4.2 units

All this is achieved when the lines meet in threes at 120° to each other. This angle is an old friend, and when more angles are added to this diagram, the honeycomb emerges. In an earlier chapter, we learned that to join three towns with a minimal road system, you must use angles of 120° — *see* Figure 7.8. The form to the left is the shortest of all possible branching patterns.

We know one of the secrets of the forest. During the summer months, trees — the leaf warriors — battle to keep their surface-area-to-volume ratio stable. Because of the *square-cube law,* the area increases as the square of the height while the volume multiplies as its cube, requiring the trees to leaf out disproportionately (*see* Chapter 1). Leaves need sunlight to perform photosynthesis. They must get to the canopy as quickly and economically as possible — branching at 120° angles allows just that. This is another secret of the trees. In the world

EXPLOSION
Length: 233.1 units
Average: 3.37 units

TOTAL LENGTH
OF OPPOSITES

BRANCHING
Length: 77.9 units
Average: 4.2 units

Figure 9.6

of the living, this pattern is a product of descent with modification, as Darwin would say. In the other world, it's the way natural forces behave, part of their economy — as with soap bubbles.

Although the spiral is a poor model for a tree transporting nutrients, on rare occasions we use it in our architecture. The best-known example is the Solomon R. Guggenheim Museum in New York City conceived by the legendary American architect Frank Lloyd Wright. Outwardly it appears like a teacup formed from ribbons of white spiraling concrete.

On the inside it's a cathedral of modern art and sculpture crowned by a dome of light illuminating the galleries below — *see* the photograph below. Whether you promenade up or down the spiral, you pass each painting once and once only. The Swiss-French architect Le Corbusier has designed several square spirals like the Museum at Ahmedabad in India. Unfortunately, Le Corbusier's work is often marred by a mystic adherence to the golden ratio as if nature knew no other number. What does 1.618 add to architecture? Nothing! What extra significance does phi give to any work? None! This distraction and its dismissal were noted in Chapter 6 (*see* page 236) — the higher phi foolishness that humans have mastered.

SPIRAL INTERIOR OF THE GUGGENHEIM MUSEUM

Spirals may be gentle things like art galleries and pumpkin tendrils, or they may be hurricanes and tornados sweeping away everything in their path. Meanders on the other hand are by definition easy-going, circuitous, like the trail of a saunterer to the Holy Land. Their strength is released slowly at every turn, as in a stream or a bend in a brook. But explosions are always violent! Their energy catapults in every direction like water from a large dog shaking itself after a swim — it's not enduring like a hurricane or gentle like a stream but immediate and everywhere.

Explosions can be discovered in unexpected places, although they are impossible to find in architecture except as the ultimate product of terrorism. But thinking in the broadest sense, they occur even in the spoken word. The created language of the Klingons of *Star Trek* fame is guttural and vigorous, full of (ex)plosive consonants like *k, c, p,* and *q*. The names of their warriors often begin with *k*, the capitol of their homeworld is Kronos, and the founder of their empire was Kahless. This awareness of language hasn't been lost on advertisers: Kodak, Compaq, Prozac, and Ex-Lax are all two-syllable words created to imply strength or immediacy. Research shows the public thinks positively toward words that *explode forcefully* from the mouth — at least they did in the past. Perhaps it was the catastrophe of 9/11 — the day everything changed — that caused the shift to softer nonexplosive sounds.

As an example of this, and on a lighter note, advertisers have now switched to three-syllable words using soft and lyrical sounds like *l* and *s, z,* and the sibilant *c*. Hence we get evocative brand names like Celebrex (think celebrate) for arthritis and Levitra (think levitate, elevate, vital) for erectile dysfunction, and so it goes.

The spiral, meander, explosion, and various branching patterns reveal much about the world; they let us peek at nature's blueprints. Two numbers summarize the properties of each shape: the total length and the average distance of any point or leaf from the center. Know these and you are close to comprehending the form and therefore the function for this triptych of patterns. Form determines function and these two numbers govern both.

In three dimensions, corkscrews and telephone cords are spirals, a tangled ball of string is a meander, and the crown of hair on your head is an explosion. But nature isn't kind to *pure forms* — she seldom allows them. Winds blow, insects chew, beavers gnaw, diseases destroy, molds grows, fungi adhere, snows burden, rains beat, and gravity warps, but the patterns persist. The universe was born in an explosion that evolved into spiral galaxies and meandering dust clouds and nebulae. In the maelstrom of life, these patterns somehow stay mounted on their design whether on earth or in the heavens above. Down through the empty halls of time, littered with the debris of countless "eternal" civilizations, these three patterns and their derivatives have always been among the immortal riders in the whirlwind of existence.

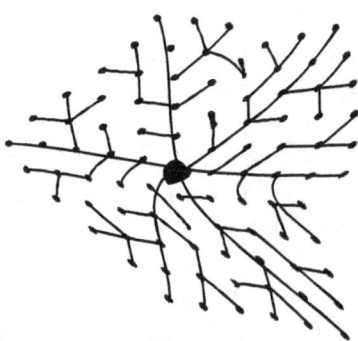

A RIDER IN THE WHIRLWIND

THE PATTERNS CREATED by nature have a permanence unknown to human creations. We all strive to leave something of a legacy in the wake of our life's course. Despots do it with grandiose buildings and monumental statues of themselves, but they will all suffer the fate of Ozymandias in Shelley's sonnet. Its ironic inscription follows:

> *And on the pedestal these words appear:*
> *"My name is Ozymandias, king of kings:*
> *Look on my works, ye Mighty, and despair!"*
> *Nothing beside remains. Round the decay*
> *Of that colossal wreck, boundless and bare*
> *The lone and level sands stretch far away.*

One day the Pyramids at Giza will be no higher than an anthill. If mountains can be weathered away, what chance do human monuments have? Yet some things humans *conceive* may, just may, last as long as our species — perhaps even longer. I'm alluding to ideas and theories. Euclid's theorems will be celebrated when the great dramas of Aeschylus are forgotten; when Shakespeare is gone from the Western canon, Einstein's theories will still be remembered. Callimachus said it best in the last two lines of *Heraclitus:*

> *Still are thy pleasant voices, thy nightingales, awake;*
> *For Death, he taketh all away, but them he cannot take.*

None of this is meant to comment on the relative value of art or science. After all, value can't be measured by longevity. Some mayflies live but an hour while bristlecone pines exist for more than 4,000 years — the pyramids were under construction when these trees were seedlings!

Many artists and scientists believe the act of creation has its core in the discovery of likenesses. We have seen the patterns of spiral, meander, and explosion in nature and the second secret of the trees: branching at 120° angles. Consider now one of the great paintings of the Renaissance, Leonardo's *The Lady with the Ermine*. This portrait shown in Figure 9.7, hangs in the Czartoryski Museum in Cracow. As in most of da Vinci's works, there has been much overpainting and restoration in the 500 years or so since its creation, but the head of the girl, her hand, and the arresting presence of the beast are surely the master's own. What do they tell us? What do they say after five centuries?

The Lady with the Ermine
by Leonardo da Vinci, about 1490

Figure 9.7

Leonardo is comparing the lady to the ermine, the human to the animal. She was Cecilia Gallerani, the 15- or 16-year-old mistress of Ludovico Sforza, usurper of Milan, and patron to da Vinci. The ermine, called a stoat in Europe, was the heraldic symbol of purity and the adopted emblem of the Sforza family. The artist was suggesting a comic parallel between her purity and that of the beast. The name *Gallerani* is reminiscent of the Greek word for ermine, *galée,* and hence a pun on the girl's name.

As Jacob Bronowski writes in *Science and Human Values,* the whole picture is in some sense a pun. Leonardo has matched the ermine to the maid. Her body is almost turned in the opposite direction to her head, as is the stoat's; all eyes are fixed on something off-canvas. Her claw-like hand strokes the beast's back and its paw seems to do the same to her arm. The flattened hair emphasizes her wide forehead like that of the animal. I believe this ermine is one of the most perfectly painted animals of the Renaissance: the sinuousness of its body, the power of its gaze, and the arresting perfection of its head are matchless. The beautiful head of the girl contrasts with the tightness of her lips and slight upward curl of her mouth lending an air of limited mental ability. Bronowski catches the parallel and states the pattern:

> As we look, the emblematic likeness springs as freshly in our minds as it did in Leonardo's when he looked at the girl and asked her to turn her head. *The Lady with a Stoat* is as much a research into man and animal, and a creation of unity, as is Darwin's *Origin of Species*.[2]

The wisely cautious reader may think I'm finding meanings and making inferences in Leonardo's masterpiece that the artist never intended. To prove this isn't so, consider what the eminent art historian Kenneth Clark has to say in his splendid book on da Vinci:

> Very often in reading the description of a picture by a man of letters we feel that what the writer takes to be a stroke of dramatic genius is an accident of which the painter was quite unaware. With Leonardo this is not the case. We know from his notebooks and his theoretical writings on art how much thought he gave to the literary presentation of his subject.[3]

Artist, scientist, and mathematician have varying degrees of freedom in the creative process. The painter and poet have a dimension of sovereignty not given to the scientist who *must* conform to the truth — the physical reality of the world. This conformity is what confers the extraordinary power of science to make and change the world, remolding it nearer to the heart's desire. Whereas the scientist conforms to the world as he finds it, the mathematician is entirely different. He bows to the world as he conceives it. The arithmetician has created numbers so large he has nothing to count with them; theories so outlandish they couldn't possibly portray the real universe. He has three geometries to describe the nature of space (Euclidean, Riemannian, Lobachevskian), only one of which can be true because each contradicts the others while agreeing with its own basic axioms. Present astronomical data from the WMAP satellite (*see* the last two pages of Chapter 8) signify that the geometry of space is flat. Consequently, scientists pick the theory whose axioms imply this, the Euclidean.

Artists often rebel against their culture in obvious ways, but they also conform in subtle ways. Scientists follow the world as it is. Mathematicians obey the axioms of their own system so that they don't contradict themselves. Let us consider examples of each:

ART: *We are such stuff as dreams are made on; and our little life is rounded with a sleep* — a cultural truth about the human condition. Normally we avoid the word *truth* when talking about painting, poetry, and music and refer to emotions and convictions.

SCIENCE: *The planets revolve around the sun* — a truth of correspondence. The truths of science evolve to increasingly more exact statements about the real world. Ptolemy said the planets travel on circles riding on larger circles about the earth. Copernicus wrote that they move in circles about the sun. Kepler measured the heavens and found they course the void in ellipses at varying speeds about the sun. Newton wrapped this in the blanket of universal gravitation and showed all orbits are conic sections. Einstein fine-tuned even this. Next?

MATHEMATICS: *You can't divide by zero* — a truth of consistency: if you do, you'll break other mathematical laws and contradict yourself. This might be all right for the comedian who says, "I never repeat myself; let me say that again," but it's unbecoming for a human activity with the highest logical standards.

These divisions are seldom as clear as I've described. Each recognizes the truths of the others, and some statements are difficult to pigeonhole. And although the truths of mathematics are considered eternal, the truths of art, science, and human society evolve over time. We wouldn't be content to accept the cultural or scientific truths of ancient Greece, still less of Egypt. "All peoples are created equal" is a moral truth foreign to every ancient society — even to the citizens of Periclean Athens. Since the Nuremberg War Crimes Trials, justifying behavior solely on the statement "I was just following orders" is not deemed a defense. The Waffen–SS officers were just following orders. Japanese commanders were just following orders. Serbian genocide squads were just following orders. This *justi*fication is a moral vacuum. Those who accept it plunge into an abyss where no one should descend. Morality evolves, and so do the truths of art and science.

Art often has visual, anatomical, and psychological patterns like Leonardo's *The Lady with the Ermine*. The truths of art have little to do with technique — an autistic child's sketch can be arresting in what it reveals. Musicians call their patterns themes, melodies, or tunes. Writers call their parallels, similes, metaphors, personifications, allegories, and the like. Science is white magic, it predicts the future: you do this, and that will happen. It's a pattern that applies over time. The truth of science is one of correspondence between statements and reality. On the other hand, mathematics writes statements about other statements concerning patterns of numbers, forms, shapes, and so on. The truth of mathematics is one of consistency. Patterns are such stuff as art, science, and mathematics are made of, and these works are rounded with our hopes. They are the truths that dreams fashion.

METHINKS IT IS LIKE A WEASEL

I wish to propose for the reader's favourable consideration a doctrine which may, I fear, appear wildly paradoxical and subversive. The doctrine in question is this: that it is undesirable to believe a proposition when there is no ground whatever for supposing it true.
Bertrand Russell, *Sceptical Essays*

MOST PATTERNS that aspire to science are false, and it's a major task to separate the wheat from the chaff. As you can well imagine, doing this is of no small importance to mankind. After all, we declare ourselves *the* rational animal — on this, we claim our distinctiveness from all other creatures. It's our evolutionary heritage to adapt by altering the environment. Every other life form adapts by changing themselves. Exceptions exist: the heat-retaining and heat-dissipating body types of Inuit and Ruandan (*see* Figure 1.7); the holes, dams, and nests of woodchucks, beavers and birds. But these are minor variations on basic survival themes. We have our song and we'd best sing it. And if we can't, we are like the owl that doesn't hear the mice under the snow or the wolf that can't run through the evergreen forests.

In the previous chapter we huddled in the cold rocks until eventually emerging to drive a lion from a wildebeest rib cage. Most such kill sites were discovered by noticing flocks of vultures circling over the carcass — a pattern in space. Consider the Inuit hunter who knows that twice a year the caribou herds come through a narrow valley in their great migrations — a pattern in time. Finding true patterns was the basis of our survival in the previous million years; finding true patterns will be important for our survival in the next 100 years. Our biology has honed us to be pattern junkies. How do we know when a trend or pattern is significant and not just an outcome from the confluence of multiple events? How do we distinguish significance from coincidence? That is the question! Well, you begin by examining the evidence. And as Russell implied in the opening quote, belief should be directly proportional to evidence.

We'll do what we've done before — strip away everything superfluous. That's easy to do. Consider the following sequence of results, supposedly from flipping a fair coin 20 times:

T H H H H T H T T H T T H H H H H T H H

Examine this sequence carefully. Do you believe it's entirely random? Are the heads (Hs) and tails (Ts) distributed as they should be? The runs of four and then five heads seem beyond what we should reasonably expect. After all, aren't Hs and Ts to occur with equal frequency in a sequence this long? Notwithstanding any of these questions, I simply wrote down the *first results* obtained from flipping the same coin 20 times. The reader is encouraged to do this and inspect the outcomes. You'll almost certainly be surprised.

Have the laws of chance become unglued? Not in the least. This is an example of the "clustering illusion," a phenomenon well known to psychologists and statisticians. It's a false intuition that random events, which often occur in clusters, are not truly random. Some mystics, like Carl Jung with his theory of synchronicity, have elevated this illusion to the status of a world-embracing concept. In real life, clusters occur: you have a run of good luck, three old school chums telephone in the space of a week, on the way to work you see two different license plates with the Antichrist's number 666.

Not only do we fail to recognize random sequences because of their naturally occurring clusters, but we're equally inept at creating them. Most people when asked to make up a sequence of 20 random Hs and Ts are incapable of doing so. If they begin HHH, say, they'll think it's time for a T to keep the outcomes balanced. If they have HTHTHT, they'll feel it's time to break the repetition of pairs with a T. When attempting to create a random sequence, *thinkers* need not apply. Coins, on the other hand, have no memory, nor do they think about the future; their entire existence is now. If a coin turns up Hs in 10 consecutive times, the next toss is still fifty-fifty — mathematicians say each toss is an *independent event*. So are rolls of a die, spins of a roulette wheel, and the gender of newborns. The coin has no prejudice, the wheel no bias, and the womb no recollection. The womb doesn't whisper to itself after the birth of seven consecutive girls that it's high time for a boy. Those that think otherwise suffer from the gambler's fallacy. Since we're pattern junkies — inherently incapable of recognizing or creating random sequences — we're ill equipped to comprehend a universe abounding in random events. We must be taught to do so. If an understanding of *independent events* were widespread, Las Vegas would still be a one-burro shack in the desert!

Expanding a pattern beyond its original parameters is often possible and always exciting, and the *linear* sequences offer such an opportunity. Consider their two-dimensional analogues shown in Figures 9.8 and 9.9. A computer randomly generated one of these; living organisms created the other. Which is which?

Look at each carefully; they're quite different. Have you made your choice? A drum roll please. . . . And the answer is *random* for the top design, *ordered* for the bottom. You might ask the same question of a few friends. Do a little survey and you'll find the preponderance of people confuse the random with the ordered, just as they did for the sequences, and for the identical reason — the inability to recognize *clustering* as natural consequence of chance. Whether our brains have difficulty with this question because of genetics or culture, I don't know, but this almost universal failing can have dangerous consequences.

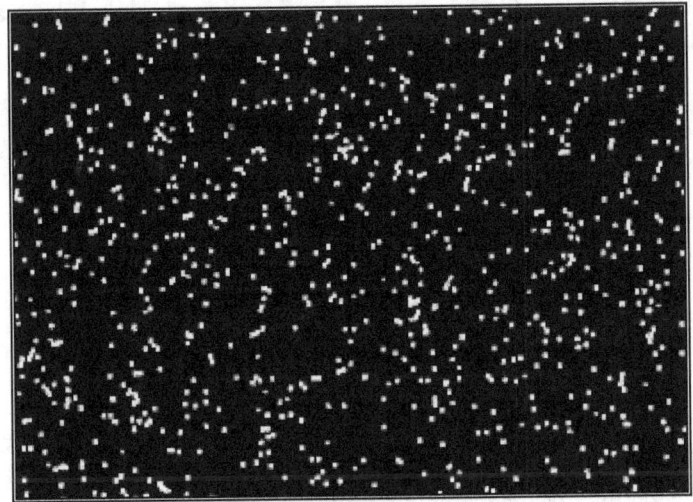

Figure 9.8

Nature offers many examples similar to the random design above. Consider the starry night over our heads, which we populate with heroes and creatures from a former Golden Age. And for one of poetry's greatest images, Omar Khayyám used this "starry" randomness in the *Rubáiyát's* final quatrain:

> *And when Thyself with shining Foot shall pass*
> *Among the Guests Star-scatter'd on the Grass,*
> *And in your joyous errand reach the Spot*
> *Where I made One — turn down an empty Glass!*

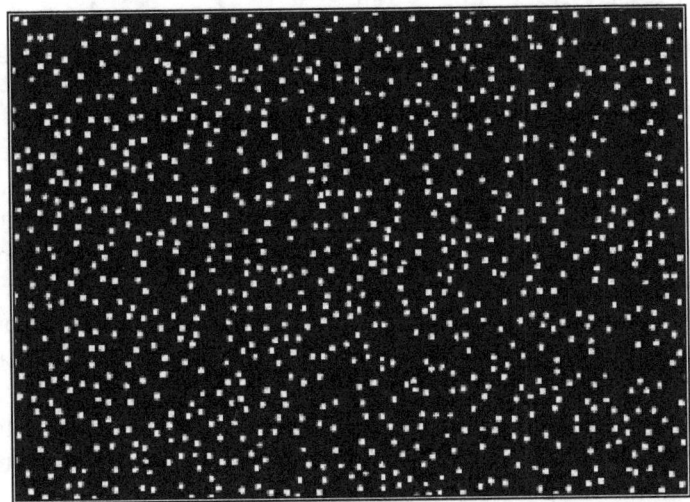

Figure 9.9

Stephen Jay Gould quotes his friend and Nobel laureate Edward Purcell on these imaginary designs we "see" in random star fields:

> What interests me more in the random field of "stars" is the overpowering impression of "features" of one sort or another. It is hard to accept the fact that any feature — be it string, clump, constellation, corridor, curved chain, lacuna — is a totally meaningless accident, having as its only cause the avidity for pattern of my eye and brain! Yet that is perfectly true in this case.[4]

Our non-random figure is spectacularly weird in its origins — no amount of conjecture could uncover its genesis. These seeming dots of light on a blue sky are really the *glowing rear ends of fly larvae* from the roof of New Zealand's Waitomo Cave. The voracious larvae, like chained dogs, form circles of avoidance around themselves thereby creating an "unlumpy" sky. If instructed to randomly put dots on a page, most of us would produce a design closer to the worms than the heavens.

Previously I alluded to the prospect that this human failing can have grave consequences. "Life," wrote G. K. Chesterton, "is full of a ceaseless shower of small coincidences . . . . It is this that lends a frightful plausibility to all false doctrines and evil fads." And what are these small coincidences but a lumpiness in the cascade of a billion daily events — a trivial coming together in time rather than space.

Search history and you will find abundant examples to support almost anything. Someone claims economics is the single and pervasive motivator of all human actions. You cite the Trojan War as a battle by the Greeks to avoid paying a tax to trade into the Black Sea. So the great deeds of Achilles, Hector, Epius, and Odysseus were about money; Homer's poetry was a penny to their pocketbook. After you have finished gathering "evidence," you write a book to tell the world. People call you Karl Marx; you change history — millions die. Someone claims to have a cure for cancer. You collect anecdotal evidence to "prove" that apricot pits eliminate malignant tumors. You write a book. People call you Everyman; you change lives — hundreds die. The patterns of Marx and Everyman aren't true riders in the whirlwind, just ghosts in the minds of true believers.

Undoubtedly the reader and this writer have at some time said, "Now that's a coincidence." You fly to Chicago's O'Hare Airport and in the terminal, or perchance in the seat beside you, is your grade five teacher, Miss Jenkins. Both of you are amazed that after three decades you should cross paths so far from home. You reminisce and then say goodbye still marveling at this chance meeting — surely a great coincidence. Or is it? As a professional class, teachers are the foremost travelers; it's often their major retirement activity. Would it have been just as amazing to run into any of your grade-five classmates at O'Hare? How about any other grades from kindergarten through university? Consider your many friends and numerous acquaintances and relatives. Would it have been just as incredible to see one of them in the terminal lounge? The answer to all these questions is a booming Yes! How about reading a newspaper or magazine and seeing one of these people mentioned? Or sitting beside someone with the same first and last name as yourself? Or perhaps they knew Miss Jenkins, or ... or .... The combinations are extremely abundant, but we didn't consider any of this. We just said, "Now that's a coincidence."

This error is tantamount to noticing *only* the five consecutive heads in the previous sequence of 20 outcomes or concentrating *only* on a random clumping in Figure 9.8 — you miss the big picture, the context, or the sample space as mathematicians call it. Without knowing this, you cannot assign a probability to an event from that space. Recall the challenge to Galileo to determine the likelihood of rolling a 10 with three dice. First, he had to find the size of the sample space, 216 (6x6x6), then the size of the event, 27. So, the probability of this happening is 27 times out of 216. You read that a major-league pitcher with 25 wins is coming to your hometown team. Great! Further on in the article you learn that these wins were over a four-year period. Not so great! Without knowing the size of the sample space for the pitching records, the dice sums, or the everyday coincidences, any element of surprise is unjustifiable.

Let's return to Miss Jenkins with this understanding of sample spaces. If you or I had *specified* — before leaving home — that we would meet our grade-five teacher at O'Hare, then that would be a terrifyingly accurate prediction, but we did no such thing. We waited until after an *unspecified* event happened and then we proclaimed it marvelous and astonishing. We're like the archer who shoots an arrow off in any direction, then runs and paints a large bull's-eye around where it landed and cries, "Hit, I've made a hit!" Coincidences are incredibly accurate when made after the fact and a penny a dozen when you look for them.

The best summary of the last few pages is a cartoon I recently saw. Some sheep are at a cocktail party all drinking martinis. The ewes are all wearing white evening gowns; the rams all have white tuxedos. Off to one side at the bar two ewes are intently engaged in conversation. One says to the other, "Now that's a coincidence, because you know I'm a follower too."

Probabilities and coincidences are rife with paradox. Often the main source of this is the noteworthy size of the event or the sample space. Recall Miss Jenkins again. I've shown that the number of possible coincidences involving a traveler at O'Hare International is amazing, yet in this and many other cases an exact count is impossible. Even in cases where we can enumerate the space, it's often unexpectedly large. Consider the following famous problem: how many students must be in a classroom, or guests at a party, to have a 50 percent likelihood that at least two have identical birthdays? The answer is very small. Would you believe only 23? With a group this size it's nonetheless possible to make 253 different pairs: {Bob, John}, {Mary, Bob}, {Sam, Jenny}, and so on. It's this large number of pairs that allows the low number of people required to have a 50 percent chance of *at least* two having identical birthdays. (No, the number of pairs needed is not half of 366.) To be 100 percent certain of a match, however, you need 367 students (remember leap year) because if you have 366 pigeonholes, and more objects, no less than two must double up.

One evening on the Johnny Carson show, a guest brought up this paradoxical birthday puzzle. Johnny, unwilling to accept it, polled his audience of 120 people for a match to *his birthday* on October 23. No one did, and the guest was unable to explain what went wrong. Although the question Johnny asked is closely related to the classic birthday problem, it has one major difference.

Carson's problem involved a *specific* rather than an *unspecific* birthday — this small difference often causes confusion. How many people did Johnny need in his audience to have a 50 percent chance of an exact match to his birthday? The answer is astonishingly, *namely* 253. In other words, he had to be in each of 253 pairs to have half a chance: {Johnny, Mary}, {Johnny, Jenny}, {Johnny, Bob}, and so on — this is the identical number of pairs needed for an *unspecified* match.

Summary: For either case — unspecified or specified — to have a 50 percent probability of a birthday match you need 253 pairs. For the unspecified, you can form these pairs from just 23 people. In the specified case, you get them from 253 people plus yourself (or whosever's birthday you wish to match). *See* the Chapter Notes for the mathematics on both birthday problems.

Meanwhile back at the cocktail party, two rams are discussing their zodiac signs. One says to the other, "So you were born in the spring too. Then we're both Aries — that's soooo... exciting." With our knowledge of the birthday problem, we might expect that when only a few people are in a room *some* identical zodiac signs are probable. The likelihood of a match among four people is 43 percent; among five it's 62 percent; among *spring lambs* it's 100 percent. So, what's your sign?

If a man look sharply and attentively, he shall see Fortune;
For though she is blind, she is not invisible.
Francis Bacon, *Of Fortune*

The ancient Romans called her Fortuna; the Greeks named her Tyche; modern Europeans and Americans dub her Lady Luck; but by whatever name, she is pictured blindfolded toward sinner and penitent, potentate and pauper. Medieval songs and poems were written to celebrate the whimsy of her ways, and contemporary composers have set many of these to music. For example, consider the opening verses of Carl Orff's spectacular choral piece **Carmina Burana:**

LATIN TEXT	TRANSLATION
O Fortuna	O Fortune
velut luna	just as the moon
statu variabilis,	you vary your state,
semper crescis	always increasing
aut descrescis;	or decreasing;

Unlike our ancestors, we often dismiss Fortuna and deceive ourselves into believing we are masters of our individual destinies. Not for us "the slings and arrows of outrageous fortune." We see our doctors, take our pills, and undergo our surgeries — cosmetic and otherwise — to stay youthful and healthy. And the ultimate rock upon which we rely for every contingency is *insurance*. Yet those that reach for the fabled golden ring of *absolute* security delude themselves. As Demosthenes said, "Nothing is so easy as to deceive oneself; for what we wish, we readily believe."

The more credulous among us interpret random groupings of natural objects as divine apparitions. They see visions in the clouds, perchance the Virgin Mary, or perhaps her visage in the wooden texture of a door, or Jesus in the linoleum. They delude themselves. With a toadying Polonius and a teasing Hamlet, Shakespeare has a gem of dialogue (Act III), mocking all those who would see only what they wish to see:

Hamlet: Do you see yonder cloud that's almost in shape of a camel?
Polonius: By the mass, and 'tis like a camel indeed.
Hamlet: Methinks it is like a weasel.
Polonius: It is back'd like a weasel.
Hamlet: Or like a whale?
Polonius: Very like a whale.

We need not be smug and believe such imaginings couldn't happen to us for they surely have.

Recall our sequence of 20 random flips of a coin:

T H H H H T H T T H T T H H H H H T H H

This sequence has 13 Hs and only 7 Ts. But aren't these outcomes to be equal in the long run? Well, yes and no. The *ratio* of Hs to Ts will grow closer and closer to 1:1, even though their absolute difference will generally grow apart. And who is to say how long a run is? There is no law of small numbers for sequences; they can be anything!

The streaks of hits and misses by basketball players are very like these sequences of Hs and Ts. Similar to a coin, a good player will sink approximately 50 percent of his shots, and coins can be biased to model other probabilities or players.

Every basketball fan knows the phenomenon of the "hot hand," streaks of successive baskets, times when the player is "in the groove" or "on fire." These are the moments when fans are on their feet screaming with excitement as the hero sinks basket after basket. The trouble with this deeply entrenched eternal verity is that it's false — totally false.

At the beginning of his splendid commonsense book *How We Know What Isn't So*[5], Thomas Gilovich conclusively demonstrates that the shooting sequences of players with the hot hand or the cold hand are *indistinguishable* from coin tossing. In other words, the random streaks of heads or tails from coins (*see* the above sequence with a run of four and then five heads) accurately model basketball statistics.

Gilovich's analysis was detailed and took into consideration the many objections of fans and coaches. He studied every basket made by the Philadelphia 76ers for more than one season. His discoveries were twofold:

- The probability of making a successful second basket did not rise following a successful first shot.
- The number of runs of successful baskets did not vary from a random coin-tossing model.

Did Gilovich's work convince fans, coaches, and players of the unreality of the hot hand? Not in the least! The phenomenon seemed so real, so emotional, that his rational analysis was ignored or discounted. Gilovich mentions that Red Auerbach — the fourth most successful coach (regular season and playoffs) in NBA history — said the following upon hearing of these results, "Who is this guy? So he makes a study. I couldn't care less." But a straightforward computer analysis of all 1,048,576 possible sequences for 20 coins flips shows that 46 percent of them have runs of four or more heads (hits), and 18 percent have five or more. This is the source of the belief in the hot hand and Auerbach's derision: the human failure to recognize *natural clustering* as just that and nothing more. (Whatever has been learned about *streaks* in basketball applies with equal veracity to completed passes in football, base hits in baseball, and goals in hockey.)

In first-year logic courses the professor will tell you that from a false premise any foolishness can follow — even more so for the clustering illusion, the gambler's fallacy (mentioned previously), and any narrow selection of data. I ask the reader to reflect beyond the harmless folly of the cluster illusion in spectator sports to dubious doctrines in politics, religion, and economics. In a discerning sentence Pascal noted, "We want to be deceived." Great crimes are often perpetrated by fanatics wedded to some "perceived universal truth" be it Communism, Nazism, Islam, or Christianity. We don't need a history lesson to recall the worst of these. Every truth should be tentative! Science is proof without certainty; faith is certainty without proof.

THE RIDERS

Great ideas come into the world as quietly as doves.
Albert Camus (1913- 60)

A DECADE AFTER WORLD WAR II, mathematicians John von Neumann and Stan Ulan invented cellular automata[6]. They envisaged these as discrete systems of dots (cells) on an infinite chessboard whose behavior was completely specified by local relations. It seems an innocent idea. Cellular automata became incredibly popular after John Horton Conway created his *game of life*.

The *game of life* (or simply *life*) isn't a game in the ordinary sense — there are no players, no winners, no losers. Once the pieces are placed in the starting position, *the rules* determine all else that follows. Nonetheless, it's full of surprises proving determinism doesn't equal dullness. Properties emerge that no inspection of the starting positions will ever disclose. Like biological life, simulated *life* is revealed as the clock ticks.

A cell can be *live* or *dead*. We will indicate a live cell by placing a black circular marker on its square; we show a dead cell by leaving its square empty or with a gray marker. Every cell on the chessboard has eight neighbors sharing edges and corners: two vertical, two horizontal, and four diagonally —

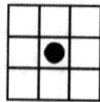

these may be live or dead. The clock moves and with each tick the overall state of the entire grid evolves to the next step according to three rules:

Rule 1: A dead cell (shaded gray) with exactly three live neighbors becomes a live cell at the next step (3 unrelated cases shown).

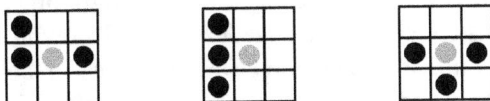

Rule 2: A live cell (two or three live neighbors), shown by the white disc, stays alive at the next step (3 cases shown).

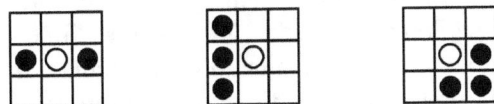

Rule 3: In all other cases a cell dies or stays dead from overcrowding or loneliness at the next step. Consider the evolution to extinction of the five markers at the left.

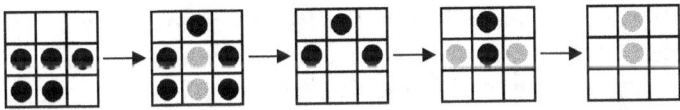

After trying many other possibilities, Conway chose these three rules. Some alternatives caused the cells to be born too quickly while other variations initiated death too quickly. The *game of life* balanced these tendencies of birth and death, making the outcome unclear and thus interesting. This game, in some sense, holds a mirror up to evolutionary processes. Life and *life* have three possible outcomes extinction, (eternal) growth, or stability.

Life is one of the simplest examples of "emergent complexity" or "self-organizing systems" — the most recent scientific discovery of pattern formation and a new horse for the riders in the whirlwind. Recall how two fundamental elements, hydrogen and oxygen, combine to create the unexpected substance water with marvelous emergent properties. In a similar manner this new topic helps us explain how elaborate designs emerge from very simple rules. It will teach us how the leopard got its spots, the zebra its stripes, and the butterfly its glorious wings.

Nature is complicated and we're usually uncertain of its rules; with *life*, however, we know them all. By examining a simple input of cells and the often incredibly elaborate output, we can learn much about general pattern formation in the world outside the study. It's impossible, however, to understand the vast intricacy and dynamism of even simple inputs with words or static diagrams. Fortunately, and as a measure of its popularity, Conway's *life* is the most often programmed computer game in existence. Curious readers should download the following freeware to watch their own versions of cellular automata in action: http://psoup.math.wisc.edu/Life32.html

Life32 is the best, fastest, and most user-friendly *life* player around; half an hour with this program will give you a deep understanding of this game. An early pattern to try is the R-pentomino (to the left) — the first design that defied Conway's attempts to simulate it by hand. Starting with these five dots and our three rules, this pattern gets complicated *very* fast, and it doesn't stabilize until 1,103 generations have passed. Many early computer programs were written to determine the fate of this small design, but with Life32 you can solve it immediately.

If you've been following the R-pentomino on the computer screen, you've noticed that some of its parts are moving. This was an exciting early discovery about *life*. These common moving patterns, called gliders (at the left), consist of just five cells — not very different from the R-pentomino. Follow this design for four steps, and it looks just like it did when it started. There's only a single difference: a shift along a diagonal. Repeat this process, and you'll have a moving object. Remember, the rules said absolutely nothing about movement — these moving patterns just appeared. This is a clear and convincing illustration of *emergent complexity*. The whole seems more than the sum of parts.

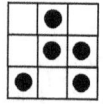

What have we learned from the *game of life?* Three things. First, from very simple rules astonishingly complex properties arise. Second, these emergent properties are in no way apparent in the initial form. Third, the outcome is influenced by local conditions only — the eight neighboring cells. It's as if you had an orchestra without a conductor or a team without a coach.

This last point is difficult for humans to accept. No unseen intelligence directs these things; no lord of creation is required for such designs to exist. Those without a clouded vision can see such designs in nature's tapestry. Clearly, a flock of starlings gracefully flying in perfect coordinated form has no leader, and a school of mackerel dancing through the sunless sea has no choreographer. The writer of Proverbs 6:6-8 said all this centuries earlier:

> *Go to the ant, you sluggard;*
> *consider its ways and be wise!*
> *It has no commander, no overseer or ruler,*
> *yet it stores its provisions in summer*
> *and gathers its food at harvest.*

Perhaps you don't have access to a computer but you would still like to better comprehend cellular automata? Consider a simple one-dimensional version that begins not with an entire chessboard of cells (two-dimensional automata), but rather a single row extending to infinity, left and right. As before, cells are colored black (alive) or white (dead). Now, however, each has a neighborhood of only two adjoining cells, one to the left the other to the right. As with Conway's game, each tick of the clock produces a new generation (horizontal strip) according to a few rules — one in this case. Our example also starts with one cell. One rule, one cell. What could be simpler?

Rule: A cell remains or becomes black (alive) in the next generation if one — but not both — of its neighbors is identical; otherwise it remains or becomes white (dead).

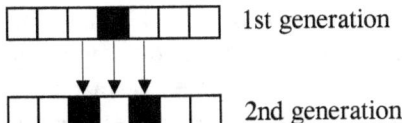

These one-dimensional cellular automata have an advantage over the *game of life:* successive generations are successive rows right under their predecessor. The result is a two-dimensional grid of cells portraying the entire evolution of the top row, which in this case is a single live cell.

We're about to press the button on the computer — the moment of creation for this program. Before turning the page, can you predict what it will look like? It's an old friend. You've seen it many times.

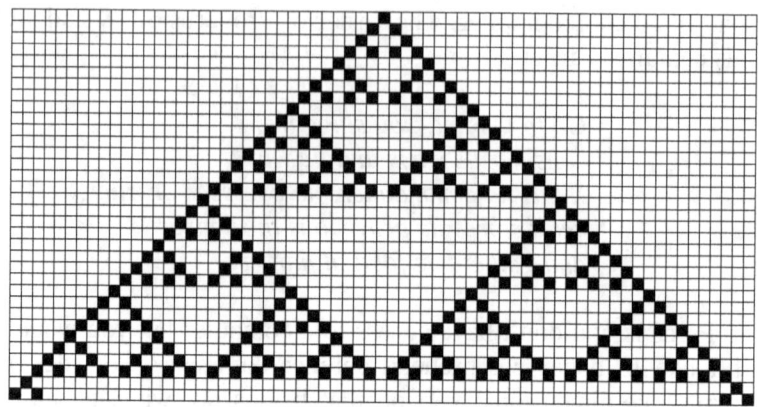

AN EMERGENT DESIGN — PASCAL'S TRIANGLE

Figure 9.10

Most people believe that to make something complicated you must do something complicated. Computers are complex, airplanes are complex, and we have separate industries dedicated to their intricate constructions. Nevertheless, the design above — an example of emergent complexity arising from one rule — shows that assumption is not universally true.

Recall that this pattern is really Pascal's triangle with every odd number replaced by a black square and every even by a white. Just so that the connection is obvious, *see* Figure 4.11, the Pascal–Sierpinski fractal. The relationship between cellular automata and Pascal's triangle is now clear for everyone to see. But how is this related to the natural world of land, air, and water? The transition from the above figure to something as simple as seashells, for example, is made possible by hormones. For pattern formation, hormones come in two varieties: activators (promoting growth) and inhibitors (preventing growth). From Chapter 4, consider again the tent olive shell shown on the next page. It's an image of Sierpinski's triangle with all its self-similar triangles plus the roughness of nature we have come to expect when forces interact.

Like the tips of a tree, shell growth occurs only on the leading outside edge. Material is deposited by a row of cells on the foremost margin of the mollusk's mantle. Through hormonal activators and inhibitors of colored pigments, the shell lays down patterns that are rarely altered thereafter. And the rule(s) that governs this activation and inhibition must be very like the cellular automata rule for the Figure 9.10. Many natural phenomena appear complex at some level, but they're really the result of simple underlying computational mechanisms that are essentially cellular automata at work. And these self-organize into complex systems based solely on local events — democracy at the cellular level if you like. The attractive pattern of triangles on a tent olive shell is a good example of this labor.

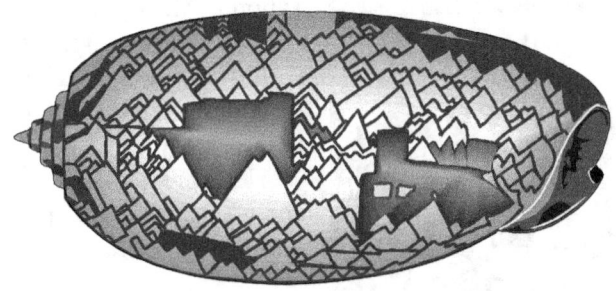

TENT OLIVE SHELL, *Oliva porphyria*

Of the many species of equids (horses, asses, and zebras) that dominated the world's grasslands for the last eight million years, only seven are left, and three of these are zebras. Stripes, it would seem, are good for survival. When zebras gather in herds, this pattern may be a way to confuse predators. In groups these stripes make it almost impossible to focus on a single animal, and when zebras run they create a *flash* pattern distorting the vision of would-be predators. However, the boldness of the design — the very opposite of camouflage — is useless for a *solitary* animal. In legend and science, many other uses for this anti-camouflage have been proposed, but this explanation seems most convincing.

Although the full range of utility for stripes may not be known, cellular automata can simulate the means of their genesis. The black and white stripes display the effects of hormonal activation and inhibition. Cells in the skin called melanophores make melanin pigments. For the melanophores to produce melanin, certain hormones must be active in the skin during embryonic development. So the zebra's coat, the one it's born with, reflects the early interaction of chemicals diffusing through the skin of the embryo.

CHAOS

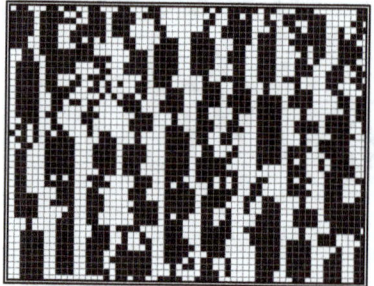

STEP 5

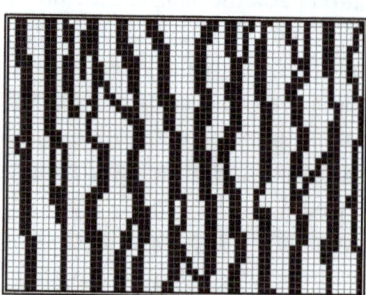

STEP 10

ZEBRA

This is a two-dimensional cellular automaton, and we begin with a random distribution of black and white cells on a chessboard grid like the top diagram opposite. Each black cell represents a certain level of pigment activation. Most coat patterns (leopard, cheetah, giraffe, and so on) are thought to develop from some such starting point in the embryo.

Here are the new rules to apply that create the real physical effect:

Rule 1: Activator hormones that are near each other reinforce their effect.

Rule 2: At the same time, activator hormones diminish the effect of other activators far away by inhibiting their ability to switch on their nearby neighbors.

As in Conway's game, each cell is black or white, on or off, and when the clock ticks the cells interact with their neighbors according to the above rules. By the 5th step some pattern appears (second from the top); by the 10th (third from the top) a definite zebra design emerges.

The first rule is a positive feedback loop: the more you get, the more you get. If left unchecked the zebra would be entirely black. The second rule, on the other hand, is a negative feedback system in which more leads to less by putting a dampening effect on the runaway processes of Rule 1. Together these rules form a delicate balance within the zebra's skin, an equilibrium that results from a long evolutionary heritage of natural selection for stripes. This balance was achieved on the African grasslands by a million year dance between lion and zebra, predator and prey.

Nature's emergent complexity results from its efficient and ubiquitous ability to self-organize. Although, as we have learned from previous chapters, not all designs arise in this manner; nevertheless, it is one of the least known but most powerful strategies for achieving pattern and order in the natural world.

WE HAVE COME A LONG WAY from the seniors' residence and our two elderly Greek friends. Once again we learned that space has an unseen structure, one that permits only an even number of odd dots. Beyond this mandatory feature is nature's toolkit, which contains three very practical templates for her expression of order: spirals, meanders, and explosions. In the world of art, Leonardo taught us that patterns could be psychological as well as visual and thereby add power to a painting. They also add structure, power, and even surprise to music and poetry. We stumbled into the garden of false patterns — illusions of our own construction. While dallying over probability paradoxes, we wondered about our difficulty in comprehending the birthday problem. Also, the phenomenon of the "hot hand" in sports emphasized our deficiencies regarding the cluster illusion. Lastly we learned about cellular automata and the *game of life*. With one rule and one square, we created Pascal's famous array. And cellular automata showed us how easily — without the aid of DNA — a leopard gets its spots, a zebra earns its stripes, and the world creates many of its patterns. We *have* come a long way.

Some fear all this science has taken us from the center of god's creation to a dark place under the stairs. But far from being constructed in his image, we are the product of an incredibly long evolutionary process. And with advancing knowledge, our often postulated differences from earth's other creatures have slipped away. We are no longer the sole toolmaker, the lone communicator, or the one creature that reasons; by a count of our DNA, we are 98 percent chimpanzee. But to quote Tennyson, "that which we are, we are," a deeply reasoning animal capable of profound understanding of the world around us and beyond. We are a part of, not apart from, nature — our differences are a matter of degree, not kind. Yet what a degree that is! Our task is to comprehend humanity's place in the universe. This is what we should celebrate in art, poetry, and music. We leave fear to those who would fall on their knees. Let's come out from the dark and ascend the stairway to the new heavens.

CHAPTER — 10

THE SUM OF ALL THINGS

"... this place is sacred — thick-set with laurel, olive, vine; and in its heart a feathered choir of nightingales makes music. So sit thee here on this unhewn stone ..."
Sophocles, *Oedipus at Colonus*

WE STRIVE to find universals: a substance that will dissolve anything (one wonders what you would store it in), the fountain of eternal youth, the origins of life, the theory of everything (TOE). All these and more have underlying elements about ultimate things. Notions such as these imply power and permanence. But age and disease weaken us, parents and friends die, epidemics spread, storms destroy, the environment degrades, governments behave little better than lunatics, and wars proliferate where ignorant armies clash by night. We seek constancy and longevity, but all around us is a deadly flux.

"EPIUS, have you ever wondered why we've lived so long?"

"Yes. Often! We have much in common, you and I, artist and scientist. Odysseus blessed you for keeping the suitors entertained and so distracting them from his wife, Penelope. As I recall, you sang so often of the Achaean heroes that Penelope herself asked you to sing a different song. And on that terrible, triumphant day after the returning hero had slaughtered everyone in the great hall except you and the priest, what did you do? You placed your lyre carefully on a table. If you were to be slain, you didn't want this instrument damaged. That's a lesson only a poet would teach. What did Odysseus do? He slew the priest and spared you saying he could not kill a man of god. In the *Odyssey* he declared:

> *All men owe honor to the poets — honor*
> *and awe, for they are dear to the Muse*
> *who puts upon their lips the ways of life."*

"Nothing in all these years," said Phemius, "has dimmed your passion. You speak and think as clearly as when you designed and built the great wooden horse. Its construction ended the war, and Odysseus blessed you for your skill. I suppose in a way we're both the blessed sons of the wandering hero. Is that what you meant when you said we have much in common?"

"That's part of it. But it's impossible to recollect everything for a full comparison. Even though I've lived three millennia, there are vast periods — occasionally whole centuries — I don't recall. How can you think about what you can't remember?"

"I know what you mean," said the poet. "I wonder if the periods we recall and the ones we don't are the same."

They each got a mug of coffee from the kitchen and went to the solarium of their seniors' residence, Happy Acres. In the morning sunlight, with coffee and oranges, they recalled their lives at leisure. The *best* remembered, sometimes the only remembered times, were ancient Greece to late Alexandria, the Renaissance, and Modern Europe and America. Their lives outside these periods were blurred, and the Dark Ages were indeed dark. Early Greece and the Renaissance, the two greatest periods of art, literature, and sculpture were also the times in which science was born, the times when the two men felt most alive.

"The biggest discovery of science is science itself," said Epius. "I mean humanity has for centuries — millennia even — looked for ways to influence nature to increase the food supply and cure disease. Before science we vainly prayed and used other forms of magic, some still do. After all, the only purpose in praying to the gods is to have influence over them."

"But why wasn't the structure of science born all at once, full-grown and in complete armor, like Athena from the head of Zeus or art in the caves of Lascaux and Chauvet?" asked the poet.

"The deductive side of science — the part we call mathematics — was born with Pythagoras, nourished on the Greek islands, reached maturity on the mainland, and found its apotheosis in Alexandria with Euclid's *Elements*. Everything in the *Elements* was meant to be a *deduction* from ten axioms. This was the Greek way and their gift to science. The second part — the induction — was born during the Renaissance with the Scientific Revolution."

Phemius asked his friend if he would explain the difference between these two faces of science. Which is more powerful, which more certain?

"How long, my friend, have we lived in North America?"

"At least a century," replied the poet.

"And during all that time we've walked and hiked over most of this continent. Even here, at Happy Acres, we stroll the grounds and the nearby roads. Throughout these decades, we must have seen thousands of red squirrels."

"Epius, what's your point? I asked you to explain deduction and induction, and you're telling me about squirrels. Where has your legendary directness gone?"

"I'm getting to a *definition* of induction. Be patient, we have the time. . . . After collecting from our random walks some information on these noisy mammals, I now propose a 'theory' about them: *All red squirrels are red.* I propose this statement because I've noticed these squirrels to always be the same rufous color. That's induction: you reason from the particular to the general. Since I haven't seen every such squirrel, the theory goes far beyond the facts of my observations, yet it seems reasonable wouldn't you say?"

"It's more than reasonable, it's true, but trivial. I need another coffee to stay awake."

"You're being impatient again. What would you say if I now told you that my theory — based on induction — is false, plainly false? I can prove it. Here's a photograph of a 'white' red squirrel, and by its eye, it isn't an albino either. So, all red squirrels are not red. One counter-example kills any theory based on induction," concluded the scientist.

"WHITE" RED SQUIRREL

Figure 10.1

"Now you're confusing me. First, you set up this induction process and then you shoot it down. Why are you attacking your passion, or bliss as I call it?"

"I'm attempting to be honest, a virtue as important to any method of seeking knowledge as it is to your art. As I said, we have much in common, you and I."

"Let me see if I understand this," reflected Phemius. "You're saying that all the theories in science — regardless of the number of confirming instances — are never, and never can be, 100 percent confirmed. And that extends even to, say, universal gravitation, so things might *fall up*."

"The short answer is yes, but that incredibly remote possibility doesn't merit consideration. All the same, the fact that the sun has always risen doesn't mean it will always rise.

"People commonly view scientists, myself included, as a little arrogant, know-it-alls as they say. Paradoxically, any sense of superiority we may have rests on induction, which, as the philosopher David Hume said, is *logically* indefensible. But it reflects the way the world is. Induction gives us the power to predict the future and therefore control it."

The poet reminded the scientist that he had asked two questions concerning deduction and induction: which is more powerful, which more certain?

"Deduction isn't about this world," Epius continued, "it's about internal consistency. If a triangle is right-angled, then the sum of the squares on the two shorter sides equals the square on the longest side. Pythagoras proved this once and forever." Epius recalled how the sage had shown him the proof when they were together on the Island of Samos. "Within the rules of mathematics, his proof is 100 percent certain. Deduction speaks of an ideal world; induction talks about the real world. The first is certain, the second is not; in some sense, the first is powerless; the second is not. The Greeks discovered the former; the Renaissance created the latter."

"Good," said Phemius, "I think I understand this now. Let's get out of here and go for a walk. We need to keep fit or the staff will transfer us to the nursing-home section of Happy Acres."

The Greek aristocracy, the only ones with enough leisure to speculate about anything other than the source of their next meal, preferred the abstract worlds of geometry, lyric poetry, and heroic sculpture. Not for them the things of this world — raw reality. Plato's *Dialogues* speak about the abstract concepts of truth, justice, the good life, and the ideal state ruled by philosopher kings. Euclid's edict to use only straight edge and compass was really an injunction against the measuring instruments of artisans and slaves. Ancient Greece was barren ground for any theory requiring detailed observation and measurement of the natural world.

Enter Leonardo da Vinci! Born an illegitimate child in the tough and tumble of the Renaissance he was the right person, at the right time. He wasn't educated in the curriculum of the upper classes of Italian society. Largely self-taught, he learned Latin only in middle age and never Greek.

"Epius, do you remember the time we both worked for Leonardo? We built those striking wooden models of the five regular solids that the master later painted for Pacioli's *De Divina Proportione* (*see* Figure 2.1)."

"Those were the great days, the best in a thousand years. I recall Pacioli afterward hired you to do the dropped-letter calligraphy for his book. And Barbari even included you as Pacioli's student in his famous portrait of the friar. Some student! But with your good looks, you seemed a mere boy. So there you are, immortalized, forever staring directly into the viewer's eyes (*see* Figure 2.2)."

"What I remember about Leonardo was his unrivaled attention to detail — for him nothing seemed too great an effort. Even as Andrea del Verrocchio's apprentice, assisting on the *Baptism of Christ,* Leonardo displayed that trait. Do you recall the story about the two angels in this painting [shown to the left]?"

"No, I don't — now it's your turn to teach me."

"Well, Epius, the apprentice painted one of these two angels. Can you point out which one? . . . Yes, yes, that's right; your artistic sense seems as well developed as your scientific. It's the one on the left. Leonardo's angel is more finely drawn."

"The other angel," the scientist noted, "appears to need the immediate close attention of a skilled oculist — I hope this judgment isn't too harsh."

Exeunt Epius and Phemius.

Detail from the *Baptism of Christ,* Uffizi, 1472-75, Andrea del Verrocchio

Figure 10.2

Jacob Bronowski commented on this:

> It is usual to say that Leonardo's angel is more human and more tender; and this is true, but it misses the point. Leonardo's pictures of children and of women are human and tender; yet the evidence is powerful that Leonardo liked neither children nor women. Why then did he paint them as if he were entering their lives? Not because he saw them as people, but because he saw them as expressive parts of nature. We do not understand the luminous and transparent affection with which Leonardo lingers on a head or a hand until we look at the equal affection with which he paints the grass and flowers in the same picture.[1]

Giorgio Vasari (1511-74), Leonardo's biographer, says Verrocchio never touched colors again, being most indignant that a boy should know more of art than he did.

Phemius pointed out that if induction needed the particulars of nature for its birth, then Leonardo was its godfather. The master distrusted all large theories; he saw that nature displays herself in detail — the small features he put into the rocks and grasses that the angels are kneeling on. Other Renaissance artists had this view, unlike the artists of the Middle Ages. But Leonardo went further. He understood that science as well as art has its expression in particulars. It's not just the devil that's buried in the details; it's the *origins of induction*. More than a century passed after Leonardo before Francis Bacon laid out the intellectual basis for induction. As previously mentioned, the greatest discovery of science is science itself.

Medieval "scientists" didn't examine nature for answers; instead they looked to Aristotle. And so, they continued to repeat his every error. One often-quoted example of this was that women have fewer teeth than men. A cursory examination by Aristotle of Mrs. Aristotle's mouth would have easily dispatched this blunder, but he never bothered, and neither did anyone in the Middle Ages. Fortunately, Leonardo couldn't read Greek.

We have insisted that induction — the scientific method — requires attention to the minutiae of the natural world. And how do we gather these minutiae? By being both *curious* and *active*, traits common to artists as well as scientists. The kind of curiosity plus activity my cousin and I had when discovering the terms of the Fibonacci sequence in the petals of flowers on the field of dreams.

I wish to put forward an idea — not original — but one most readers will consider outrageous, preposterous even. The proposition is that the very greatest scientists are as creative with the invention of their theories as Shakespeare was in the writing of his plays. The comparison will be between genius and genius, not engineer and journalist — the best with the best. Before you dismiss me out of hand, consider the avalanche of ideas we've encountered in the previous chapters: everything from multi-dimensional regular solids to pattern formation by shattering the symmetry of the cosmic egg to the weird world of cellular automata and a hundred other thoughts and theories along the way. Each of these ideas had threads common to art and science, often stitched together through the medium of nature. These were not anonymous inventions but imaginative ideas by known scientific geniuses.

A sign of the difference between excellent artisanship and great art is that the first is anonymous while the second is not. The Bayeux Tapestry of 1077 depicting the defeat of Harold II, King of England, by William the Conqueror, is unsigned. There is an ongoing controversy as to who wove it, monk or maid. And the names of the architects and freemasons who built the medieval cathedrals are unknown. Michelangelo signed only one of his works, the *Pietà* — when he was a young unknown. A signature on the Sistine Chapel ceiling would have been as ludicrous as unnecessary.

The story of this signature is curious and instructive. Shortly after the unsigned *Pietà* was placed in Saint Peter's Basilica, Michelangelo overheard a pilgrim remark that it was done by Cristoforo Solari, nicknamed the Hunchback of Milan. That night in a fit of anger Michelangelo took hammer and chisel and placed the following inscription in lapidary letters on the prominent band running diagonally across Mary's breast: MICHEL ANGELUS BONAROTUS FLORENT FACIBAT (Michelangelo Buonarroti, Florentine, made this). Most signatures appeared in modest locations, sometimes at the feet of revered figures or on ornamental elements in the margins. The placement of his signature and the word "FACIBAT" are powerful statements of pride — implying creativity is personal. Genius demands its rightful due. Our modern age rewards originality in the arts and sciences with the annual Nobel prizes. The relative anonymity of an age is a testament to its sterility: we know the names of more ancient Athenians than all the poor wretches of the Dark Ages combined.

Science looks to the future by building on the past. Yet some scientists, such as Richard Feynman, refused to study the work of predecessors, fearing it would impair their originality. Influenced by science, art is now expected to be forward-looking as well. This search for what is new and different has linked the artist and the scientist in the mind of the public at large: they often fear and dislike the way both see the world. The unconventional lifestyle of many artists and their often-outrageous creations repel the politically correct "nice" people. Many of these same individuals feel science dabbles in things best left alone: atomic power, genetics, and stem cell research.

These two points of agreement between art and science — lack of anonymity and a general public distrust — don't get to the core of the problem: the uncreative, unimaginative essence that most of us still think is at the cold heart of science. We believe art creates what it wishes; science describes what it must. After all isn't there a solid reality "out there" that science merely reports? This is exactly what G. H. Hardy thought of mathematics when he wrote (*see* introduction to Chapter 5), "That which we describe grandiloquently as our 'creations,' are simply the notes of our observations." What Hardy meant might be true of mathematics but even that doesn't diminish its creativity. With respect to the creativity of science, the question has two answers: yes and no.

Yes. Certain things are discovered the way the Vikings discovered Newfoundland. Science does find things that were already there: the planet Uranus, the frequency of red light, the DNA molecule. If Watson and Crick hadn't discovered the molecule of life, someone else would have — and quickly too. Yet no dichotomy exists between discovery and creation; rather there is a continuum as one blends into the other.

No. Scientists have created some theories that are as profoundly original as the greatest art. Much of modern science revolves around the theory of light, its nature, speed, and properties. One theory of light holds that it's composed of photons, incredibly small packets of energy. An alternative theory considers light to be a wave phenomenon. Both have powerful supporting evidence; both are "true," meaning each fits the data of sensation — physicists jointly call these two analogies wave-particle duality. Each theory is the different creative vision of different scientists, and we cannot choose between them. So much for an *ultimate* solid reality "out there."

Parts of reality are so far removed from our daily experience that we cannot picture them; even our analogies and poetic metaphors fall short. Some of these parts are close at hand. Can you imagine your car's pistons turning over at 6,000 rpm (100 per second)? Can you comprehend your radio and TV capturing unseen and unfelt signals from the air? Can you imagine your kitchen table as a physicist might — all space and no substance? Whether or not you can, all these things are part of reality.

Our lives are circumscribed not only in time and space but also in scale. We aren't small enough to comprehend what's "out there" on the microcosmic level. Even though the bodies of insects are immense compared to anything on the atomic scale; nonetheless, their lives are still challenging to understand. Recall the almost instant heat loss by an insect flying from sunlight to shade as outlined in Chapter 1. And gravity is meaningless to a midge.

Let's exit the weird worlds of photons and insects and stroll about in the macrocosm we know. Scientists develop differing visions of identical phenomena. We have already met a famous example involving Newton and Maupertuis. Newton did more than create a theory; he discovered infinitesimal calculus and applied it brilliantly. Like Michelangelo, when he didn't have the tools he needed, he invented them. One part of Newton's creative vision came in the form of his three laws of motion that compute the paths of projectiles and planets. His method requires a right-angled coordinate system and a calculation of *all* the acting forces. Maupertuis' elegant principle of least action (*see* pages 260–61) works in any frame of reference and requires only the projectile's or planet's mass, velocity, and total time of flight. Two men, two visions, but one subject.

In an oft-quoted remark, Newton said that if he had seen further than others it was because he stood on the backs of giants. The principle of inertia (Newton's first law) states that a body at rest remains at rest and a body in motion *remains in motion* at a constant velocity as long as no outside forces are involved. This was a reformulation and clarification of a motion law by Galileo who was one of Newton's giants.

Newton's first law banished to rest the angels from constantly having *to push* the heavenly bodies in their orbits according to the medieval cosmology of the Catholic Church. Like modern spacecraft, the whole ship keeps moving, perchance forever, when the engines are shut down. It's a long way from angels to Galileo to the law of inertia. And Sir Isaac's broad shoulders supported Einstein's vision. Galileo, Newton, Einstein: different men, different times, and different visions. It's not just science that requires a context for its creation: Shakespeare lived in the age of Elizabethan dramatists — he could not have written as he did without them. Christopher Marlowe, Robert Greene, and Thomas Lodge were a few of these. No man is an island.

We'll slip from the shoulders of giants and descend to two boys roaming fields like foxes searching for whatever they can find. Here a quartz rock, there an emerald snake, and over there a woodcock's nest. In Chapter 6, I recounted how my cousin and I had stumbled upon some terms in the Fibonacci sequence by counting flower petals. We didn't, however, discover the simple method of finding the next term by adding the preceding two. You could say, perhaps, that we had no vision of this famous sequence. But you would be wrong. Be it ever so humble we did have a theory of sorts — an idea about what the next term might be before we actually found it. We noticed something curious concerning these numbers: two odds (O) were "always" followed by an even (E). We were generalizing beyond the evidence. See for yourself:

1,	1,	2,	3,	5,	8,	13,	21,	34,	55,	89,	...
↓	↓	↓	↓	↓	↓	↓	↓	↓	↓	↓	
O,	O,	E,	O,	O,	E,	O,	O,	E,	O,	O,	...

On the occasions when we applied it, this knowledge of odds and evens was confirmed. In effect, we were predicting the future based on the past — something science has always done. Our little theory (*see* Chapter Notes) was as much a creation as a sonnet by a fledgling poet.

We have seen a few examples of different creative visions on identical subjects about what's "out there." Are these visions fixed, eternal truths chiseled on the halls of science? Of course not! The last is a minuscule part of an ever-expanding body of knowledge reported regularly in the *Fibonacci Quarterly*. Newton's three laws and Maupertuis' principle are now part of relativity theory. And who can know where the future lies for quantum-photon weirdness? We say the outcome from tossing a die is a random event, but it isn't. If all the forces were known (upward velocity, rate of roll, and so on), we could correctly predict the outcome. Some magicians have such precise motor control that they can flip a coin to land heads or tails at will. But whenever the forces are too complicated to calculate, we say it's a random event. Perhaps this is the situation in the world of the quantum; perhaps one day we'll know.

"SEEING IS BELIEVING" is a maxim of everyday existence. But even this truth must be subject to analysis. Shakespeare has Macbeth face this dilemma in Act II:

Is this a dagger which I see before me,
The handle toward my hand? Come, let me clutch thee.
I have thee not, and yet I see thee still.
Art thou not, fatal vision, sensible
To feeling as to sight? or art thou but
A dagger of the mind, a false creation,
Proceeding from the heat-oppressed brain?

Macbeth played the perfect scientist. He subjected his vision to the touch test and concluded it was a false creation existing only in his mind. This is what science must do with all its theories and hypotheses: test them against the outside world.

Often visual tests aren't enough to confirm objectivity. Eyes are wonderfully intricate, but they're not passive recorders of the outside world. Rather, they're like political spin-doctors putting their particular perspective on the facts. We see with greater acuity than the rods and cones should allow. It's as if our brains, like Adobe Photoshop, were running a sharpening program on the retinal image. Moreover, when you swing your head from side to side, only a small fraction of all the data points are recorded — or can be — but your brain integrates these into a smooth sequence like a motion picture running at 24 frames per second. And then there's the classic problem of the viewer affecting the thing seen. For example, anthropologists recording the daily lives of a less complex people forget all the while that their presence changes the natives' behavior.

Or recall the wolf and the fox with their caught-in-the-headlights look (*see page 250*) triggered by the camera's flash.

I am *not* a camera, nor are you. A camera is more objective than the human eye. The remarkable photograph at the left shows two hands; the right is one foot (30.5 cm) and the left, two feet (61 cm) from the man's eyes. This is how the camera sees them. This is how they are. Why? When an object is twice as far away, it appears to be *half* the height and *quarter* the area by the inverse-square law. Human brains, on the other hand, automatically compensate by making the more distant hand look larger. Try this with your own hands and you'll be amazed. Personally, my more distant hand appears slightly larger — bizarre! This illusion affects everything we "see." Artists naturally incorporate this abiding false impression into their works so we can view the world as evolution has blindly directed us to.

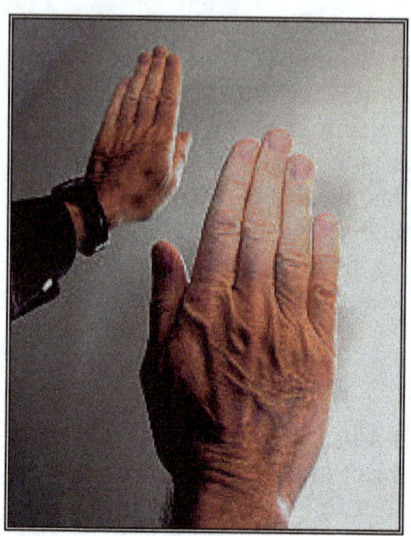

THE TWO-HAND ILLUSION

Figure 10.3

Concerning the late-Renaissance painter El Greco, there exists an enigma regarding what he actually saw. Critics have often commented that the painter had a problem with his eyesight. Witness, they would say, the strange upward elongation of his

St. Jerome, Cardinal, by El Greco, 1590, Frick Collection, New York

Figure 10.4

subjects and the surrounding buildings and vegetation. His portrait of St. Jerome, the cleric who translated the Bible into Latin, exemplifies this. Let's assume his eyesight was defective: how would El Greco "see" his own distorted figures? Well, since they're already elongated, his presumed ocular problems would further twist and distort the figures into mere caricatures. But this conclusion is absurd and so the painter's eyesight was fine. The agonizing subjects of his portraits with their melancholy features are a reflection of his times and talent intended to elicit a heightened sense of religiosity during the terrible years of the Spanish Inquisition. Again, every painting is a picture and a vision; El Greco's vision is clear.

Beyond all the difficulties in seeing — hallucinations, sharpening and filling in the images, changing them, distorting their proportions — there are further problems. Surely, one of the major dilemmas is seeing only what we *expect to see*. The present is seen with images stored from the past. The brain expects and accepts the most probable answer. This is the overriding reason why the creation of new ideas and patterns, by artists or scientists, is rare and to be treasured. What happens when we see something entirely new? Simple. We don't see it! Consider the following five examples.

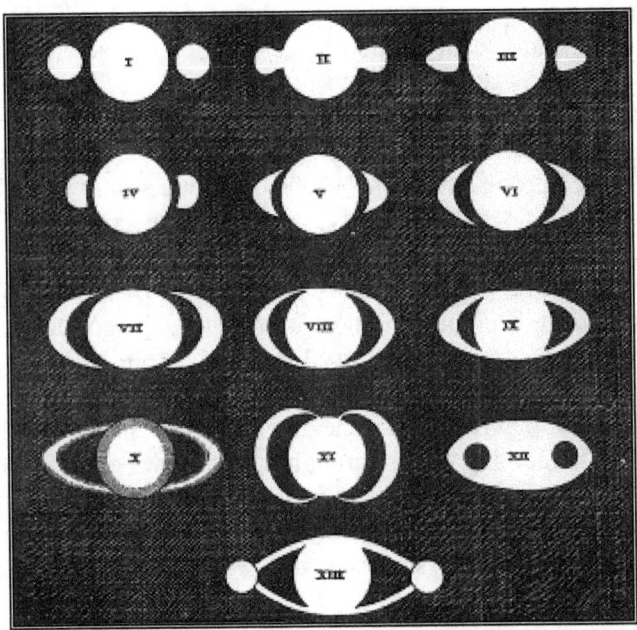

COMPOSITE FIGURE FROM
HUYGENS'S *SYSTEMA SATURNIUM*

Once upon a time in lands far away lived two great men, Galileo Galilei and Christiaan Huygens. The former was the first earthling to see the rings of distant Saturn. Never having encountered anything like this before, he thought the planet had ears or perhaps two satellites that never moved. A few years later, Huygens also saw the rings and made many curious sketches of them (*see* previous page), again seeing only what he expected to see. Unsatisfied with these images, Huygens went on a vision quest where he discovered the true nature of the rings — the modern vision we have and spacecraft confirm. These two heroes of upper earth will always be remembered as the true "Lords of the Rings."

The red planet Mars has been the home of all hopes for extraterrestrial life for at least two centuries. The American astronomer Percival Lowell drew very elaborate maps of the canals he "saw" on the Martian surface through his telescope. Many of his contemporaries claimed to have seen them as well. Lowell thought these canals were used to conduct water from the polar ice caps to the equatorial regions of a dying planet. But when NASA sent the two Viking spacecraft to the planet in 1977 the canals vanished. Perhaps the old maxim should be reversed to "believing is seeing" for all the canals were in Lowell's head. He wanted to see them and so he did. Viking brought a sharp reality although the question of bacterial life on Mars is open — so far.

On an earthly level — at least in higher latitudes and elevations — is the ubiquitous snowflake. No evidence exists that anyone in Europe before the 1600s had noticed that snowflakes are hexagons. Almost certainly, they were expecting them to be the more common shapes — circles or squares. Not until Kepler in 1611 published his *On the Six-Cornered Snowflake* did its form become common knowledge. If no one points something out, it may go a very long time before being noticed.

Twenty years ago Douglas R. Hofstadter, author of *Gödel, Escher, Bach,* wrote an article for *Scientific American* on the perils of perceiving things through the windows of expectation. The little diagram on the next page is an old example of what he called "default assumption," or seeing what we expect to see. What does this sentence say? Are you positive? *Look again.*

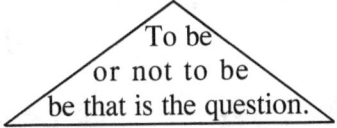

Even the most experienced editors misread this sentence because they *know* what they're supposed to read.

The issue of gender correctness bedevils current writers. We struggle between gender sensitivity and grammatical awkwardness. A trend is developing to refer to any person in authority of either gender as "sir" — a practice adopted in the *Star Trek* TV series. Consider the issue of using the pronoun "he" to stand for "he and she." In some contexts this can be confusing. And think about the old riddle of a father and son in a terrible car accident. The father is killed; the son in the seat beside him survives but is badly injured. He is rushed to the hospital, where the distraught surgeon mutters, "I can't operate on this boy, he's my son." What's the answer to the riddle? We'll wait.

. . . .

After five minutes, some people — of both genders — will still be scratching their heads. If you haven't solved this puzzle, you've made the default assumption that all surgeons are men.

PERHAPS YOU CHUCKLED at the great Galileo mistaking the rings around Saturn for ears or two small satellites. Other astronomers of his time had similar problems perceiving their true nature. Now we can't see anything else, and we'll never see the canals on Mars again. Perhaps also you immediately spotted the extra word in the quotation from *Hamlet*. And maybe you even knew the surgeon was a woman. So, you've passed these initial tests; now it's time for the mother of all perceptual problems. I guarantee the test below will perplex you.

The American artist Adelbert Ames created bizarre perspective puzzles; the one on the next page is his best. When the room is viewed through a peephole at point A, binocular depth perception becomes impossible. Now your brain has to make a perceptual choice: either the people inside the room shrink (or grow) to impossible sizes or else the room isn't the normal cubical shape. Our brains opt for an ordinary *cubical room* —

THE AMES ROOM WITH IDENTICAL TWINS

a surprising choice. Yet, the left corner is twice as far away as the right (*see* diagram below). This quite properly causes our minds to see the right twin as taller than the left. If they changed positions, they would also change sizes. So, it's the severely distorted room and our perception of it that's at the root of this illusion. The Ames Room appears normal only from point A. Shift your position and the paradox vanishes. It's difficult to accept, although we must, that seeing isn't always believing.

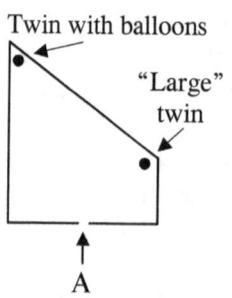

This tendency to see what we want to see — to gaze through the veil of our prejudices and beliefs — has disturbing implications. There is danger here! The danger of being in an arrested state of development, tethered to the post of our preconceptions. How do we untie these cords and free ourselves? To cut this Gordian knot we need the feathered edge of creative imagination tested against the real world. Although he had solved many mysteries of nature, Galileo failed, where Huygens succeeded by first imagining that Saturn had rings and then seeing that the data fit this view. Peter Medawar, in his *Advice to a Young Scientist,* rejects the literary notion that scientific discoveries are made by "just looking around":

I myself believe it to be a fallacy that any discoveries are made in this way. I think that Pasteur and Fontenelle would have agreed that the mind must already be on the right wavelength, another way of saying that all such discoveries begin as covert hypotheses — that is, as imaginative preconceptions or expectations about the nature of the world and never merely by passive assimilation of the evidence of the senses. . . . The truth is *not* in nature, waiting to declare itself . . . every discovery, every enlargement of the understanding begins as an imaginative preconception of what the truth might be.

THROUGHOUT THIS BOOK runs a minor theme in a major key: Pascal's triangle. We heard it in Chapter 2 in English and Chinese versions. It appeared in Chapter 4 singing only two notes: odd and even forming the Pascal-Sierpinski fractal. In Chapters 5 and 6, it encored, producing the Fibonacci numbers — those counters of flower petals, sunflower spirals, and the sacred guardians of the golden number. Ultimately it took a bow in Chapter 9 in connection with cellular automata. This is a short list of its many recitals. Is there a new aria, one that throughout the centuries no one has ever heard? Yes, and to discover it you need a creative vision — one missed for a millennium. I first heard this new song in Bruce Ewen's article "Pascal's Triangle Is Upside Down" from the February 1970 issue of the *Mathematics Teacher*. That's right, for all these centuries we've been looking at the triangle *upside down,* on its head so to speak. What the data alone didn't imply, a creative imagination saw.

⋮	⋮	⋮	⋮	⋮	⋮	⋮	⋮
1	3	3	1	0	0	0	...
1	2	1	0	0	0	0	...
1	1	0	0	0	0	0	...
1	0	0	0	0	0	0	...
1	−1	1	−1	1	−1	1	...
1	−2	3	−4	5	−6	7	...
1	−3	6	−10	15	−21	28	...
⋮	⋮	⋮	⋮	⋮	⋮	⋮	⋮

PASCAL'S TRIANGLE UPSIDE DOWN

Figure 10.5

Every entry in the above array is the sum of the *two numbers below* (e.g. $-10+15=5$ and $2+1=3$ in bold). Our standard Pascal's triangle is in the upper part, while there's a once-hidden sideways version in the lower shaded region. The table goes to infinity in three directions. These previous hidden entries hold new mathematical treasures beyond the scope of this book. Subtle things are often close at hand.

In some respects artists and scientists are blind. None of us stands astride Mount Olympus seeing and knowing all. Copernicus and Galileo were certain that the planets rotated about the sun in perfect circles at constant velocities, but they were certainly wrong. And Kepler would still be buried in data and calculation if he hadn't first imaginatively seen that planets might move in ellipses at varying velocities. Tyco Brahe's astronomical data on Mars — in Kepler's possession — later confirmed this. Would anyone disagree that if El Greco had painted the subhuman practices of the Spanish Inquisition, he would have been morally superior to the El Greco who painted cardinals and saints? Are we to admire Keats' knowledge of history when in his stirring sonnet "On First Looking into Chapman's Homer" he mistook Balboa for Cortez?

Science and art are very human forms of knowledge — venturing forward, taking chances, often in error. A myriad of maxims show how to correct errors, but Oliver Cromwell's underlies any hope of improvement: "I beseech you in the bowels of Christ, think it possible you may be mistaken." If you want fixed, eternal "truths," don't choose the arts or sciences, but rather wade into the dogma of the church, synagogue, mosque, temple, or Capitol. It's the ultimate sign of our humanity that we can *change our minds* on the presentation of new evidence. To do less is dehumanizing. Never mistake reasonableness as a failing. As Bronowski wrote in *The Ascent of Man*, "Science is a tribute to what we can know although we are fallible."

Jacob Bronowski was a Polish Jew. In the chapter "Knowledge or Certainty," he pays tribute to his murdered friends and family by returning to Auschwitz. Here he dissects the essence of evil with some of the most powerfully written sentences of the 20th century:

> It is said that science will dehumanize people and turn them into numbers. That is false, tragically false. Look for yourself.

This is the concentration camp and crematorium at Auschwitz. This is where people were turned into numbers. Into this pond were flushed the ashes of some four million people. And that was not done by gas. It was done by arrogance. It was done by dogma. It was done by ignorance. When people believe they have absolute knowledge, with no test in reality, this is how they behave. This is what men do when they aspire to the knowledge of gods.[2]

"Epius, Epius, are you awake?"

"I am now."

"Why are we living in Happy Acres?"

"Phemius, I thought you liked it here — the cleanliness, the meals, the facilities, and above all the quiet."

"It was good at first. Now I find it too clean, too quiet. It's a perfect little world. Can't you *hear* the difference?"

"I don't hear anything. Go back to sleep."

"That's just it, Epius. It's September and in this place you can't even hear the crickets!"

"OK, OK, I'm wide-awake now. What do you want us to do? We're getting old you know."

"We've been getting old for centuries. It's this place. It's devouring our spirit, our joy of life. Don't you feel that? If you get excited, they give you a pill; if you get depressed, they give you a pill. The inmates here just talk about the past. You and I have the longest past of anyone, but we don't dwell on it. Do you know what I miss most, beyond the crickets and the mental stimulation?"

"I can't imagine."

"Dogs, dogs, dogs. We can't keep pets here. I remember old Argus; he was a pup when Odysseus went off to Troy, and the first to recognize him upon his return. Wagging his tail at his master's homecoming, he died for joy. That was when I fell in love with animals. Let's get a dog again, Epius, a big dog."

"Then we'll have to move. Maybe I'm feeling a little too comfortable here myself. In a way you're right — nothing here seems memorable. Our joys aren't as high or our pains as deep — this may be another of those periods we don't remember. Perhaps it's time we left. Where would you like to go?"

"Greece! Let's go home to the islands: Samos, Delos, Naxos, Rhodes. Yes, Rhodes! These were our Happy Isles.

Since vacationing on Frackinos two years ago I've grown restless. Epius, let's go home again."

"We'll let the staff know in the morning. Now go back to sleep."

"We can't do that. I've been telling you they're going to put us in the nursing-home section of this place. And if we say we're leaving, they'll transfer us in the clothes we're standing in. We'll be labeled 'runners'; they'll lock those wristbands on us that sound the alarms when you go through an exit. Let's pack and leave now, right now, before sunrise."

"You're serious. Well, I've never yet turned away from a challenge, so let's do it. I'll get ready. The only things I want to take are a few keepsakes, remembrances of times past."

"What?"

"The latch to the door on the Trojan horse, the wooden truncated icosahedron [see frontispiece] I built for Leonardo, and my calculator. You'll take your lyre?"

"And my copies of Homer and the Greek poets. We're traveling light today."

A week later the poet and the scientist are on the Island of Rhodes in the ancient town of Lindos — the birthplace of Chares. Their tiny white-stucco house sits on the side of a cliff going down to the dazzling blue Aegean. The air is full of flowers, bird song, and the smell of the sea.

Early one morning they walked to a height of land for a clear view of the eastern horizon. Rosy-fingered Dawn, called Aurora by the Romans, was at her resplendent best. And as golden Helios climbed out of the sea, he was accompanied by his two great sundogs.

Epius murmured, "I was at Achilles' side the day he slew Aurora's son, Memnon, and the ground flashed as bright red as this morning's sky."

With teardrops of dew all around, and as if not to offend the sacred calm, his friend whispered, "A much worst fate awaited her husband Tithonus. Dawn asked Zeus to grant him immortality, but she forgot to also ask for eternal youth. As Tithonus aged more and more, he shriveled up into a chirping grasshopper. I fear that may yet happen to us."

"Phemius, Phemius, let's take the bus up to Rhodes this morning and see the harbor where the Colossus stood. We can have lunch and be back by supper. You'll feel better then."

"It's all gone, you know, every last piece of it — the remains were sold as scrap 1000 years ago. And there will come a time when even its memory will be forgotten. You can't go there, no one can. You'll only be saddened by what you don't see. Let's stay here and discuss ideas as we have always done, and then we'll go swimming in the afternoon with our new lab puppy."

"The statue itself may have vanished, but there are two immortal things remaining: your *inscription to freedom* and my *square-cube law*. These will never die. What's the poem you've often recited that expresses an astonishing longevity for the mind's creations?"

"You mean *Heraclitus,* the one Callimachus taught me":

They told me, Heraclitus, they told me you were dead,
They brought me bitter news to hear and bitter tears to shed.
I wept as I remembered how often you and I
Had tired the sun with talking and sent him down the sky.

And now that thou art lying, my dear old Carian guest,
A handful of gray ashes, long, long ago at rest,
Still are thy pleasant voices, thy nightingales, awake;
For Death, he taketh all away, but them he cannot take.

THEY WERE SILENT. Sitting in full sunlight on the hillside overlooking the sea, with laurel and olive trees all around, these two legends out of time reflected on their "nightingales." After a while, Epius put his protective arm around his friend who then looked up and smiled. Their thoughts glided like ships over a sunlit sea. The things they remembered down through their long history were imaginative ideas: those that weave facts into science and those that compose color, marble, and sound into art — trying to make sense of existence. They knew then, as they had always known, that ideas are the true immortals.

The stars are fixed, but the planets wander. The forest is still, but the wolves run. The woods are quiet, but the nightingale sings.

Chapter Notes

All the following entries are referenced in the main text by the words *Chapter Notes* or by a numerical superscript.

Chapter — 1

[P. 14] You could ask a hundred people what the word *gematria* means before getting a correct answer. Even if someone gave the right reply, it is doubtful they could explain gematria's connection to the Bible (*see* Rev. 13:18). So here's a dictionary definition: *gematria* is the ancient practice of using the numerical values of Greek, Hebrew, or Latin letters to change words into numbers. From public school, you may recall that only seven Latin letters (I, V, X, L, C, D, and M) have a numerical value. On the other hand, in the Greek and Hebrew alphabets *every symbol* was both a letter and a number. Any Webster's dictionary under "Special Signs and Symbols" will provide a full list of the Greek letters with their numerical values. The Hebrew number/letter code can be found on the Internet.

1. Stephen Jay Gould, *Ever Since Darwin* (New York: W. W. Norton & Company, Inc., 1977), p. 179.
2. Ibid., p. 181.
3. Ibid., pp. 196 ff.
4. Richard F. Burton, *Biology by numbers* (United Kingdom: Cambridge University Press, 1998), pp. 161-63.

Chapter — 2

[P. 42] The next row of the polytopes of the cube:
64   192   240   160   60   12   1

The text method for finding new table entries (double the number above and add the number to its left) is clearly reminiscent of the rule for Pascal's triangle. The reason for this similarity can be seen when Figure 2.13 is rewritten as below — focus on the bold numbers.

Let n = number of dimensions and r = number of polytopes of varying types. Also, let $_nT_r$ be the table entry in the nth row and rth column. With this notation, we have the following non-recursive formula, $_nT_r = 2^{n-r} \binom{n}{r}$.

For example, $_4T_3 = 2^{4-3} \binom{4}{3} = 2^1 \times 4 = 8$.

n \ r	0	1	2	3	4	5
0	1(2^0)	0	0	0	0	0
1	1(2^1)	1(2^0)	0	0	0	0
2	1(2^2)	2(2^1)	1(2^0)	0	0	0
3	1(2^3)	3(2^2)	3(2^1)	1(2^0)	0	0
4	1(2^4)	4(2^3)	6(2^2)	4(2^1)	1(2^0)	0
5	1(2^5)	5(2^4)	10(2^3)	10(2^2)	5(2^1)	1(2^0)
⋮	⋮	⋮	⋮	⋮	⋮	⋮

There is yet a third method of finding table entries. Let n be the number of dimensions, as above, then the coefficients of $(2x+1)^n$ will give the entire nth row. Here for example is the second row: $(2x+1)^2 = 4x^2 + 4x + 1$.

[P. 42] Description of the extra (6th) regular polytope in four dimensions:

- No equivalent in other dimensions
- 24 octahedral cells, 96 triangular faces, 96 edges, 24 vertices
- Self-dual
- 3 octahedra meet at an edge

[P. 48] A manuscript on which an earlier text has been effaced and the vellum or parchment reused is called a palimpsest. In medieval church circles, it was a common practice to erase an earlier piece of writing by means of scraping or washing in order to prepare it for a new text. Motives were twofold: economic and the desire "to convert" pagan Greek texts by overlaying them with the word of god. Modern historians, more interested in older writings, have employed infrared and digital enhancement techniques to recover the erased texts, often with outstanding results.

A single reclaimed palimpsest is the only known copy of Archimedes' important *On the Method of Mechanical Theorems* and the original Greek version of *On Floating Bodies*. It also contains copies of Archimedes' *On the Measurement of the Circle, On the Sphere and the Cylinder, On Spiral Lines,* and *On the Equilibrium of Planes.* Previously, these had been identified only from secondary sources.

[P. 52] By definition, the circumference of any disc that passes through a sphere's center is a "great circle." And any arc of this great circle is a geodesic (i.e. the shortest distance on the sphere) for the arc's end points.

[P. 52] The Swiss mathematician Leonhard Euler (1707-83) is probably the most prolific member of his craft. Among his many discoveries is the well-known formula $F - E + V = 2$: F is the number of faces, E the number of edges, and V the number of vertices. This formula is true for all the three-dimensional convex polyhedra. We will use it to prove that 12 pentagons, neither more nor less, are required to *close* a polyhedron while the hexagons can be any number. The proof is as follows:

Let the number of pentagons be P.
Let the number of hexagons be H.
So, the total number of faces is $P + H$.
Therefore, the number of edges is $E = \dfrac{5P + 6H}{2}$
since each edge is shared by two polygons.

And the number of vertices is $V = \dfrac{5P + 6H}{3}$

because each vertex must be shared by three polygons.
Now substitute these into Euler's formula.

$$P + H - \left[\dfrac{5P + 6H}{2}\right] + \left[\dfrac{5P + 6H}{3}\right] = 2$$

Multiplying both sides by 6 clears the fractions.
6P + 6H − [15P + 18H] + [10P + 12H] = 12.
Clear brackets and collect like terms.
This gives P = 12 with H canceling out of the equation.
Therefore, the number of pentagons is always 12, while the number of hexagons may be any whole number.

[P. 58] The next row of Pascal's triangle:
1 7 21 35 35 21 7 1

1 From a paper written in 1798 by Immanuel Kant titled "On the First Ground of the Distinction of Regions of Space".
2 Martin Gardner, *The Ambidextrous Universe* (New York: Charles Scribner's Sons, 1979), p. 141.
3 Ibid., p. 148.
4 Peter S. Stevens, *Patterns in Nature* (New York: Atlantic-Little, Brown Books, 1974), p. 14.
5 D'Arcy Wentworth Thompson, *On Growth and Form* — Abridged ed./ Edited by John Tyler Bonner (Great Britain: Cambridge University Press, 1984), p. 152.
6 Ibid., p. 167.
7 Georges Ifrah, *From One to Zero* (New York: Penguin Books, 1985 English translation from the French), p. 466.
8 Oystein Ore, *Cardano: The Gambling Scholar* (New York: Dover Publications, 1958), p. vii.

CHAPTER — 3

[P. 69] The difficulty in determining the motion of three celestial objects or bodies moving under no influence other than that of their mutual gravitation is called the three-body problem.

No solution of this dilemma (or the general problem involving more than three bodies) is possible.

[P. 71] The sum of the first hundred numbers is 5050. Consider this series: $1+2+3+ \ldots + 98+99+100$. We believe Gauss added $1+100$ to get 101 and $2+99$ to again get 101, and so on. By this approach, he got 50 sums each of 101, so their total must be 50 x 101 or 5050. This method can be generalized.

[P. 75] Program for Figure 3.2:

```
SCREEN 12: REM This screen provides the best resolution.
pi = 4 * ATN(1): REM pi = 3.1415926...
PAINT (0,0): REM Creates a white background.

CIRCLE (283, 230), 3, 0: REM  These six lines put three
PAINT (283, 230), 0:      REM  dots on the horizontal axis.
CIRCLE (pi / 2 * 60, 230), 3, 0
PAINT (pi / 2 * 60, 230), 0
CIRCLE (5 * pi / 2 * 60, 230), 3, 0
PAINT (5 * pi / 2 * 60, 230), 0

FOR N= pi/2  TO   5 * pi / 2   STEP   pi /1000
FOR c = -2  TO  5   STEP  1: REM These are the constants.

    IF c > 2 THEN
        r = c - 2 * SIN(N)
        v = SQR(r)
        PSET (N * 60, v * 60 + 230), 0
        PSET (N * 60, -v * 60 + 230), 0
    ELSEIF c > 2 * SIN(N) THEN
        r = c - 2 * SIN(N)
        v = SQR(r)
        PSET (N * 60, v * 60 + 230), 0
        PSET (N * 60, -v * 60 + 230), 0

    END IF
    IF INKEY$ = CHR$(27) THEN STOP
    REM Press the "esc" key any time to stop.
```

NEXT c
NEXT N

[P. 90] To explore cobweb diagrams in a dynamic fashion visit www.lboro.ac.uk/departments/ma/gallery/doubling/.

[P. 115] President Garfield's proof of the Pythagorean theorem:

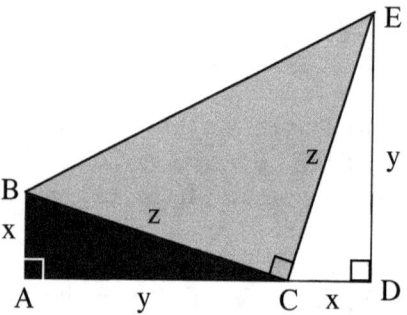

The area of the entire figure, trapezoid ABED, is the product of its base, $x + y$, and the average of its height.

That is $\left[(x+y) \dfrac{(x+y)}{2} \right]$.

But the area of the trapezoid is also equal to the sum of the areas of its three right $\triangle$s.

The large isosceles $\triangle$ has area $\dfrac{zz}{2}$.

The other two $\triangle$s each have area $\dfrac{xy}{2}$.

Therefore, the area of the trapezoid is $\dfrac{z^2}{2} + \dfrac{2(xy)}{2}$.

Equate these areas and simplify:

$$\dfrac{(x+y)(x+y)}{2} = \dfrac{z^2}{2} + \dfrac{2(xy)}{2}$$

and so, $x^2 + 2xy + y^2 = z^2 + 2xy$

$x^2 + y^2 = z^2$ Q.E.D.

[P. 117] What follows is an iterative version of the "division statement" we learned in secondary school. We can rewrite this as an equation: (divisor)(quotient) + remainder = dividend. All the factors and terms in this equation are positive whole numbers. Let us condense the formula by a few letter substitutions, and manipulate it so it becomes the equation below.

$$\text{divisor}\overset{\text{quotient}}{\overline{)\text{dividend}}}$$
$$\vdots$$
$$\overline{\text{remainder}}$$

$$\begin{array}{ll} \text{divisor} = d & \text{quotient} = q \\ \text{remainder} = r & \text{dividend} = D \end{array} \longrightarrow q = \frac{D - r}{d}$$

We would like q to represent a single digit in the decimal expansion of rational numbers. And so, the above equation

$$q_n = \left[\frac{10r_{n-1} - r_n}{d} \right] \quad \begin{array}{c} 0 \leq r_{n-1}, r_n < d \\ \text{and} \\ q_n \text{ is a digit in the period.} \end{array}$$

must stand for a single division in what can be a very long process — recall William Shanks and 1/17389. The new dividend, D, equals 10 times the r_{n-1} remainder. Therefore, the nth term (a digit) of the quotient is q_n. The [ ] brackets indicate this is the "greatest integer function." For instance: [2.6] = 2, [7.99] = 7, and [3.1] = 3.

For example, remember 1/7 = 0.142857 142857 1428.... So our divisor is d = 7, and choose r_{n-1} from 1,2,3,4,5,6, say 3 as the initial value. Substitute these into the formula and evaluate.

$$q_1 = \left[\frac{10 \times 3 - r_1}{7} \right] \longrightarrow q_1 = \left[4\frac{2}{7} - \frac{r_1}{7} \right] \longrightarrow q_1 = 4 \text{ and } r_1 = 2$$

To show this formula is iterative, substitute $r_1 = 2$:

$$q_2 = \left[\frac{10 \times 2 - r_2}{7} \right] \longrightarrow q_2 = \left[2\frac{6}{7} - \frac{r_2}{7} \right] \longrightarrow q_2 = 2 \text{ and } r_2 = 6$$

Do it again:

$$q_3 = \left[\frac{10 \times 6 - r_3}{7} \right] \longrightarrow q_3 = \left[8\frac{4}{7} - \frac{r_3}{7} \right] \longrightarrow q_3 = 8 \text{ and } r_3 = 4$$

By continuing this iterative process beyond **428**, the full period of 1/7 can be constructed.

[P. 119] Newton–Raphson method:

The equation must be put into the form f(x) = 0, where $f^1(x)$ is the derivative.

The formula: $x_{n+1} = x_n - \dfrac{f(x_n)}{f^1(x_n)}$.

Example: Find one root of $x^5 = 37$.

$f(x) = x^5 - 37 = 0$, and $f^1(x) = 5x^4$.

Therefore, $x_{n+1} = x_n - \dfrac{(x_n)^5 - 37}{5(x_n)^4}$.

Let your first guess be $x_0 = 3$:

$$x_1 = 3 - \dfrac{3^5 - 37}{5 \times 3^4} = 2.49135802469$$

Plug this in as x_n in the formula and get the next guess:

2.18516863143
next guess:
2.07269258474
next guess:
2.05910584704
next guess:
2.05892416855
next guess:
2.05892413648

which is accurate to 10 decimal places.

1 Ian Stewart, *Does God Play Dice?* (United Kingdom: Penguin Books, 1990), p. 18.
2 Jacob Bronowski, *The Ascent of Man* (United States: Little, Brown and Company, 1973), p. 180.
3 Eric Temple Bell, *Men of Mathematics* (United States: Simon and Schuster, 1937), pp. 221-22.
4 Ian Stewart, *Does God Play Dice?* (United Kingdom: Penguin Books, 1990), pp. 85-6.
5 James Gleick, *Chaos: Making a New Science* (United States: Penguin Books, 1987), p. 16.

6 Plutarch, *The Lives of the Noble Grecians and Romans*, translated by John Dryden and revised by Arthur Hugh Clough (United States: The Modern Library, Random House, Inc. undated), p. 801.
7 Bertrand Russell, *History of Western Philosophy* (London: George Allen and Unwin Ltd., 1946), p. 762.
8 Martin Gardner, *Order and Surprise* (United States: Prometheus Books, 1983), pp. 151-55.

Chapter — 4

[P. 135] What *fractal dimension* means has undergone many revisions, and no universal definition seems available. And even what it means *to be a fractal* is undecided like the words *beautiful* and *ugly*. Mandelbrot now believes we would be better off without a definition.

[P. 139] Computer program for Figures 4.6 and 4.15:

```
OPTION BASE 0
SCREEN 12
RANDOMIZE TIMER

xmin = -1.1: xmax = 1.1
ymin = -1.1: ymax = 1.1
dx = xmax - xmin: dy = ymax - ymin
pi = 4 * ATN(1)
LOCATE 10, 15
PRINT "esc key must be pressed to stop program"

LOCATE 15, 25:  REM row, column
INPUT "enter # vertices :", nvert
NV = nvert - 1
M = 1 / (nvert / 3 + 1)
CLS
PAINT (0, 0)
```

```
FOR IV = 0 TO NV:  REM establish polygon vertex coordinates
      angle = 2 * pi * IV / nvert
      IF nvert MOD 2 = 0 THEN angle = angle + 2 * pi / nvert / 2
      X0(IV) = -SIN(angle)
      Y0(IV) = COS(angle)
NEXT IV

DO
      I = INT((NV + 1) * RND)
      x = M * x + (1 - M) * X0(I)
      y = M * y + (1 - M) * Y0(I)
      GOSUB draw0
      IF INKEY$ = CHR$(27) THEN STOP
LOOP
END

draw0:
      xq = 640 * (x - xmin) / dx
      yq = 480 * (ymax - y) / dy
      PSET (xq, yq), I
RETURN
```

[P. 147] The fourth level:

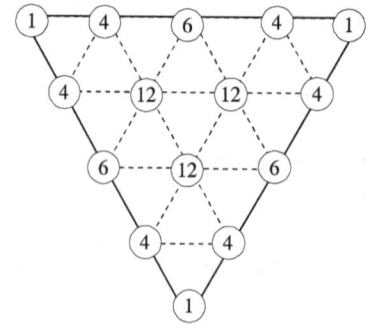

[P. 151] Same program as above for [P. 139].

[P. 156] Here is another definition of the Mandelbrot man: it is the set of all points in the complex plane that remain prisoners for every iteration of $z_{n+1} = z_n^2 + c$ when the initial value of z_n is 0, and c is a constant.

1 Benoit B. Mandelbrot, *Fractals: Form, Chance, and Dimension* (United States: W. H. Freeman and Company, 1977), pp. 48-9.
2 Ian Stewart, *Does God Play Dice?* (United Kingdom: Penguin Books, 1990), p. 242.

CHAPTER — 5

[P. 173] Reverse triangle numbers:

55	55
66	66
153	**351**
171	171
595	595
666	666
3003	3003
5995	5995
8778	8778
15051	15051
17578	**87571**
66066	66066
185745	**547581**
617716	617716
828828	828828
1269621	1269621
1461195	**5911641**
1680861	1680861
3544453	3544453
5073705	5073705
5676765	5676765
6295926	6295926
. . .	. . .

Reversible triangle numbers can be divided into two groups:

- Palindromes, e.g. 55, 3003
- Non-palindromes, e.g. 153, 17578, 185745, and 1461195.

[P. 173] Truncated triangle numbers:

Readers with a home computer and some knowledge of programming can prove that only 6 truncated triangle numbers exist: i.e. 15, 36, 66, 105, 153, 666. After you have written and run your program, the print-out will be a list of these half-dozen numbers. But how do we know this inventory is complete? Reason as follows. To have a truncated triangle number with five digits implies you had one with four digits. But, if your program is correct, then there are none with four digits. Accordingly, there cannot be any with five. And reasoning in a similar fashion, there can be none with six digits or seven or eight, and so on.

[P. 175] *Googolplex:* a googol is 1 followed by a hundred zeros (i.e. 10^{100}), and a googolplex is 1 followed by a googol of zeros (i.e. 10^{googol}). That's quite large!

[P. 183] What Dr. Matrix didn't tell us was that this property of powers works for any two consecutive numbers.

1   James R. Newman, *Volume Four of The World of Mathematics,* (New York: Simon and Schuster, 1956), p. 2,348.
2   Edward Kasner and James Newman, *Mathematics and the Imagination* (New York: Simon and Schuster, 1940), p. 359.
3   Martin Gardner, *The Night is Large* (New York: St. Martin's Press, 1996), p. 268.
4   The Newsletter of the Mathematical Association of America v. 14, no. 6, Dec. 1994.
5   David Ulansey, "The Mithraic Mysteries," *Scientific American,* Dec. 1989, p. 133.
6   St. Augustine, *City of God* (United Kingdom: Penguin Books, 1984), p. 790.
7   G. H. Hardy, *A Mathematician's Apology* (United Kingdom: Cambridge University Press, 2001), p. 105.

Chapter — 6

[P. 215] Computer program to draw Figure 6.11 — the spirals on a sunflower's head:

```
SCREEN 12
    PAINT (0, 0)
    pi = 4 * ATN(1)
    t = (137.507764# * pi) / 180: REM t is in radians
    r = 0
    a = 0

FOR k = 1 TO 500: REM Play with this variable.
    r = r + 0.4
    a = a + t
    x = r * COS(a)
    y = r * SIN(a)
    CIRCLE (x + 320, y + 232), 3, 0
    PAINT (x + 320, y + 232), 0
    IF INKEY$ = CHR$(27) THEN STOP
NEXT k
```

[P. 225] Let's work out the implications of Herodotus' report that the area of each triangular face is equal to the square of its height. Consider this figure:

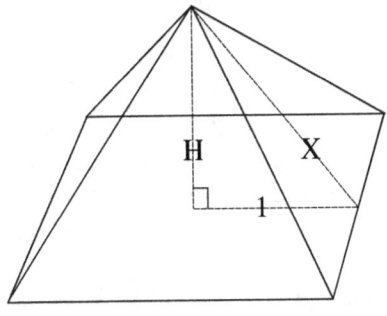

GOLDEN PYRAMID

Let X be the slant height of any triangular face,
and assume the base is 2 units long. In no way does this
assumption affect the generality of the conclusion.

Therefore the face area $= \left(\frac{1}{2}\right)$(base)(height)
$$= \left(\frac{1}{2}\right)(2)(X)$$
$$= X.$$

Now the slant height is the hypotenuse of a right-angled triangle whose legs are 1 (half the base) and H (the pyramid's height).

By the Pythagorean theorem, $X^2 = 1^2 + H^2$
and so $H^2 = X^2 - 1$.

When the area of the face equals the square of the height, we have the following equation:

$$X = X^2 - 1 \text{ or } X^2 - X - 1 = 0.$$

This is easily solved by the quadratic formula to give the positive root:

$$X = \frac{(1 + \sqrt{5})}{2}$$
$$= 1.618...$$
$$= \text{phi}.$$

Because $H^2 = X^2 - 1$,
therefore $H = \sqrt{X^2 - 1}$
$$= \sqrt{(\text{phi})^2 - 1} \text{ (see page 213)}$$
$$= \sqrt{\text{phi}}.$$

So the slant height, X, is phi and the perpendicular height, H, is the square root of phi. This is the golden triangle/pyramid of myth and legend.

1. Joseph M. Jauch, *Are Quanta Real?: A Galilean Dialogue* (IN: Indiana University Press, 1973), pp. 63-4.
2. Ian Stewart, *Nature's Numbers* (New York: BasicBooks, 1995), p. 139.
3. Ibid., p. 146.
4. Henry David Thoreau, *Thoreau: Walden and Other Writings* (New York: Bantam Books, 1962), p. 148.

5 Howard Eves, *Historical Topics for the Mathematics Classroom* (Washington, DC: National Council of Teachers of Mathematics, 1969), p. 206.
6 Herodotus: *The Persian Wars*, translated by George Rawlinson (New York: Modern Library Book, Random House, 1942), p.179.
7 David Bergamini, *Mathematics* (New York: Time-Life Books, 1963), p. 94.
8 Bertrand Russell, *History of Western Philosophy* (London: George Allen and Unwin Ltd., 1946), p. 48.
9 James Harrison, *The Pattern & The Prophecy: God's Great Code* (Toronto: Isaiah Publications, 1995), pp. 103-111.
10 Jacob Bronowski, *The Ascent of Man* (United States: Little, Brown and Company, 1973), p. 56.

Chapter — 7

[P. 260] The least-action principle:

$$action = mass \times velocity \times distance = m \times v \times d$$

By using kinetic energy, $E = \frac{1}{2}mv^2$, and $d = vt$, we can manipulate the above as follows:

$$\begin{aligned} action &= mvd \\ &= mv(vt) \quad (\text{recall, } d = vt) \\ &= mv^2 t \\ &= 2Et \quad (\text{recall, } E = \tfrac{1}{2}mv^2) \\ &= 2 \times energy \times time \end{aligned}$$

1 Bernd Heinrich, *Bumblebee Economics* (United States: Harvard University Press, 1979), pp. 148-49.
2 Michael Grant, *Cicero — On the Good Life* (New York: Penguin Books, 1971), pp. 86-7.
3 Robert Weinstock, *Calculus of Variations: With Applications to Physics and Engineering* (New York: Dover Publications Inc., 1974), pp. 67-8.

4 Richard Feynman, *QED: The Strange Theory of Light and Matter* (New Jersey: Princeton University Press, 1985).
5 Thomas A. Moore, *Macmillan Encyclopedia of Physic*, "Least-Action Principle" (New York: Simon & Schuster, Macmillan Publishing, 1996), p. 840.
6 Daniel Dennett, *Darwin's Dangerous Idea* (New York: Touchstone Books, 1996), p. 21.
7 William Paley, *Natural Theology: Selections* (USA: The Bobbs-Merrill Company Inc., 1963), p. 4.
8 Richard Dawkins, *The Blind Watchmaker* (U.K.: Penguin Books Ltd., 1988), p. 5.
9 Peter S. Stevens, *Patterns in Nature* (New York: Atlantic-Little, Brown Books, 1974), p. 196.

CHAPTER — 8

[P. 308] Duals of regular solids:

REGULAR SOLIDS	FACES	VERTICES	EDGES
Tetrahedron	4 ←——→	4	6
Cube	6 ↘	↗ 8	12
Octahedron	8 ↗	↘ 6	12
Dodecahedron	12 ↘	↗ 20	30
Icosahedron	20 ↗	↘ 12	30

1 Martin Gardner, *The Ambidextrous Universe* (New York: Charles Scribner's Sons, 1979), pp. 6-7.
2 Ian Stewart and Martin Golubitsky, *Fearful Symmetry: Is God a Geometer?* (USA: Blackwell Publishers, 1992), pp. 27-8.
3 Leopold Infeld, *Whom the Gods Love* (USA: The National Council of Teachers of Mathematics, 1978).
4 Hermann Weyl, *Symmetry* (USA: Princeton University Press, 1982), pp. 66-7.
5 Carl Sagan and Ann Druyan, *Shadows of Forgotten Ancestors* (USA: Random House, Inc., 1992), p. 15.

CHAPTER — 9

[P. 330] The number of odd nodes in a network is always even:

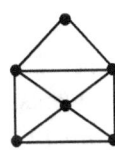
HOUSE

A node (vertex) is a point from where roads (lines) emerge or converge. A road connects nodes and has two ends. If R is the number of roads, 2R is the number of ends — an even number. Let K_n be the number of nodes of the nth order. So in the HOUSE $K_2 = 1$, $K_3 = 2$, and $K_4 = 3$. Because a node of the nth order is one at which n roads emerge there are n ends there. Now we can construct a second expression for the total number of ends and set it equal to the first (that is, 2R):

$$2K_2 + 3K_3 + 4K_4 = 2R.$$

$2K_2$ and $4K_4$ are even and may be dropped without changing the parity. So, $3K_3$ = an even number.

$3K_3 = 2K_3 + 1K_3$ = an even number.

Also $2K_3$ may be dropped without affecting the parity, implying K_3 is an even number — in this case 2.

This entire argument generalizes to any network of nodes and roads. Hence, the number of odd nodes is always even.

[P. 355] Classic birthday problem — *unspecified day:*

To prove just 23 people are sufficient to have a 50 percent chance of at *least* one birthday match, proceed as follows. Work indirectly by finding the probability the dates are *all different* and then subtract this from 1 (or 100 percent if you're dealing with percentages). The 23 dates can be chosen in 365 x 365 x 365 x . . . x 365 ways (23 repetitions). If all must to be different, the possibilities are 365 x 364 x 363 x . . . x 343 (23 repetitions). Now divide these:

$$P(\text{all different}) = \frac{365 \times 364 \times 363 \times \ldots \times 343}{365^{23}}$$

$$P(\text{at least two the same}) = 1 - \frac{365 \times 364 \times 363 \times \ldots \times 343}{365^{23}}$$

$$= 51\%$$

Johnny Carson's problem — *specified day:*

The probability of someone's birthday not being October 23, say, is 364/365, and because birthdays are independent events, the probability of two people not having being born on this specified date is 364/365 x 364/365. For the product to equal one half or 50 percent, you require 253 people.

$$P(\text{none are October 23}) = \frac{364 \times 364 \times 364 \times \ldots \times 364}{365^{253}}$$

$$P(\text{at least one is October 23}) = 1 - \frac{364^{253}}{365^{253}}$$

$$= 50\%$$

To have a match for his birthday at the 50 percent confidence level, Johnny needed an audience of 253 people.

The reader can confirm both answers with a hand calculator.

1. Carl Sagan, *The Cosmic Connection: An Extraterrestrial Perspective* (New York: Doubleday, 1973), pp. 189-90.
2. Jacob Bronowski, *Science and Human Values* (New York: Harper & Row, 1972), p. 28.
3. Kenneth Clark, *Leonardo da Vinci* (England: Penguin Books, 1975), p. 95.
4. Stephen Jay Gould, *Bully for Brontosaurus: Reflections in Natural History* (New York: W.W. Norton & Company, 1992), p. 268.

5 Thomas Gilovich, *How We Know What Isn't So* (New York: The Free Press, a division of Simon & Schuster Inc., 1991), pp. 11-6.
6 Scott Camazine, "Patterns in Nature," *Natural History* magazine, June 2003, pp. 36-8.

CHAPTER — 10

[P. 376] Even and odd numbers in the Fibonacci sequence:

+	E	O
E	E	O
O	O	E

The distribution of odds (O) and evens (E) depends on the two starting numbers and the addition table at the left. The result is more a matter of mathematics than science.

1 Jacob Bronowski, "The Creative Process," *Scientific American* magazine, September, 1958, p. 62.
2 Jacob Bronowski, *The Ascent of Man* (United States: Little, Brown and Company, 1973), p. 374.

INDEX

A

Abel, Niels Hendrik, 302
Achilles, 83, 352, 386
acrostics, 168
activators and inhibitors, 146, 362-364
Adam, 172, 231-32, 236, 237, 289
Aeschylus, 160, 344
Agrippa of Nettesheim, 231, 233, 236
air pressure in soap bubbles, 280-82
Alexander the Great, 1, 102, 332, 333
Alexandria, 48, 248, 368
Alhambra, the, 309-10, 311, 312
Alice (of Wonderland), 131, 248, 290
Allah, 182, 183, 188
Allen's rule, 24-25, 335
Ames, Adelbert, 380-81
Andromeda galaxy, 166, 254
angles of 109.5°, 276, 280
angle of incidence, 247-48, 296
angles of 120°, 272
    cracks, honeycomb, etc., and, 269, 272, 341
    minimizing distance with, 54, 268-69, 341
    soap bubbles and, 268-69, 271, 276, 284
    tree branching at, 338, 341, 344
    see also three-way corners
animals:
    bear, 83, 205, 241, 256, 281
    bison, 215, 240-41
    elephant, 3, 11, 22, 26, 94, 95, 131-32
    fox, 89, 246, 250, 286-87, 290, 378, *see also* Reynard
    hippopotamus, 11
    leopard, 21, 359, 364, 365
    lion, 240, 314, 349, 365
    lynx, 89
    mouse, 20-21, 22, 209
    musk ox, 256
    red squirrel, 286, 369
    vole, 243, 286, 287
    wolf, 21, 205, 214, 243, 250, 256, 262, 290, 315, 366, 378
    zebra, 359, 363-65
Antichrist's number, 172, 174, 350
    see also numbers supernatural: 666
Apollo (Greek god), 5, 292
Apollo (moon rocket), 197, 249
Archimedes, 36, 141, 182, 300
    death of, 50
    rhombicuboctahedron of, 34-35
    streaking by, 71, 294
    tomb of, 244-46, 259
    truncated icosahedron of, 48-49, 50
Argus (Odysseus' dog), 385
Aristotle, 36, 79, 147, 277, 372
Arnold, Matthew, 299-300
Ars Magna (Cardano), 301, 302
astrology, 61, 167
asymmetry, 290, 297, 320
Athena (Greek goddess), 83, 228, 368
Athens, 3, 79, 160, 228, 236, 347
attractor(s): 93, 94, 95, 116, 122, 164
    period-two, 90
    point, 86-87
    see also strange attractors
Auerbach, Red, 357
Augustine of Hippo, 170-71, 178
Aurora (Roman goddess), 386

B

Babylonians, 113, 301
Bacon, Francis, 355, 372
bacteria, 84, 109-10, 294, 380
Barbari, Jacopo de, 33, 371
Barbier's Theorem, 193
Barnsley, Michael, 125, 137-39, 143, 149-51, 162
Bartholdi, Federic Auguste, 1
Bayeux Tapestry, 373

beauty and bilateral symmetry,
    291, 292, 303, 320
Belloc, Hilaire, 263, 264
bellybutton, 231, 235
    see also navel
Bergmann's rule, 24-25
Bhagavad–Gita, 327
Bible, 67, 178, 379
    153 Fishes in the Net (John
        21:11), 169-71, 176, 196
    Adam's navel and the, 232
    equidistant-letter sequences (ELS)
        in the, 186
    gematria and the, 172, 185
    Moses' horns in, 332-33
    pi's (π) value in the, 195-96
    triangle numbers in the, 171-73
    see also Scriptures, Holy
bifurcation diagram, the, 95-96, 99,
    237, 317
Big Bang, 325-26
billiards (and bouncing), 247-49
birds:
    black-billed cuckoo, 27
    black-capped chickadee, 25
    chimney swift, 325
    golden-crowned kinglet, 25
    hummingbird, 72
    owl, 89, 281, 322, 349
    penguin, 23
birthday problem:
    specified (Johnny Carson's
        problem), 354, 355
    unspecified, 354, 355, 365
    see also pages 405-6 of the
        Chapter Notes
Blake, William, 7, 12-15, 63-65,
    127
Borwein, Peter, 204
Botticelli, Sandro, 221
Brahe, Tyco, 384
brain coral, 334
Bronowski, Jacob, 70, 113, 242,
    345-46, 372, 384-85
Browne, Sir Thomas, 232, 273
bucky ball, 50, 52, 309
    see also truncated icosahedron
bumblebees, 22, 205, 209-10, 244,
    246, 285
Burgess Shale, 109, 122
Bush, George W., 163

butterflies, 293, 359
butterfly effect, 99-102, 104-5,
    106, 107, 109-10, 119, 122

C

calculus, 68, 70, 245, 253, 258,
    270, 375
Callimachus (Greek poet), 160,
    344, 387
Camus, Albert, 358
Cantor, Georg, 91, 219
Cardano, Gerolamo, 60-61, 301-2
    see also the Preface
Carroll, Lewis, 29, 95-96, 247,
    276
Carson, Johnny, 354, 355
Carthaginians, 50, 255
Catholic Church, 266, 273, 376
cellular automata, *see* game of life
chaos game:
    with six points, 150-51
    with three points, 137-42
chaos, definitions of, 68
Chardin, Jean *(A Boy Blowing a
    Soap Bubble),* 266, 276
Chares of Lindos, 2, 3, 4, 291, 386
Chauvet cave, 240, 241, 297, 368
Chesterton, G. K., 352
Christ, 168-69, 175, 266, 384
    as the fish symbol, 178
    Corpus Hypercubus, 44-45
    *Soldiers casting Lots for Christ's
        Garments,* 63-65
    The Last Supper, 70
Christianity, 44, 107, 236, 358
Churchill, Winston, 223
Cicero, 106, 245-46
Clark, Kenneth, 346
Clarke, Arthur C., 47, 80
clockwork universe, 76, 122
clustering illusion, 349, 350-51,
    357-58, 365
cobweb diagrams, 87-88, 89-90,
    95, 122
Coleridge, Samuel Taylor, 157
Colossus of Rhodes, 1, 2, 3, 4,
    291, 387
    collapse of, 4
complex plane, the, 120-21, 153,
    154, 155
constellations, 162, 187, 188, 352

contingent history, 109, 111-12, 140
continued fractions, 203-4, 213-14
Conway, John Horton, 311-12, 358-359, 360, 361, 364
Copernicus, Nicolas, 8, 347, 384
cowboy's first dilemma, 251
cowboy's second dilemma, 252
Coxeter, H.S.M., 29, 211
Cromwell, Oliver, 266, 384
Crucifixion in painting, the, 44-45
cube:
    cube-corners and bouncing, 249, 250
    projection of a, 277-78
    unfolding a, 43, 44
Curie point, 325
Curie, Pierre, 315, 316, 318, 321
Curie's Principle, 315-16, 318
curves of constant width, 190-93, 194

D

D'Alembert, Jean le Rond, 56
Da Vinci Code, The (Dan Brown), 236
da Vinci, *see* Leonardo
Dali, Salvador:
    Colossus of Rhodes, 2
    Corpus Hypercubus, 44-45, 279
    The Sacrament of the Last Supper, 235
Daniel, Book of, 14, 105, 332
Dante Alighieri, 123, 137, 189, 327
    see also, Inferno, The
Dark Ages, 14, 48, 230, 297, 368, 374
    see also Middle Ages
Dawkins, Richard, 274
de Camp, L. Sprague, 2
De Divina Proportione (Luca Pacioli), 30, 35, 49, 52, 371
de Morgan, Augustus, 130, 198
deduction, 118, 368-70
    see also induction
deductive reasoning, 114-15
del Ferro, Scipio, 301
Delian puzzle, 5-7
Delos, Island of, 5, 385
Delphi, 5, 8
delta (δ), the constant, 189, 212, 237-39, 242
    see also Feigenbaum, Mitchell
Democritus, 12, 13
Demosthenes, 106, 356
Dennett, Daniel, 261
Descartes, René, 284
descent with modification, 22, 109, 262, 341
    see also natural selection
dice:
    frequencies for any number of cubical dice, 62-63
    handedness of, 65
    non-cubical, 63
    perfect polyhedral, 60, 61, 92, 166, 353
    Soldiers casting Lots for Christ's Garments, 63-65
Dido, Queen, 255, 256, 257
differential equations, 68, 69, 84, 103
dinosaurs, 25, 110
divergence angle, 217, 219
    see also golden angle
divine proportion, 213, 236
    see also phi
    see also golden number
DNA, 194, 220, 289, 294, 365, 366, 374
Dobzhansky, Theodosius, 244
dodecahedron, Etruscan, 32
Doré, Gustave, 170
double printing (by animals), 243, 285
doubling the cube, *see* Delian puzzle
Dr. Irving Joshua Matrix, 183-84
    see also page 400 of the Chapter Notes
drawings, reversible, 299
Drosnin, Michael, 184-86, 201
duality and the Platonic solids, 234, 307-8, 321
    see also page 404 of the Chapter Notes
Dürer, Albrecht, 33, 235

E

e (base of the natural logarithms), 118
earth, 8, 67, 79
    area/volume ratio, 17

associated with a cube, 33
coldest place on, 23
destiny of the, 15, 317
distance to the moon, 249
impacts, 110
Kepler's association, 36
life on, 16, 83, 107, 109, 112, 319
magnetic field of, 324
moon and the, 71
orbit of, 347
plate tectonics of, 18
spherical, 277, 282
wobbles of, 169
Ecclesiastes, 97, 123
Eden, 12, 43, 47, 237, 242, 289
Einstein, Albert, 40-41, 67, 117, 266, 347, 376
  equations of, 134
  longevity of his theories, 344
  meandering stream model of, 194, 333
  on light, 253
  reflections on the face of, 296-97
El Greco, 378-79, 384
electromagnetic radiation, 16, 253
electrons, 16, 167, 276
Elements, Euclid's, 300
  deductive proofs of, 118, 368
  on phi, 228
  on the regular solids, 32, 33, 38, 321
  structure of the, 115
  see also Euclid
emergent complexity, 359, 360, 362, 365
  see also emergent properties
emergent properties, 90, 221, 359, 360
  see also emergent complexity
Empedocles, 261
enantiomorphs, 40, 293-95, 296-97, 320-21
energy:
  kinetic (K.E.), 73-75, 132, 133, 194
  least-action principle and, 260-61
  potential (P.E.), 73-75, 133, 194, 265
  size and, 8-9
Epeios, *see* Epius

Epidaurus, theater at, 229-30, 236
Epius, 3-4, 56-57, 79-83, 185-86, 247, 249-50, 291-92, 327, 329-30, 352, 367-71, 385-87
  see also Epos
Epos, 159-60
  see also Epius
equidistant-letter sequences (ELS), 185-86
Escher, Maurits, 309, 312
  Drawing Hands, 294
  Horsemen on a Möebius frieze, 310
  Limit Circle III, 129-30
Eskimos, *see* people: Inuit
Euclid, 41, 67, 162, 212, 248, 255, 284, 300, 338, 344
  deductive reasoning, 118, 368
  beauty and, 161, 337
  on phi, 228
  on the regular solids, 32, 33, 38, 321
  structure of the *Elements,* 115, 370
  see also Elements, Euclid's
Euclidean geometry, 40-41, 128
Euclidean space, 346-47
Euclidean tools, 5, 198-200, 370
Euler, Leonhard, 330
Eve (Heva), 231-232, 237, 289
evolution, 23, 27, 220, 244, 365, 378
  Charles Darwin and, 8, 22, 96, 108, 109, 261-62, 266, 273, 274, 285, 341, 346
  distant origins of, 336
  honeycomb and, 272
  orthogenesis and, 109, 122, 140
  size and form in, 9, 19
Ewen, Bruce, 383
Exodus, 168, 333
expressway or roadway problem, 274-75
  for three towns, 268
  for four towns, 270-71
  for five towns, 271
extraterrestrials, 65, 110, 166, 380

F

feedback loops, 84-85, 92, 364-65
Feigenbaum, Mitchell, 99

fig tree of, 99, 119, 152, 156, 159, 237-39
  constant of, 237, 238-39
  see also delta (δ)
Femos, 159-160
  see also Phemius
Fermat, Pierre de, 253, 255, 260
Ferrari, Ludovico, 302
Feynman, Richard, 223, 254, 374
Fibonacci:
  drone bumblebees and, 210
  flower petals and, 205, 241-42, 373
  numbers, 210, 213-14, 217, 221, 383
  sequence, 207, 211, 212, 214, 229, 235, 236, 376
  spirals, 208-9, 216, 218, 220-223
Fibonacci sequence of odd and even numbers, 376
  see also page 407 of the Chapter Notes
fiddleheads, 331, 336
Fishes in the Net (153), 177-78, 196
Flatland, 29, 38-39, 43, 44, 47, 277
flowers:
  jewelweed (touch-me-not), 337
  Queen Anne's lace, 336-37, 340
four dimensions, 43-44, 46, 47, 277
  see also hypercube
Frackinos Island, 125, 127, 128, 130, 131, 137, 146, 151, 159, 386
fractal dimension, 133-36, 140-41
  see also page 397 of the Chapter Notes
fractions, improper, 79, 120
Francesca, Piero della, 49
Franklin, Benjamin, 104, 115
frieze patterns, 310, 311-12, 321
Frost, Robert, 317
Fuller, Buckminster, 50-51

G

galaxies, 65, 162, 331, 336, 343
Galileo Galilei, 10-11, 36, 61, 63, 67, 266, 353, 375, 376, 384
  Dialogue on the Great World Systems by, 201, 266, 286
  pendulum and, 71-73, 76
  on Saturn's rings, 379-80, 381, 382
  The Discorsi of, 10
Galois, Évariste, 300-1, 302, 304, 321
gambler's fallacy, 350, 357
game of life, 358-61, 365
Garden of Eden, 12, 43
Gardner, Martin, 39, 40, 108, 165, 166, 183, 292-93, 295
Garfield, President James, 115
  see also page 394 of the Chapter Notes
Gates, Bill, 163
Gauss, Karl Friedrich, 120, 181, 189
  as a child prodigy, 71
  see also page 393 of the Chapter Notes
gematria, 14, 172, 177, 183, 185, 186
  see also page 389 of the Chapter Notes
generative spiral, the, 216, 217, 221
genes, 214, 285, 336
Genesis, 67, 79, 122, 172, 259
geodesic domes and spheres, 50, 51-52
  pentagons and, 52
  see also pages 391-92 of the Chapter Notes
Gilovich, Thomas (*How We Know What Isn't So*), 357
Gleick, James (*Chaos*), 100
Goat of Mendes, 231
Gogh, Vincent van, 221-23, 254, 299-300, 331
golden angle, 217, 218-20
golden number or ratio, 118, 383
  art and, 235-36, 342
  curious properties of, 213-14
  golden angle and, 220
  Pacioli on, 234
  Parthenon and, 229
  pentagram and, 230-31
  Phidias and, 228
  see also phi
golden rectangles, 234, 235
Goliath of Gath, 9-10
googolplex, 175
  see also page 400 of the Chapter Notes

Gould, Stephen Jay, 8-9, 18, 55, 109, 110-11, 122, 140, 352
Grand Gallery of the Great Pyramid, 224, 228
grasses, growth of, 215
great circle (geodesic), 52
  see also page 391 of the Chapter Notes
Gregory, James, 202
groups, cyclic, 306-7
groups, mathematical, 302-3, 304-6, 307-8, 309-10, 312, 314, 315, 319, 321-22

H

Haeckel, Ernst, 53, 54-55, 279-80
Haldane, J.B.S., 25, 27
Hamlet (Shakespeare), 327, 356, 381
handedness, 38-40, 46, 292-95, 333
  of dice, 65
Hanno the Carthaginian, 255
Hardy, G. H., 163, 165, 180, 181, 374
Harrison, James *(The Pattern & The Prophecy)*, 178, 232
Heinlein, Robert, 42-43, 44
Heinrich, Bernd, 244, 322
heliocentrism, 36, 262, 266
Helios, 2, 386
Heraclitus (philosopher), 105
Heraclitus (poet), 105, 289, 344, 387, *see also* the Preface
Heraclitus (poem), 105, 344, 387
  see also the Preface
Hero of Alexandria, 248, 249, 253, 255
Herodotus, 255
  cube pyramid of, 225, 226-28
  phi pyramid and, 224-28
  see also pages 401-2 of the Chapter Notes
Hesiod, 79
Hilbert, David, 91
Hofstadter, Douglas R., 380
Hollywood, 11, 22, 108, 123, 253
Homer, 4, 82, 186, 292, 352, 386
hominids, 19, 110, 113
Homo erectus, 110-11
Homo sapiens, 8, 19, 110-11, 164, 202, 301

honeycomb, 272, 273, 323, 341
horns (as symbols of power), 332-33
horoscopes, 61, 167
hot hand (or cold hand), 357, 365
Hubble Space Telescope (HST), 162, 331
human hand, the, 241, 242
humanoid, 65, 189
Hume, David, 370
Huram of Tyre, 195
Huygens, Christiaan, 76, 379-80, 382
hypercube, 41-43, 277-79, 308
  see also tesseract

I

Iliad (Homer), 3, 8, 160
imaginary unit (i), 120-21
independent events, 350
induction, method of, 368-70, 372
  see also deduction
Infeld, Leopold *(Whom the Gods Love),* 301
Inferno, The, 123, *see also* Dante
infinity, 29, 75, 90, 127, 148, 155, 201, 204, 237, 384
insects, 27, 83, 223, 289
  bumblebee, 22, 205, 209-10, 244, 246, 285
  dragonfly, 269-70
  eyes of, 272
  heat retention in, 21-22
  logistic function and, 84-85, 93
  size of, 22-23, 375
  surface tension and, 263
inverse-square law, 12, 334, 378
Ishmael, 188
Islam, 59, 182, 184, 309, 358
isoperimetric problem, 256, 274, 275
Iterson, G. Van, 217
Ithaca, Island of, 83, 159, 250

J

Jauch, J.M., 201
Jesus, 44-45, 70, 168-69, 173, 174, 175, 177-78, 356
John, Gospel of, 169, 196
Job, 169, 263
Josephus, Flavius, 10

Julia sets, 152-55, 156, 157
Julia, Gaston, 152, 156
Jung, Carl, 349-50
Jupiter, 17, 36, 112
Jurassic Park, 25, 123

K

Kaaba, the, 182, 188
Kaczynski, Ted (Unabomber), 186
Kant, Immanuel, 39, 40
Kasner, Edward, 164, 165
Keats, John, 384
Kennedy, President John, 186
Kepler, Johannes, 212, 260
  laws (rules) of, 67, 133, 347, 384
  molecule of, 38, 50, 309
  snowflakes and, 380
  System for the Heavens of, 35-37
Khalifa, Dr. Rashad, 183-84
Khufu (King), 224-25
King Lear (Shakespeare), 188
I Kings, 196
Koch, Helge von, 125
Koran, 182, 183, 185, 186, 188
  number 19 and the, 183-84
  see also page 400 of the Chapter Notes
Krishna, 327
Kronecker, Leopold, 189, 200
"Kubla Khan" (Coleridge), 157
kudu (antelope), 231, 332, 333

L

Lady Luck, 355
Lagrange, Joseph-Louis, 302
Lambert, Johann Heinrich, 199, 203-4
Laplace, Pierre Simon marquis de, 67, 87
Law of Universal Gravitation, 12, 14, 133-34, 347, 370
laws of motion, 14, 316, 375, 376-377
Lazarus, Emma, 1
Le Corbusier, 235, 342
least-action principle, 260-61
  see also Maupertuis'
  see also page 403 of the Chapter Notes
least distance, 252-53
least time, 253, 255, 260

Leibniz, Gottfried Wilhelm, 32, 70, 202
Leonardo of Pisa (Fibonacci), 207
Leonardo da Vinci, 30, 35, 38, 49, 52
  Baptism of Christ, 371-72
  Lady with the Ermine, 344-46, 348, 365, 371-72, 386
  The Last Supper, 70, 189, 190, 233, 235, 236, 290, 300, 331, 339
  Vitruvian Man, 233, 236, 290
Lincoln, Abraham, 115, 186
Lindemann, Ferdinand von, 200
Lindos, town of, 2, 3, 386
Lorenz, Edward, 99-103, 105, 122
Louvre, 54, 236, 240
Lowell, Percival, 380
Lucas sequence, 211-12, 214, 221
Lucas, Édouard, 211

M

magnets and magnetic fields, 324-25
mammals, 8, 20, 22, 24, 25, 84, 110, 112, 285, 293, 369
Mandelbrot set or man, 157, 160-161, 237
Mandelbrot, Benoit, 78, 87, 127-129, 130-32, 136, 156-57
  almond tree of, 152, 156, 160, 237
Marcellus (Roman general), 50, 244
Mars, 17-18, 36, 189, 334, 380, 381, 384
Marx, Karl, 106, 352
Matthew, Gospel of, 168
Maupertuis' *principle of least action,* 260-61, 265, 285, 338, 375, 376-77
Maupertuis, Pierre-Louis Moreau de, 261, 375
McDermott, Jeanne, 158
Medawar, Peter, 382-83
Melville, Herman, 186
Memnon (Ethiopian king), 83, 386
Menander (Greek poet), 301
Menger sponge, 148
Mercury (planet), 17, 36
metabolic rate, 20-22

methane molecule, 276, 309
Michelangelo Buonarroti, 189,
    232, 300, 332, 333, 373, 375
  Pietà, 373
Middle Ages, 33, 35, 38, 189, 225,
    372
  see also Dark Ages
Millay, Edna St. Vincent, 161, 337
Millennium, the, 176, 179
Milton, John, 13, 14, 16, 67, 91
mirror(s), 40, 256, 290
  group theory and, 304
  images and nature of, 294
  parallel, 131
  reflection in, 248-49, 251
  reversals in, 292-93, 295-96,
    297, 320-21
Miss Jenkins, 353, 354
Möebius, August Ferdinand, 40, 310
Mohammed, the prophet, 182, 188
moon, 18, 47, 71, 162, 249, 254,
    277, 312, 335
More, Henry, 46
Moses, 332-33
Moslem(s), 178, 186
Mr. T. Square, 46, 277-78

N

Nasca lines, 142
natural and unnatural selection, 19,
    22, 214, 243, 244, 262, 274,
    285, 331, 365
  see also descent with modification
navel, 94
  Adam's, 231-32, 236
  center of the earth and the, 8
  phi and the, 233, 236
  see also bellybutton
Nazis, 127, 185, 306, 358
neolithic, 31-32, 37, 51
Newman, James R., 164, 165
Newton, Isaac, 7, 36, 37, 67, 70,
    83, 118, 194, 243, 245, 253,
    259, 266, 285, 347
  –Raphson method, 119, 120, 121-
    122
  apple of, 71, 181
  Kepler's laws and, 133-34
  Maupertuis' principle and, 260-26
  scientific rationalism of, 12-15

vision of, 375-76
Newton–Raphson method, 119,
    120-122
  see also page 396 of the
    Chapter Notes
nightingales, 130, 344, 367, 387
nonlinear differential equations, 69,
    71, 99, 102, 103, 151
non-periodic decimals, see
    numbers: irrational (surds)
number of completeness (seven),
    186-188
number systems, different, 165
numbers natural:
  algebraic, 199-200, 213
  complex, 79, 120
  dullest, 181-82
  imaginary, 79, 120
  irrational (surds), 5, 79, 91, 94,
    117-19, 120, 128, 198-200,
    203 213, 214, 219, 239
  rational, 91, 94, 102, 115-17,
    118, 119, 120, 198-99, 203-
    204, 213, 220
  real, 79, 91, 120, 152-53, 198, 199
  transcendental, 199-200, 203,
    213
numbers supernatural:
  153 (Fishes in the Net), 80
    127, 133, 169-71, 172, 173,
    174-78, 179-81, 183, 196-97
  666 (number of the beast), 163,
    172, 173, 174-75, 176, 181,
    183, 350
  see also Antichrist's number

O

Odysseus, 3, 82, 83, 159, 187, 250,
    300, 352, 367-68, 385
Odyssey (Homer), 3, 82, 367
On Growth and Form (Darcy
    Wentworth Thompson), 53,
    146, 313
Oppenheimer, Robert, 327
Opus Dei, 236
Orff, Carl, 355
orthogenesis, 108, 109
Orwell, George, 166
oxymorons, 16, 91, 239, 274
"Ozymandias" (Shelley), 344

P

Pacioli, Luca, 33-34, 35, 49, 213, 234, 371
Pagels, Heinz *(Perfect Symmetry)*, 326
Paley, William *(Natural Theology)*, 273-74
palindromes, 184, 289
Pappus (Greek mathematician), 48
parabola, 88, 90-91, 237, 287
Paradise Lost, see John Milton
Parthenon, the, 228-29, 236
Pascal, Blaise, 58, 60, 145, 237, 291, 298, 358
Pascal's triangle, 57-60, 63, 365
   cellular automata and, 361-63
   Chinese version of, 59
   Fibonacci sequence and, 210-11
   Omar Khayyám and, 58
   Pascal's pyramid and, 146-48
   Sierpinski and, 144-45, 147, 149, 383
   triangle numbers and, 172
   upside down version of, 383-84
Pasteur, Louis, 294, 321, 383
Patroclus, 83, 160, 292
Penelope, 367
pentadactylism, 180
pentagram, 209, 230-31, 232-33, 236
people:
   Cheyenne Indians, 8
   Inuit, 8, 19-20, 25, 236, 313, 348, 349
   Jews, 8, 59, 127
   Neanderthals, 19, 84, 110, 241
   Ruandans, 19-20, 25, 236, 313, 348
period doubling, 88, 95, 98, 237, 239
periodic decimals, *see* numbers: rational
Persian Wars, The (Herodotus), 225
phase portraits, 69, 72-73, 74-75, 77-78
Phemius, 1, 3-4, 56-57, 79-83, 184-186, 247, 249-50, 291, 327, 329-330, 368-372, 385-87
   see also Femos
phi (φ), 118, 189, 212-14, 217, 220-221, 223, 225, 237, 239, 242
   art and, 235
   foolishness and, 236, 342
   Khufu's pyramid and, 226-28
   Parthenon and, 228-30
   pentagram and, 231, 233
   navel and, 232-33
   regular solids and, 234
   see also divine proportion
photon(s), 16, 250-51, 252, 374, 375, 377
pi (π), 212, 237, 239, 242
   355/113 (Chinese value) for, 197, 199
   as a continued fraction, 203-4
   as an infinite sequence, 201-2
   as an irrational number, 199
   as a transcendental number, 200
   Bible's value for, 195-96
   circle squarers and, 5, 198
   curves of constant width and, 193
   definition of, 189
   legislating the value of, 197-98
   meandering streams and, 194-195, 333
   Rhind Papyrus' value of, 195
   to 38 decimal places and beyond, 118-19
pine cones, 209, 210, 215, 236
pineapples, 208, 209
Pisces, 167, 169
planet(s), 221, 298, 366, 375
   earth's place among, 327
   extra-solar, 112
   heliocentrism and the, 347
   Kepler's system of the, 36-37
   least-action principle and, 261, 285
   origins of stable, 316-17
   paths of, 380
   plate tectonics and, 17-18
   rings around, 384
Plateau, Joseph, 284
Plato, 33, 38, 61, 162, 370
Platonic non-solids, 149
Platonic solids:
   complete symmetry groups of the, 307
   duals of the, 234, 307-8, 321
   relative volumes of the, 259
   see also page 404 of the Chapter Notes

Pliny the Elder, 2
Plutarch, 50, 105-6
Pluto (planet), 112, 157
Poincaré, Henri, 69, 87
point of accumulation, 95, 117, 239
pointillism, 241
Polyclitus (Greek architect), 229
polytopes, 40, 41, 42, 308
   see also pages 389-90 of the Chapter Notes
precession of the equinoxes, 169
prehistory, 31, 107, 331
primordia, 216-18, 220, 221
probabilities in coin tossing, 56-58, 60, 92, 318, 349-50, 356-57, 377
Protagoras, 6, 166
protons, 16, 167, 193
Proverbs, 361
Ptolemy (Alexander's general), 1
Ptolemy, Claudius, 253, 347
Purcell, Edward (physicist), 352
Pyramid at Giza, 191, 224-25, 236, 293, 344
   Herodotus' cube and, 224-27
   Jehovah's Witnesses and, 224
   Masonic Order on, 225
   phi and, 225-28
   Thoreau on, 224
   see also pages 401-2 of the Chapter Notes
Pythagoras, 33, 36, 37, 198, 209, 212, 227, 228, 245
   Bertrand Russell on, 230
   deduction and, 368, 370
   five-pointed star of, 230-31, 236
   flaw in the philosophy of, 117, 118
   look-see proof of the theorem of, 114
   mystical and rational sides of, 113-114
   philosophy of, 115-16
Pythagorean theorem, 189, 212, 244
   proof of the, 114-15
   see also page 394 of the Chapter Notes
Paul, the apostle, 246
Pythagoreans, 29, 32, 117, 118, 230, 245
Pythia (priestess), 5, 6

Q

QBasic programs, 75, 92, 139, 151, 215
   see also pages 393, 397, 401 of the Chapter Notes
quadratic formula or equation, 69, 212, 301, 302
quantum theory, mechanics, or field, 68, 123, 253-54, 261, 377

R

Rabin, Yitzhak, 185, 186
radiolaria, 53-55, 279-80, 309
Raleigh jets, 323
Ramanujan, Srinivasa, 181, 184, 204
Raphson, John, 119
rational angles, 219
regular solids (proof only five exist), 32, 303
relativity theory, 68, 123, 261, 377
Rembrandt, 295
Resurrection Property, the, 174, 179
retina(s), 250, 254, 272, 285
Reuleaux, Franz, 191-92
Revelation, Book of, 14, 129, 172, 187, 188, 332
Reynard, 287, 243-44, 285-86, 287
   see also animals: fox
Rhind Papyrus, 195
Rhodes, 1, 3, 4, 385, 386
Roman numerals, 172
Rome, 61, 168, 187, 230
Rubáiyát (Omar Khayyám), 58-59, 104, 108, 351
Russell, Bertrand, 58, 111, 165-66, 184
   on belief, 348, 349
   on deductive reasoning, 114
   on Georg Cantor, 91
   on mathematics, 161
   on Oswald Spengler, 106
   on Pythagoras, 113, 117, 230
Russell, Charles Taze, 224

S

Sagan, Carl, 18, 67, 201, 316, 336
Samos, Island of, 113, 370, 385
sample space, 353, 354, 355
I Samuel, 10

Saturn, 36, 112, 379-81, 382
Schliemann, Heinrich, 159
science versus the humanities, 141, 160
scientific creationism, 274, 324, 331
scientific method, 107, 123, 372
Scriptures, the Holy, 14, 172, 177, 333
  see also Bible
Seahorse Canyon, 157, 161
seashells, 83
  enantiomorphic, 295, 321
  handedness of, 46, 295
  tent olive, 145-46, 362-63
self-similarity, 79, 98, 122, 148, 156
  chaos game and, 150
  fig tree's, 238-39
  fractal dimension and, 134-36, 140-41
  Koch snowflake and, 125-26, 150
  Lorenz attractor and, 103-4
  natural world and, 131, 161-62
  Pascal–Sierpinski triangle's, 144-145, 362
  pi (π) and, 201
  tent olive shell and, 145-46, 362-63
Seurat, George, 241
Seven Bridges of Königsberg, 329-331
  see also page 405 of the Chapter Notes
Seven Wonders of the World, 1, 2, 187, 228
Sforza, Ludovico, 35, 345
Shakespeare, William, 67, 188, 285, 344, 356, 373, 376, 377
Shanidar man, 84
Shanks, William, 116
Shelley, Percy Bysshe, 262, 290, 344
Siamese twins, 293, 296, 321
Sierpinski triangle, 137, 140-42, 144, 145, 148, 150, 151, 362
  deterministic construction of, 143-44
  generalizations of, 149, 150
Simeon Stylites, 116, 226
Sistine Chapel, 232, 242, 373
skepticism, 12, 46, 266
Slade, Henry, 46
Snell's law, 253

snowflake, 48, 123, 319, 380
solar system, 17, 112, 316-17
solipsism, 164, 166
Solomon, King, 169
Solomon's Temple, 195
Sophocles, 160, 367
  see also the Preface
space-filling polyhedra, 148
specified versus unspecified events, 353, 354-55
Spengler, Oswald, 106, 107
square-cube law, 3, 4, 5, 6-7, 9, 10, 11, 12, 16, 18, 20, 22, 24, 26, 28, 231, 341, 387
St. Jerome, 333, 379
stability and instability, 317, 321, 323, 325
  definition of, 316
  ferromagnets and, 324-25
  galaxy formation and, 318, 325-326
  logistic function and, 317
  solar system and, 316-17
  spider's web and, 323-24
  symmetry and, 318
Star of David, 125
Star Trek, 65, 253, 381
Statue of Liberty, 1
Steiner, Jacob, 270
Stevens, Peter S., 48, 285, 337
Stewart, Ian, 68, 76, 79, 158, 218, 221, 303, 318
Stoppard, Tom, 123-24
strange attractor(s), 78, 103-4, 105, 119, 122, 139
  Julia sets as, 154-55
  Lorenz, 103-4
  Sierpinski, 139-40
  see also attractor
Strølum, Hans-Henrik, 195
sun, 15-17, 36, 105, 167, 169, 317, 336, 347, 384
sundogs, 4, 386
sunflower(s), 209, 210, 223
  Lucas, 215
  number of spirals on, 215-16, 383
  pairs of spirals on, 208, 211
supernovas, 335-36
surface tension, 55, 269, 313, 315, 323, 324

definition of, 264-65
drop of mercury's, 265
drop of water's, 265
soap bubbles and, 54, 267, 280
swastika, 306-7
Swift, Jonathan, 130
Sylvester, James Joseph, 151
symmetries of the equilateral triangle, 304
symmetry, definition of, 289, 303, 319

T

tapetum and nocturnal vision, 250
Tartaglia (Niccolo Fontana), 301
Tennyson, Alfred Lord, 15, 212, 250, 366
tent olive shell, 144-45, 362-63
tesseract, 44, 134, 277
  see also hypercube
Theaetetus, 32, 38, 65, 307
theory of everything (TOE), 123, 367
Thomas, David (physicist), 185-86
Thompson, Darcy Wentworth, 53, 54, 146, 313
Thoreau, Henry David, 15, 224, 254, 279
three-body problem, 69
  see also pages 392-93 of the Chapter Notes
three-space, 39, 44, 47, 102, 166
three-way corners, 54, 269
  see also angles of 120°
Timaeus (Plato), 33, 35
Tithonus, 386
Torah, 185
Toynbee, Arnold, 105, 106-7
transpiration, 28
Tree of Knowledge, 47, 207, 237
trees, 10, 25-29, 146, 162, 210, 236, 298, 331, 340, 341, 344, 387
triangle numbers, 171-73, 174, 181, 196
  reversible, 173
  truncated, 173
  see also pages 399-400 of the Chapter Notes
Trinity Function, 174-77, 179, 180
Trojan horse, 3, 159, 250, 368, 386

Troy, 48, 83, 159, 250, 385
truncated icosahedron, 48-49, 50, 52, 309, 386
  see also Bucky Ball
truth:
  cultural, 347-48
  of consistency, 347-48
  of correspondence, 347-48
Tsu Ch'ung-chih, 199
Twain, Mark, 167, 184
Tycho, lunar crater, 334-35

U

UFO reports, 276
"Ulysses," *see* Tennyson
Ulysses (James Joyce), 232
universe, 104, 108, 123, 164, 166-67, 188, 193, 201, 204, 221, 244, 254-55, 327, 346, 350
  beginning of, 67, 68
  Big Bang's background radiation in, 325-26
  center of, 8
  close-packing in, 148
  dodecahedron and, 33
  fourth dimension and, 44
  fractals in, 162
  god circumscribing, 13
  human beings and, 335-36, 343
  humanity's place in, 366
  instability in, 317-18
  Kant's empty, 39
  light in, 252
  parsimonious, 285
  Pythagoras and, 114, 115, 118
  symbol of the clockwork, 76, 122
urban legends, 25-26

V

Vasari, Giorgio, 372
Venus, 17, 36
Verrocchio, Andrea del, 371-72
Virgil, 123
viruses, 105, 110-11, 141, 281, 294, 309, 331

W

Waitomo Cave, 351-52
Wantzel, Pierre, 5
water or whirligig beetle, 263, 264

wave-particle duality, 374-75
West, Benjamin (*Cicero Discovering the Tomb of Archimedes*), 246
Weyl, Hermann, 289, 303, 306-7
whales, 24, 87, 294
Wheeler, John, 158
whirlpools, 157, 162, 332
White, Dr. Leslie Alvin, 164, 165, 167, 178
white-eye in animals, 250-51
Whitehead, Alfred North, 50, 78, 329
Wilkinson Microwave Anisotropy Probe (WMAP), 326, 347
Wright, Frank Lloyd, 341-42

Y

Yevtushenko, Yevgeny, 111

Z

Zechariah, 175
Zeus, 8, 368, 386
zodiac, 33, 167, 169, 355
Zöllner, Johann Carl Friedrich, 46

www.ingramcontent.com/pod-product-compliance
Lightning Source LLC
Chambersburg PA
CBHW060912300426
44112CB00011B/1426